DENDRIMERS FOR DRUG DELIVERY

DENDRIMERS FOR DRUG DELIVERY

Edited by
Anil K. Sharma, MPharm., PhD
Raj K. Keservani, MPharm

AAP APPLE ACADEMIC PRESS

Apple Academic Press Inc.	Apple Academic Press Inc.
3333 Mistwell Crescent	9 Spinnaker Way
Oakville, ON L6L 0A2	Waretown, NJ 08758
Canada	USA

© 2019 by Apple Academic Press, Inc.

First issued in paperback 2021

Exclusive worldwide distribution by CRC Press, a member of Taylor & Francis Group

No claim to original U.S. Government works

ISBN-13: 978-1-77463-147-8 (pbk)

ISBN-13: 978-1-77188-662-8 (hbk)

Library and Archives Canada Cataloguing in Publication

Dendrimers for drug delivery / edited by Anil K. Sharma, MPharm, Raj K. Keservani, MPharm.

Includes bibliographical references and index.
Issued in print and electronic formats.
ISBN 978-1-77188-662-8 (hardcover).--ISBN 978-0-203-71315-0 (PDF)

1. Dendrimers in medicine. 2. Drug delivery systems.
I. Sharma, Anil K., 1980-, editor II. Keservani, Raj K., 1981-, editor

| R857.D36D46 2018 | 615.1'9 | C2018-903320-7 | C2018-903321-5 |

CIP data on file with US Library of Congress

Apple Academic Press also publishes its books in a variety of electronic formats. Some content that appears in print may not be available in electronic format. For information about Apple Academic Press products, visit our website at **www.appleacademicpress.com** and the CRC Press website at **www.crcpress.com**

DEDICATION

The present book is dedicated to
our beloved

Aashna,
Anika
Atharva
&
Vihan

CONTENTS

ABOUT THE EDITORS

Anil K. Sharma, MPharm, PhD
Delhi Institute of Pharmaceutical Sciences & Research, University of Delhi, India

Anil K. Sharma, MPharm, PhD, is a lecturer at the Delhi Institute of Pharmaceutical Sciences & Research, University of Delhi, India. He has published 28 peer-reviewed papers in the field of pharmaceutical sciences in national and international reputed journals as well as 15 book chapters and 10 edited books. His research interests include nutraceutical and functional foods, novel drug delivery systems (NDDs), drug delivery, nanotechnology, health science / life science, and biology / cancer biology / neurobiology. He graduated with a degree in pharmacy from the University of Rajasthan, Jaipur, India, and received a Master of Pharmacy (MPharm) from the School of Pharmaceutical Sciences, Rajiv Gandhi Proudyogiki Vishwavidyalaya, Bhopal, India, with a specialization in pharmaceutics. He earned his doctorate (PhD) from the University of Delhi, India.

Raj K. Keservani, MPharm
Faculty of Bachelor Pharmacy, CSM Group of Institutions, Allahabad, India

Raj K. Keservani, MPharm, is an associate professor at the Faculty of Bachelor Pharmacy, CSM Group of Institutions, Allahabad, India. He has more than 10 years of academic (teaching) experience from various institutes in India in pharmaceutical education. He has published 35 peer-reviewed papers in the field of pharmaceutical sciences in national and international journals. He has also published 16 book chapters, two co-authored books, and 10 edited books. He is also active as a reviewer for several international scientific journals. Mr. Keservani graduated with a pharmacy degree from the Department of Pharmacy, Kumaun University, Nainital (UA), India. He received his Master of Pharmacy (MPharm) (specialization in pharmaceutics) from the School of Pharmaceutical Sciences, Rajiv Gandhi Proudyogiki

Vishwavidyalaya, Bhopal, India. His research interests include nutraceutical and functional foods, novel drug delivery systems (NDDS), transdermal drug delivery / drug delivery, health science, cancer biology, and neurobiology.

LIST OF CONTRIBUTORS

Juliana Palma Abriata
School of Pharmaceutical Sciences of Ribeirão Preto, University of São Paulo, Ribeirão Preto, São Paulo, Brazil

Patrícia Mazureki Campos
School of Pharmaceutical Sciences of Ribeirão Preto, University of São Paulo, Ribeirão Preto, São Paulo, Brazil

Abhay S. Chauhan
School of Medicine and Public Health, University of Wisconsin–Madison, Madison, WI 53705, USA; and School of Pharmacy, Concordia University Wisconsin, Mequon, WI–53097, USA

Nirupam Das
Pharmaceutical Chemistry Laboratory, Department of Pharmaceutical Sciences, Assam University, Silchar–788011, India; and Department of Pharmaceutics, Indian Institute of Technology, Banaras Hindu University, Varanasi–221005, India

Sudipta Das
Department of Pharmaceutics,Netaji Subhas Chandra Bose Institute of Pharmacy, Nadia–741222, India

Michał Gorzkiewicz
Department of General Biophysics, Faculty of Biology and Environmental Protection, University of Lodz, 141/143 Pomorska Street, 90–236, Lodz, Poland

Swarnali Goswami
Delhi Institute of Pharmaceutical Sciences and Research, New Delhi–110017, India

Sougata Jana
Department of Pharmaceutics, Gupta College of Technological Sciences, Ashram More, GT Road, Asansol–713301, India

Nilofer Jasmin
Department of Pharmaceutics, Gupta College of Technological Sciences, Ashram More, GT Road, Asansol–713301, India

Ramadoss Karthikeyan
Vignan Pharmacy College, Vadlamudi–522213, India

Gurpreet Kaur
Department of Biotechnology, CT Group of Institutions, Shahpur Campus, Jalandhar–144020, India

Raj K. Keservani
Faculty of B.Pharmacy, CSM Group of Institutions, Allahabad, India

Barbara Klajnert-Maculewicz
Department of General Biophysics, Faculty of Biology and Environmental Protection, University of Lodz, 141/143 Pomorska Street, 90–236, Lodz, Poland; and Leibniz Institute of Polymer Research Dresden, Hohe Str. 6, 01069 Dresden, Germany

Oruganti Sai Koushik
Vignan Pharmacy College, Vadlamudi–522213, India

Palanirajan Vijayaraj Kumar
Faculty of Pharmaceutical Sciences, No. 1, Jalan Menara Gading, UCSI University (South Campus), Taman Connaught, Cheras 56000, Kuala Lumpur, Malaysia

Bibek Laha
Department of Pharmaceutics, Gupta College of Technological Sciences, Ashram More, GT Road, Asansol–713301, India

Phung Ngan Le
Department of Materials and Pharmaceutical Chemistry, Vietnam Academy of Science and Technology, HCMC70000, Vietnam

Robert Lee
Ohio State University, Columbus, Ohio, USA

Rahul Maheshwari
Department of Pharmaceutics, BM College of Pharmaceutical Education and Research, Khandwa Road, Indore, Madhya Pradesh–452 009, India

Juliana Maldonado Marchetti
School of Pharmaceutical Sciences of Ribeirão Preto, University of São Paulo, Ribeirão Preto, São Paulo, Brazil

Cuu Khoa Nguyen
Department of Materials and Pharmaceutical Chemistry, Vietnam Academy of Science and Technology, HCMC70000, Vietnam

Dai Hai Nguyen
Department of Materials and Pharmaceutical Chemistry, Vietnam Academy of Science and Technology, HCMC70000, Vietnam

Abayomi Tolulope Ogunjimi
School of Pharmaceutical Sciences of Ribeirão Preto, University of São Paulo, Ribeirão Preto, São Paulo, Brazil

Joyita Roy
Department of Pharmaceutics, Gupta College of Technological Sciences, Ashram More, GT Road, Asansol–713301, India

Kumar Sandeep
Department of Preventive Oncology, AIIMS, New Delhi–110029, India

Ankit Seth
Department of Pharmaceutics, Indian Institute of Technology, Banaras Hindu University, Varanasi–221005, India

Anil K. Sharma
Delhi Institute of Pharmaceutical Sciences and Research, University of Delhi, New Delhi–110017, India

Piyoosh A. Sharma
Department of Pharmaceutics, Indian Institute of Technology, Banaras Hindu University, Varanasi–221005, India

Sushant K. Shrivastava
Department of Pharmaceutics, Indian Institute of Technology, Banaras Hindu University, Varanasi–221005, India

Shivani Srivastava
Department of Medicinal Chemistry, IMS, BHU, Varanasi–221005, India

Muktika Tekade
TIT College of Pharmacy, Technocrats Institute of Technology, Anand Nagar, Bhopal–462021, India

Rakesh K. Tekade
National Institute of Pharmaceutical Education and Research (NIPER)–Ahmedabad, Opposite Air Force Station Palaj, Gandhinagar–382355, India

Ngoc Quyen Tran
Department of Materials and Pharmaceutical Chemistry, Vietnam Academy of Science and Technology, HCMC70000, Vietnam

Yamini Bhusan Tripathi
Department of Medicinal Chemistry, IMS, BHU, Varanasi–221005, India

Durgavati Yadav
Department of Medicinal Chemistry, IMS, BHU, Varanasi–221005, India

LIST OF ABBREVIATIONS

AAF	aminoacetamido fluorescein
AAV	adeno-associated virus
ABC	ATP-binding cassette
AC	a-cyano-4- hydroxycinnamic acid
ALL	acute lymphoblastic leukemia
APC	antigen presenting cells
ARMD	age-related macular degeneration
BBB	blood–brain barrier
BCRP	breast cancer resistance protein
BCS	biopharmaceutical classification system
CAT1	cationic amino acid transporter
CBPs	choline-binding proteins
CD	circular dichroism
CHT	choline transporter
CNT2	concentrative nucleoside transporter
CR	controlled release
CRDDS	controlled release drug delivery systems
CV	cyclic voltammetry
DCMN	dendrimers-coated magnetic nanoparticle
DDS	drug-delivery system
DEX	dexamethasone
DHB	dihydroxybenzoic acid
DIT	dithranol
DLS	dynamic light scattering
DOPE	dioleoyl phosphatidylethanolamine
DOX	doxorubicin
DR	diabetic retinopathy
DS	dielectric spectroscopy
DSC	differential scanning calorimetry
EDA	ethylenediamine
EGFR	epidermal growth factor receptor
EM	erythromycin
EPR	electron paramagnetic resonance

EPR	enhanced permeability and retention
FA	folic acid
FGF	fibroblast growth factor
FITC	fluorescein isothiocyanate
FMDV	foot-and-mouth disease virus
GE	gel electrophoresis
GI	gastrointestinal
GIT	gastrointestinal tract
GLUT1	glucose transporter
HA	hemagglutinin
HER2	human epidermal growth factor receptor 2
HIV	human immunodeficiency virus
HMGB1	high mobility group box 1
HP-b-CD-PEI	hydroxypropyl-b-cyclodextrin-grafted polyethylenimine
HPA	hydroxypicolinic acid
HSV	herpes simplex virus
IR	immediate release
LAT1	large neutral amino acid transporter
LCST	low temperature solution temperature
LDHN	lipid dendrimer hybrid nanoparticle
LLS	laser light scattering
LPS	lipopolysaccharide
MAP	multiple antigen peptide
MCT1	monocarboxylic acid transporter
MDNP	modified dendrimer nanoparticle
MMLV	moloney-murine leukemia virus
MRI	magnetic resonance imaging
MRP	multidrug resistance protein
MRSA	methicillin-resistant *Staphylococcus aureus*
MS	mass spectrometry
MTD	maximum tolerated dose
MTX	methotrexate
NBT	nucleobase transporter
NCI	National Cancer Institute
NIR	near-infrared
NMR	nuclear magnetic resonance
NRTKs	non-receptor tyrosine kinases
OR	oral route

OS	organosilicon
OSR	oral sustained release
PAMAM	poly-amidoamine
PAMAMOS	poly-amidoamine-organosilicon
PB	probucol
PDGF	platelet-derived growth factor
PDT	photodynamic therapy
PEG	olyethylene glycol
PEI	polyethylenamine
PEPE	polyether-copolyester
PI	propidium iodide
PLL	poly-L-lysine
PMB	polymyxin B
PNIPAM	poly-nisopropylacrylamide
PPI	poly-propyleneimine
RES	reticuloendothelial system
ROS	reactive oxygen species
RTKs	receptor tyrosine kinases
SANS	small-angle neutron scattering
SAXS	small-angle X-ray scattering
SEC	size exclusion chromatography
siRNA	small interference RNA
SIV	simian immunodeficiency virus
SLN	solid lipid nanoparticles
SMV	simvastatin
SPSJ	Society of Polymer Science, Japan
SSD	site-specific delivery
STDs	sexually transmitted diseases
SV-40	Simian virus-40
TAM	tamoxifen
TfR	transferrin receptor
THAP	trihydroxyacetophenone
TOB	tobramycin
VEGF	vascular endothelial growth factor
VRE	vancomycin-resistant *Enterococcus*
WGA	wheat germ agglutinin
XRD	x-ray diffraction

PREFACE

We all know that science is witnessing an everlasting phase of evolution. In particular, nanotechnology-based products have been developed, and many of them have been demonstrated to be of great potential for mankind. The applications of nanotechnology are well-spread, encompassing material science, engineering, medical, dentistry, drug delivery, etc. Pharmaceutical manufacturers, healthcare personnel, and researchers are concerned about nanotechnology-based product development. In recent times, enormous developments have taken place with respect to delivering active pharmaceutical ingredients to the target sites; thereby sparing the normal functioning biological systems. One such nanometric architectures are dendrimers, which constitute a class of highly branched polymers symmetrically organized around the core, with hyperbranched arms as a tree.

The present book strives to provide the requisite information to its readers pertaining to dendrimers. The text of this book is written by highly skilled, experienced, and renowned scientists and researchers around the globe with up-to-date information to offer drug delivery knowledge to readers, researchers, academicians, scientists, and industrialists worldwide.

The book *Dendrimers for Drug Delivery* is comprised of 11 chapters that describe the introduction of dendrimers, physicochemical characteristics, and applications dealing with drug delivery as core of all. The materials used, synthesis, and characterization have been presented in elaborate fashion.

Chapter 1, *Dendrimers: Branched Nanoarchitectures and Drug Delivery,* written by Raj K. Keservani and colleagues, presents an introduction to dendrimers. In addition, it presents an overview of diverse applications of dendrimers. The authors have summarized the dendrimer applications and the striking features of dendrimers.

The details of usual properties, structure, classification, methods of synthesis have been presented in Chapter 2, *Dendrimers: A Tool for Advanced Drug Delivery*, written by Patrícia Mazureki Campos and associates. The authors have discussed the dendrimer structure and properties with suitable instances. In addition, there is a discussion of synthesis methodology and applications in various fields. The authors have provided information about toxicity of dendrimers too.

Chapter 3, *Dendrimers: General Features and Applications,* written by Durgavati Yadav and colleagues, gives a general account of dendrimers. It begins with definitions, describing synthesis methods and ends with a detailed application part.

The structure activity relationship in dendrimers is described by Chapter 4, *Computational Approach to Elucidate Dendrimers,* written by Gurpreet Kaur. The author has provided a general description of dendrimers, embracing classification, synthesis, and nomenclature with simulation using principles of computational science. This evokes a fascination for structure modifications suited to need of delivery system.

Chapter 5, *An Overview of Dendrimers and Their Biomedical Applications,* written by Nirupam Das and associates, deals with issues related to applications relevant to biological systems in particular. The authors have endeavored to provide exhaustive information about dendrimers covering almost every aspect. Further, the biocompatibility and toxicity aspects are also provided to render the information more useful to its readers.

The description of dendrimers having predesigned drug delivery potential has been provided by Chapter 6, *Dendrimers for Controlled Release Drug Delivery,* written by Phung Ngan Le and colleagues. Beginning with an introduction to dendrimers the chapter moves forward to discussion of smart dendrimers that possess the ability to respond to alteration in pH, temperature, light etc. Further, it gives details of formulations made to release the drug in a controlled manner using dendrimers.

Chapter 7, *Dendrimers in Targeted Drug Delivery,* written by Ankit Seth and associates, gives a focused view of dendrimers explored for drug delivery to specific tissue/organ systems. The authors have presented the current status of dendrimers in targeting brain, eyes, liver and other organs. Similarly, the ailments like tumors, and infections addressed by targeting approaches by dendrimers have been discussed.

The uses of dendrimers for oral drug delivery have been discussed in Chapter 8, *Dendrimers in Oral Drug Delivery,* written by Sougata Jana and colleagues. The authors have provided information relevant to bio-distribution and epithelial permeability of dendrimers followed by oral delivery of a number of drugs exploiting dendrimers. The applications are summarized in tabular form too.

Chapter 9, *Dendrimers in Gene Delivery,* written by Piyoosh A. Sharma and associates, has addressed the different aspect of dendrimer mediated gene delivery. The authors have provided an introduction to methods of gene delivery, different vectors used, and obstacles faced by the researchers in effective gene delivery. Thereafter, the gene delivery assisted by dendrimers have been discussed.

The customized applications of dendrimers focused on cancer treatment are given by Chapter 10, *Dendrimers as Nanocarriers for Anticancer Drugs,* written by Michał Gorzkiewicz and Barbara Klajnert-Maculewicz. The chapter begins with a general description of cancer and its prevalence, and the hindrances in the way to effective control of tumors, nanotechnology-based approaches for mitigation and cure of cancer. Afterwards dendrimer applications in delivery of antineoplastic agents have been discussed in detail.

Chapter 11, *Dendrimeric Architecture for Effective Antimicrobial Therapy,* written by Ramadoss Karthikeyan and colleagues, described dendrimer-based drug delivery against various infections. The diseases caused by bacterial, viral, and other microorganism have been discussed with limitations of conventional drug delivery systems. The emergence of nanotechnology-based products for microbe controls have been provided with challenges faced by such products. Subsequent to this, an introduction to dendrimers has been given, followed by their applications as drug carriers in antimicrobial therapeutics.

PART I

PREAMBLE OF NANOMETRIC ARTIFACTS: DENDRIMERS

CHAPTER 1

DENDRIMERS: BRANCHED NANOARCHITECTURES AND DRUG DELIVERY

RAJ K. KESERVANI,[1] SWARNALI GOSWAMI,[2] and ANIL K. SHARMA[2]

[1]*Faculty of B.Pharmacy, CSM Group of Institutions, 8 Milestone, Rewa Road, Iradatganj, Allahabad-212110, India.*
E-mail: rajksops@gmail.com

[2]*Delhi Institute of Pharmaceutical Sciences and Research, University of Delhi, New Delhi–110017, India*

CONTENTS

ABSTRACT

Nanotechnology has appeared to be a boon to humanity in several aspects. This is substantiated by the surge of nanotechnology-based products. For instance, the drugs entrapped within nanoparticles have been found to offer improved targeting with reduction in side effects dose, thereby showing improvement in patient compliance. Since their first-ever mention by Vögtle in 1978, the dendrimers, which are branched nanometric polymeric constructs, have been in focus of researchers. The ability to house both hydrophilic and hydrophobic drugs has rendered dendrimers the preference over other nanocarriers. The current chapter presents general features, synthesis methods, and applications of dendrimers in various domains.

1.1 INTRODUCTION

The word dendrom is Greek in origin and is made up of two parts: "dendron" meaning tree and "moros" meaning part. The term Dendrimer is internationally accepted, and "arborols" and "cascade molecules" are some other synonyms. Vögtle (1978) mentioned the first-ever dendrimers (Bhuleier et al., 1978). Dendrimer chemistry is a specialized research field, which possesses its own terms. Additionally, the varied chemical events taking place on the surface of the dendrimer are signified by a concise set of rules for structural nomenclature. A new class of dendritic polymers called "Dendrigrafts" are like dendrimers and can be constructed; they are monodisperse and have a well-defined molecular structure (Tomalia et al., 1991). Dendrimer chemistry finds a place in between the domains of polymer chemistry and molecular chemistry. They resemble molecular chemistry by stepwise carefully controlled synthesis, and they mirror the polymer world as they are composed of repeated monomer units linked together (Bosman et al., 1999; Majoral et al., 1999; Frechet et al., 2001; Newkome et al., 2001).

In the present context of architectural chemistry, considerable efforts have been dedicated to the development of polymer systems. The advent of the 20th century brought with it remarkable innovations in polymer synthesis, especially designing of polymeric macromolecules that are biodegradable. The dendrimers are the result of these advances and innovations in the field of polymer science. Dendrimers were, for the first time, synthesized during 1970–1990 by two different groups: Buhleier et al. and Tomalia et al. In contrast to linear polymers, dendrimers developed by these two groups have precisely controlled architecture with tailor-made surface groups, which could be finely tuned (Buhleier et al., 1978; Tomalia et al., 1985).

The characteristic architecture of dendrimers provides a distinct branched structure having a globular shape, which possesses numerous surface groups. It is possible to alter these surface groups according to the requirement, so that it can serve as a vehicle for drug delivery (Tomalia et al., 1990). A cationic polymer, polyethylenamine (PEI), is highly water soluble and comprises repeated units of CH_2-CH_2-NH_2. It has been proved that PEI has great potential in gene delivery as a nonviral vector (Godbey et al., 2001) and also as a carrier in the pulmonary drug delivery system (Yang et al., 2006) by virtue of the electrostatic interactions between the positively charged PEI and the negatively charged phosphate groups of DNA or cell membranes. PEI modified with sugar molecules has been shown to interact with plasma proteins, and this conjugate was less toxic as well. Therefore, it could be further explored for nanoparticle synthesis for the treatment of a variety of ailments (Wrobel et al., 2017).

There are several ways by which a dendrimer interacts with various organic drug substances; such interactions are classified into two categories, namely, the physical interactions and the chemical interactions. These interactions may be hydrophobic interaction, simple encapsulation, or hydrogen bonding (Madaan et al., 2014). The combination types of drug and dendrimer are depicted in Figure 1.1.

The routes of administration of drugs can be mainly classified as enteral, parenteral, or topical. Each of these routes has their own advantages and disadvantages. Enteral routes are oral, sublingual, or rectal. The parenteral routes are intramuscular, subcutaneous, intravascular, and inhalation administration. Topical delivery is achieved through skin and mucosal membranes (COD, 1997; http://www.network-ed.com.au/pdf/samples/9780443103315_sample, 2012; Jevon, 2010).

There has been a rapid increase in the number of biomedical applications of dendrimers from 1990 to the present. The new term of "dendrimer space" concept was defined as a new cluster, which is included in the vast volume of chemical space (Mignani et al., 2013). Zhang et al. studied the improvement of pulmonary absorption of poorly absorbable macromolecules by hydroxy-propyl-b-cyclodextrin-grafted polyethylenimine (HP-b-CD-PEI) in rats and concluded that HP-b-CD-PEI is a safe absorption enhancer for improving absorption of hydrophilic macromolecules such as peptide and protein drugs by pulmonary delivery (Zhang et al., 2015).

1.2 DENDRIMERS IN SYNTHESIS OF NANOPARTICLES

Traditionally, the three macromolecular architectural classes that are recognized are cross-linked, linear, and branched and are capable of generating polydisperse products of varied molecular weights. Conversely,

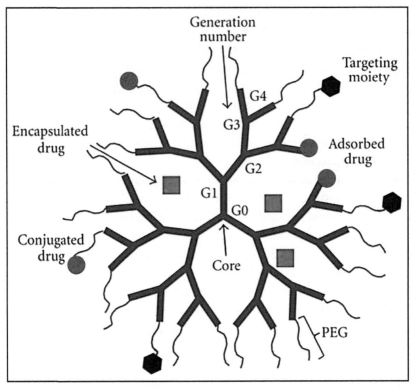

FIGURE 1.1 Methods for combining drug and dendrimer (Reprinted from Copyright © 2012 Jun H. Lee and Anjan Nan. Lee, J., & Nan, A., (2012). Combination drug delivery approaches in metastatic breast cancer. J. Drug Deliv. vol., Article ID 915375, pp. 17, 2012. doi:10. 1155/2012/915375.https://creativecommons.org/licenses/by/3.0/).

monodisperse, structure-controlled macromolecular dendrimer formulation, resembling those observed in biological systems, is possible through synthesis (Tomalia et al., 1986; Tomalia, 1995, 2004). Dendritic polymers are synthesized via a step-by-step repetitive reaction (Tomalia et al., 1990) having a hyperbranched topology which is almost perfect, radiating outwards from a core and developed generation after generation. Nearly complete control can be exerted on the parameters of critical molecular design (e.g., size, surface interior chemistry, flexibility shape, and topology) of the procedures of synthesis developed for dendrimer, which make them useful in the practical applications (El-Khouly et al., 2008, 2009).

Ligand- or polymer-stabilized colloidal noble metals have been preferred as catalysts for several years for the hydrogenation of unsaturated organic molecules (Bergbreiter and Liu, 1997). Moreover, the development of "green" methodologies for catalyzing organic reactions in aqueous solutions is gaining popularity and interest. Accordingly, the homogeneous catalytic hydrogenation of alkenes in aqueous solutions was investigated using dendrimer-encapsulated nanoparticles (Zhao, 1999).

Dendrimers are macromolecules with highly branched three-dimensional architecture and possess various special properties by virtue of being shaped like a globe and the free-void volume in their internal cavities. Due to the free-void volume located within the interior structure of the dendrimer, it can encapsulate guest molecules in its macromolecular interior, thus coating the nanoparticles and potentially preventing them from further oxidation and aggregation (Jansen et al., 1995; Hou et al., 2005). There have been many new classes of dendrimers since the synthesis of poly(propyleneimine) (PPI) dendrimers by Vögtle et al. in 1978. Starburst poly(amidoamine) (PAMAM) dendrimers by Tomalia (Tomalia et al., 1985, 1990) are the arborols of Newkome and co-workers (Newkome et al.,1985), and the polyether dendrimers described by Hawker and Fréchet et al. are some examples (Hawker and Fréchet, 1990). For preparing nanomaterial templates such as ceramic, monolithic and polymeric ones have also been used. For example, in alumina or polymeric filtration, membranes with well-defined pores can be employed to define the geometrical and chemical properties of metal, semiconductor, and polymeric nanomaterials. The template can also be removed chemically or thermally, where the naked nanomaterial will remain. The prominent advantage of this technique is that preparation of highly monodisperse particles with a variety of shapes, sizes, and chemical compositions is possible (Scho et al., 1997; Martin et al., 1999; Shelimov and Moskovits, 1999). An interesting approach used to prepare inorganic NPs is through the use of poly(amidoamine) (PAMAM) dendrimers as templates or stabilizers. Dendrimers are a novel class of polymers with a close to spherical shape and a narrow size distribution that facilitates their use as templates or stabilizers to form relatively monodispersed organic/inorganic hybrid NPs. Copper sulfide and cadmium sulfide can be formed using PAMAM dendrimers as templates or stabilizers (Xiangyang et al., 2006). Some potential applications of monodisperse nanoparticles are in the fields of catalysis, optoelectronics, magnetic, and chemical sensing. Stabilization of metal nanoparticles by chemical routes typically involves the reduction of metal salts and capping of the resulting nanoparticles with polymers or surfactants to retard growth and avoid agglomeration (Bonnemann and Richards, 2001; Roucoux et al., 2002). These templates can also be used for the synthesis of nanoparticles consisting of pre-defined shapes and sizes, namely, porous alumina, reverse micelles, and track-etched membranes (Antonietti et al., 1996; Hulteen and Martin, 1998; Pileni, 1998).

Owing to the dearth of satisfactory procedures for preparation, the phenomenon of chirality in metal nanoparticles has been sporadically explored essentially. Synthesis of DNA-templated Ag nanoparticles using both single- and double-stranded DNA has been achieved (Shemer et al., 2006). Schaaff and Whetten in an extensive study have successfully separated a series of gold–glutathione cluster compounds through gel electrophoresis on polyacrylamide (Schaaff and Whetten, 2000). PPI and PAMAM dendrimers differ in their properties. PPI have sufficient stability at very high temperatures (the onset of weight loss for G4 PPI is 470°C), whereas PAMAM dendrimers exhibit retro-Michael addition at temperatures higher than 100°C. Finally, both PAMAM and PPI dendrimers contain a statistical distribution of defects, mostly missing branches and loops (Bosman et al., 1999; Crooks et al., 2000). However, it was inferred from the measurements made by the investigation team that the polydispersity of these nanoparticles was not appreciably affected by the template imperfections. To make the NPS environmentally sustainable, these thin polymer films are embedded or coated onto the surfaces of these microparticles. The electrostatic layer-by-layer (LbL) assembly approach of preparing nanoparticulate thin films using oppositely charged polymers and NPs has recently garnered a lot of attention. This is because by this method, the thickness and composition of the films can be controlled by changing the number of alternating deposition cycles and the charged species, respectively (Decher, 1997).

1.3   DENDRIMERS IN DRUG DELIVERY

The cavities that arise between the branches of dendrimers due to the symmetric branching are occupied by either the covalent or end groups, which are backfolded. In some cases, guest molecules inhabit these cavities (Ceroni and Juris, 2003). Three factors govern the size and shape of the dendrimer molecule. They are the generation, the nature of the repeating units, and the number of branches at the core of the dendrimer. These cavities aid in the encapsulation of smaller molecules within the dendrimer framework. Therefore, dendrimers show promise as supramolecular hosts. Another avenue where extensive research is being carried out is exploiting the ability of dendrimers to act as hosts within the biological system (Klajnert and Bryszewska, 2001; Beezer et al., 2003; Boas and Heegaard, 2004).

Studies have been carried out to investigate how the size, generation, and surface functional moiety of PANAM dendrimers affect their aqueous solubility. Hence, investigation of the bioavailability of nifedipine was carried out. It was discovered that the solubility of nifedipine was higher for ester-terminated dendrimers than for the amino-terminated ones possessing the same number of surface groups. Predictably, as the size of the dendrimers increased, so did the solubility of nifedipine (Devarakonda et al., 2004). The commercialized/pipeline dendrimer preparations are summarized in Table 1.1.

Sialodendrimers or sialylated dendrimers display the exclusive feature of being effective blockers of the hemagglutination of human erythrocytes through influenza viruses. The foremost stage in the infection of a cell by influenza virus is the adhesion of the virion to the cell membrane, which occurs by the interaction of a virus receptor hemagglutinin with sialic acid groups present on the surface of the cell (Sigal et al., 1996).

In this context, the real benefit of the structural attributes of dendrimers can be realized by the designing of effective drug delivery systems that

TABLE 1.1 List of Commercialized/Pipeline Dendrimer Formulations

Brand Name	Company	Uses	Status
Priostar®	Starpharma	Water treatment	Marketed
Starburst®	Starpharma	Transfection agent	Marketed
Astramol®	Starpharma	Transfection agent	Marketed
Polylysine	Starpharma	Anticancer therapeutics	Marketed
Stratus CS®	Dade behring	Cardiac assay diagnostic	Marketed
Superfect®	Qiagen	Transfection agent	Marketed
Priofect™	Starpharma	Transfection agent	Marketed
Alert ticket™	US Army Lab	Anthrax-detecting agent	Marketed
Vivagel®	Starpharma	Prevention of HIV and STDs transmission	Clinical (Phase-III)
Dendrimer-docetaxel	Starpharma	Breast cancer treatment	Preclinical
Dendrimer-oxaliplatin	Starpharma	Colon cancer treatment	Preclinical

(Source: Madeen et. al, 2014)

could be used for the drugs developed in the past but failed to be brought into the market due to some control of their release and delivery potential. By using dendrimers, novel drug delivery systems could be formulated that are capable of enhancing the effectiveness of the overall drug delivery by ensuring its sustained release in gradual time and at a specific target (Liu and Fréchet, 1999). Some other salient features of dendrimers are their enhanced retention effects and permeability that also allow targeting of only the tumor cells in preference over the normal cells (Malik et al., 1999).

1.3.1   DENDRIMERS IN OCULAR DRUG DELIVERY

Dendrimers are used as ophthalmic vehicles for the delivery of drug to the eye. Dendrimers as nanocarriers offer the unique advantages of increased penetration through the cornea and sustained drug release. Vandamme and Brobeck investigated the development of surface-modified PAMAM dendrimers (Vandamme et al., 2005). The G1.5, G2 OH, and G4 OH PAMAM dendrimers in aqueous solution showed increase in the residence time and activities of both hosts and guests (encapsulation) pilocarpine nitrate (parasympathomimetic alkaloid) and tropicamide (pyridinylmethyl-benzeneacetamide) in miotic and mydriatic activity tests, respectively. Very recently, dendrimer-fluocinolone acetonide-G4-OH PAMAM conjugate has been prepared for the attenuation of neuroinflammation in the retina (Lezzi et al., 2012). Bioactive delivery to the eye was investigated. Vandamme and Brobeck found the PAMAM dendrimers with carboxyl- or hydroxyl-end functionalities to increase the retention of pilocarpine within the eyes. Thus, this study supports the applicability of dendrimers via the ocular route (Bai et al., 2007).

Drug targets the posterior section of the eye to treat diseases such as glaucoma, diabetic retinopathy (DR), age-related macular degeneration (ARMD), and genetically linked diseases such as retinitis pigmentosaposes, as a greater challenge. The posterior section of the eye poses a challenge for drug delivery due to the vitreous being acellular in nature and the diffusion distance toward the retina being long. Hence, the drug reaches its target incompletely, and its bioavailability also gets diminished. Practically, the amount of drug that reaches the retina is merely 5% of the amount administered to the eye (Yavuz et al., 2013; Mishra and Jain, 2014). In an attempt to increase the corneal residence time, some nonionic hydrosoluble polymers, for example, hemi-synthetic cellulose derivatives, vinyl polymers, and dextran (Buri and Gurtler, 1995; Ooteghem et al., 1995), have been used and marketed. Greaves and Wilson (1993) reported the use of solutions containing mucus glycoproteins that help to increase precorneal residence time. In order to repair the corneal wounds, a variety of dendrimer adhesives consisting of generations 1, 2, and 3 (G1, G2, and G3, respectively) associated with PEG, glycerol, and succinic acid were developed (Carnahan et al., 2002; Grinstaff, 2002; Velazquez et al., 2004). Hydrogel composed of PEGylated dendrimers that contain ocular drug molecules attached to the dendrimers efficiently deliver the drugs to the eye (Yang and Kao, 2006).

1.3.2   DENDRIMERS IN TRANSDERMAL DRUG DELIVERY

In transdermal drug delivery systems, dendrimers have been extensively used. The structure of bioactive drugs has hydrophobic moieties and low water solubility; thus, dendrimers are an efficient delivery system (Cheng et al., 2007). Transdermal route of drug delivery has the advantage of being noninvasive, and the delivery of the drug occurs through the skin for both local and systemic therapies; however, it is very selective (Naik et al.,2000). The transdermal revolution opened several avenues for the pharmaceutical industry to treat diseases, especially bone diseases (Ramachandran and Fleisher, 2000). The local immunosuppressants like cyclosporine A-loaded nanofibers for cell-based therapy have also been administered via this route (Holana et al., 2011).

Till date, PANAM dendrimers and PEGylated polyglycerol amine dendrimers have been employed for transdermal bioactive molecules delivery. This is usually accompanied by a co-treatment or a pre-treatment of the bioactives with different vehicles, for example, water, chloroform, or their mixtures; isopropyl myristrate; and so on. Three different possible mechanisms were very recently reviewed and discussed by Zhao et al. (Sun et al., 2012). Ketoprofen and Diflunisal were associated with G5 PAMAM dendrimer and presented 3.4- and 3.2-fold more permeation. Greater bioavailability of PAMAM dendrimers by employing indomethacin as the model drug in transdermal drug application was stated to be fruitful (Cheng, 2008, Jevprasesphant et al., 2003). Chauhan et al. also assessed transdermal capability of PAMAM dendrimers by exploiting indomethacin as the model drug for investigation. The findings of *in vitro* permeation experiments exhibited rise in the steady-state flux with surge in amount of all three types G4-NH$_2$, G4-OH, and G-4.5 PAMAM dendrimers. For the *in vivo* pharmacokinetic and pharmacodynamic studies, indomethacin and dendrimer preparations were applied to the abdominal skin of the Wistar rats, and blood withdrawn from the tail vein at the predetermined intervals. The indomethacin content was suggestively more with PAMAM dendrimers in comparison with the pristine drug dispersion. The inferences suggested that effective concentration could be sustained for 24 h in the blood with the G4 dendrimer–indomethacin preparation. Hence, data recommended that the dendrimer–indomethacin-based transdermal delivery system was active and might be an innocuous and efficient approach for treating different ailments (Chauhan et al., 2003). A cream formulation of PAMAM dendrimers containing resveratrol, an antioxidant, and anti-aging drug, was found to permeate more, and solubility of resveratrol was also improved. This dendrimer-drug complex offers an eco-friendly alternative to other topical formulations (Pentek et al.,2017).

1.3.3   DENDRIMERS IN TARGETED DRUG DELIVERY

Presently, general cancer chemotherapeutics are reported to be less effective in curing tumors because highly potent drugs often have nonselective action and result in side effect related to dose limiting. Therefore, drug carriers target the tumor cells specifically and are being strongly preferred for the treatment of cancer. Targeted drug delivery offers higher therapeutic index and diminishes drug resistance. For an effective targeting drug-delivery system, the base is the that is uniform and can amalgamate the different components such as targeting molecule, drug, and cancer-imaging agent (Thomas et al., 2004). Dendrimers have emerged as a versatile carrier in this regard because of their well-defined architecture, monodispersity, and tailor-made surface groups. These properties make dendrimers as effective-targeted delivery systems of bioactives. The most popular example in this context is folate-conjugated dendrimers for targeting anticancer bioactives to tumor. As the folate receptors are overexpressed on the surface of different types of cancer cells such as ovarian cancer, breast cancer, and so on, the folate-conjugated dendrimers can efficiently target anticancer bioactives to cancer cells (Quintana et al., 2002; Agrawal et al., 2008). DNA-assembled nanoclusters that help in detecting tumor cell-specific binding and internalization were evaluated *in vitro*. These DNA-assembled dendrimer conjugates may allow the combination of different drugs with different targeting and imaging agents; thus, they find an application in combinatorial therapeutics (Choi et al., 2005).

According to Patri et al., when a drug is complexed with a dendrimer as an inclusion complex, its solubility in water is improved. A dendrimer conjugate, which is cleavable and yet covalently linked, is preferred as targeted drug delivery for the premature release of the drug into the body that can be prevented. They reported minimum cytotoxic effect with the covalently linked dendrimer (Patri et al., 2005). Dendrimers have several properties that make them suitable for application in targeted drug delivery system. Folic acid has been one of the cell-specific targeting agents that have been delivered by dendrimers most successfully. PAMAM dendrimers modified with carboxymethyl PEG5000 surface chains possessed reasonable drug loading, a reduced release rate, and reduced hemolytic toxicity compared with the non-PEGylated dendrimer (Kolhe et al., 2003; Hawker, 2006; Mohammad and Antony, 2006).

1.3.4 DENDRIMERS IN ORAL DRUG DELIVERY

Oral drug delivery is the most traditional and desirable route for polymeric drug carriers and has been employed for the administration of many drugs. However, due to the large size and molecular weight, the oral bioavailability of polymeric drug carriers is often limited. Owing to the structural features, for example, molecular weight, geometry, and hydrodynamic volume, the variation in the mechanism and net permeability of polymeric drug carriers has been observed (Tajarobi et al., 2001). Oral delivery of different drug-loaded systems is impeded by issues of absorption and distribution. A plethora of different strategies have been discussed to overcome these limitations (Kojima et al.,2000). Specially, fabricated highly water-soluble and biocompatible dendrimers show improved drug properties such as solubility of orally administered drugs and their stability in biological environments. Dendrimers when attached to low-penetrating drug molecules facilitate their penetration across the intestinal epithelium, thus enhancing their absorption. Other benefits that were reported are longer half-life, higher concentration at the site action, and reduced nonspecific toxicity of loaded drugs (Duncan, 2006). The researchers concluded that PAMAM dendrimers and surface-modified dendrimers with lauryl groups could efficiently traverse epithelial monolayers via paracellular and transcellular pathways (Jevprasesphant et al., 2003). Later, propranolol–PAMAM dendrimer conjugate was investigated for transport across Caco-2 cell monolayers, and it was observed that the conjugate could reduce the effect of P-glycoprotein on the intestinal absorption of propranolol. Hence, it could be concluded that dendrimers can bypass P-glycoprotein efflux transporter and can facilitate the oral administration of drugs (D'Emanuele et al., 2004). Some examples of PAMAM-based dendrimers are as follows: conjugates PAMAM with propranolol and prodrug of naproxen. PAMAM dendrimers when assessed for pulmonary absorption of enoxaparin proved to be the vehicles for pulmonary delivery of therapeutics (Grinstaff, 2007; Kang et al., 2009).

1.3.5 DENDRIMERS IN CONTROLLED RELEASE DRUG DELIVERY

In early studies, Lemere and coworkers observed a boosting effect with intranasal dendrimeric Aβ1-15 (16 copies of Aβ1-15 on a lysine tree) but not Aβ1-15 peptide affording immune response following a single injection of Aβ1-40/42 in heterozygous APP-tg mice (Seabrook et al., 2006). A pharmacodynamic investigation of the dendritic preparation was carried out on carrageenan-triggered paw edema model. Results exhibited a 75% blockade at the 4th hour that was sustained above 50% till the 8th hour. The dendritic preparation showed 2–3 times rise in mean residence time and terminal half-life against free drugs (Asthana et al., 2005).

1.3.6 DENDRIMERS IN GENE DELIVERY

Gene delivery of drugs is more efficient than most drug delivery routes and also has lesser side effects (Pezzoli et al., 2012; Mellott et al., 2013). Dendrimers possessing high structural flexibility and partially degraded high-generation or hyperbranched dendrimers are preferred for certain gene delivery operations than intact high-generation symmetrical dendrimers (Barbara and Maria, 2001; Hinkemeyer et al., 2002; Christine et al., 2005). Dendrimers, due to their nanometric size and molecular weight, are highly popular vehicles for drug and gene delivery. Moreover, the functional groups at their surface can be altered depending upon their application. The ultimate shape of the dendrimers is globular. The surface functional groups interact with the environment. The peripheral groups are attached to the drug moieties and control their toxicity and solubility. While designing novel gene vectors, commercially available and easily functionalized poly(propylene imine) dendrimers were selected. It could be predicted with their structure that they would express during the proton sponge effect. However, toxicity, exhibited due to the presence of primary amino groups at their surface (Malik et al., 2000), should be also addressed and decreased. Studies have demonstrated that polypropylene imine (PPI) generation 2 dendrimers attach proficiently to DNA, possess low cytotoxicity to cells, and augment *in vitro* gene transfer (Zimselmeyer et al., 2002).

In order to maintain the functioning of DNA on dehydration, the dendrimer/DNA complexes were entrapped in a water-soluble polymer and then placed on or sandwiched in functional polymer films with a rapid degradation propensity to intercede gene transfection. PAMAM dendrimer/DNA complexes were utilized to entrap functional biodegradable polymer films by this procedure for substrate-arbitrated gene delivery. Research has revealed that the swiftly degrading functional polymer has prodigious ability for localized transfection (Fu et al., 2007, 2008; Tathagata et al., 2008). Kumara et al. investigated and improved gene transfection efficacy by polyamidoamine (PAMAM) dendrimers altered with ornithine residues and observed that the ornithine-conjugated dendrimers have the capability to be a novel gene carrier (Kumara et al., 2010).

1.3.7 DENDRIMERS IN CANCER DRUG DELIVERY

Cancer is the second-most common cause of death for both men and women in the United States, second only to heart diseases (Kochanek et al., 2014). Apart from DNA, several other small molecule pharmaceuticals have been carried by dendritic carriers. Upon encapsulation of cisplatin, the following effects were observed in an anticancer drug within PANAM dendrimers: slower release, higher accumulation in tumors, and lower toxicity than those of free drug (Malik et al., 1999).

Doxorubicin (DOX) is one of the most effective anticancer therapeutics available in the clinic today (Blum and Carter, 1974) and has been widely used alone or in combination to treat a variety of cancers including lung cancers (Koukourakis et al., 1999; Otterson et al., 2010; Zhong et al., 2016). DOX induces the apoptosis of cancer cells by intercalating itself to DNA double helix and thus inhibiting the progression of the enzyme topoisomerase II. Other mechanisms include the production of high-level reactive oxygen species (ROS) and cellular membrane disruption (Thorn et al., 2011).

The epidermal growth factor receptor present in human gliomas is a vital mark for the delivery of therapeutic agents to brain tumors. The chimeric monoclonal antibody cetuximab (IMC-C225) is used as a boron delivery carrier for neutron capture therapy. A profoundly boronated fifth-generation PAMAM dendrimer (G5-B1100) attached to C225 and boron neutron capture therapy has been tested, and it is being evaluated for the treatment of intracerebral brain tumors (Wu et al., 2004).

1.3.8   DENDRIMERS AS ANTIMICROBIALS

The antimicrobial activity exhibited by nanoparticles has been attributed to their relatively smaller sizes and high amount of surface area-to-volume ratio that allows nanoparticles to interact closely with membranes of viruses, fungi, or bacteria rather than the normal mechanism of releasing metal ions in solution (Morones et al., 2005). It is well known that due to the high surface area-to-volume ratio, microscopic amounts of metal nanoparticles can provide antimicrobial effects to hundreds of square meters of its host material. Many studies have shown that Ag, Cu, Ni, and Co (and their oxides) exhibit good antimicrobial activity against various bacteria (Ruparelia et al., 2008; Zhang et al., 2008; Ren et al., 2009; Ravikumar et al., 2012). Recently, PEGylated dendrimers containing silicon phthalocyanine Pc 4 against candidal infections have been formulated and demonstrated to have effective activity by killing drug-resistant *Candida albicans* (Hutnick et al., 2017). Efforts were made to improve the solubility of fusidic acid by synthesizing G2 and G3 polyester amphiphilic dendrimers (GMOA-G2-OH, GMOA-G3-OH, GMS-G2-OH, and GMS-G3-OH). This not only increased solubility but also exhibited good antibacterial activity against *Staphylococcus aureus* and methicillin-resistant *Staphylococcus aureus* (MRSA) (Sikwal et al., 2017).

1.3.9   DENDRIMERS AS BIOMIMICS

Surface topology of dendrimers is the most diverse and extensive among all biological systems. At the meter scale, it can be observed in tree branching and roots. At the centimeter and millimeter scales, lung, liver, kidney, and spleen exhibit such topology. At the nanometric scale, it can be observed in molecules of glycogen and amylopectin that act as the storehouse of energy in plants and animals. The highly branched architecture of dendrites offers special interfacial and functional advantages (Svenson and Tomalia, 2005).

1.3.10 DENDRIMERS IN MEDICAL IMAGING

Paramagnetic metal chelates such as Gd(III)-N,NV,NW,Nj-tetracarboxy-methyl-1,4,7,10-tetraazacyclododecane (Gd(III)-DOTA), Gd(III)-diethylene-triamine pentaacetic acid (Gd(III)-DTPA), and their derivatives are exploited as contrast agents for magnetic resonance imaging (MRI) as they prolong the relaxation rate of adjacent water protons (Doubrovin et al., 2004; Hay et al., 2004). Gold nanoparticles (AuNP) possess properties such as high conductivity and excellent biocompatibility. They facilitate the imaging process by acting as bimolecular nanoscopic wires having large electrode surface that help in orienting the DNA molecules for successful immobilization. These have been used to chemisorb thiolated DNA onto electrode surface (Katz et al., 2003; Liu et al., 2003; Daniels and Astruc, 2004; Lucarelli et al., 2005; Shulga and Kirchhoff, 2007; Willner et al., 2007; Pingarrón et al., 2008). One of the earliest discovered applications of dendrimers was their use in the form of carriers to fasten and improve the overall efficacy of MRI contrast agents (Velazquez et al., 2004; Wathier et al., 2004). Dendritic polymers, being analogous to protein, enzymes, and viruses, have advantage in biomedical applications and can easily be utilized for several applications. Dendrimers and other molecules can either be attached to the periphery or can be encapsulated in their interior voids (Patel and Patel, 2013). A modern day example is the use of different types of dendritic polymers as potential blood substitutes, for example, polyamidoamine dendrimers (Ruth and Lorella, 2005).

The rapid advancement of nanotechnology has given rise to several potential applications of dendrimers, such as detection and bioremediation of an extensive range of environmental pollutants. By virtue of amalgamation of analysis procedures and nanotechnology, the development of miniaturized, brisk ultrasensitive, and low-cost procedures has become conceivable to in situ and environmental monitoring devices. As has been reported, nano-material-based biosensors have been utilized for environmental monitoring (Undre et al., 2016).

Photodynamic therapy (PDT) bank on the initiation of a photosensitizing agent with visible or near-infrared (NIR) light. Upon excitation, an extremely energetic state is achieved, which, subsequent to reaction with oxygen, generate a highly reactive singlet oxygen that is capable of promoting necrosis and apoptosis in tumor cells. Dendritic delivery of PDT agents has been explored in the past few years in order to advance in tumor selectivity, retention, and pharmacokinetics (Battah et al., 2001; Chatterjee and Mahata, 2002; Nishiyama et al., 2003; Zhang et al., 2003; Triesscheijn et al., 2006; Rajakumar et al., 2010).

1.4 CONCLUSIONS

Dendrimer-encapsulated nanoparticles provide a unique advantage. These nanoparticles provide a combination of benefits such as the distinctive dendrimer structure, its three-dimensional aspect, terminal group chemistry, the presence of certain endoreceptors within the dendrimers, and radial distribution of function of polymer density. From the initial days of dendrimer synthesis, where the focus of investigation was on synthetic procedure and dendrimer properties, the interest in dendrimer chemistry has escalated. In conclusion, with respect to its versatile structure and extraordinary properties, dendrimers find their application in a plethora of fields, and consistently, further research is being pursued in the field of dendrimer technology.

KEYWORDS

- drug delivery
- PAMAM dendrimers
- polymers
- radio imaging agents
- synthesis
- therapeutics

REFERENCES

Agrawal, A., Asthana, A., Gupta, U., & Jain, N. K., (2008). Tumor and dendrimers: a review on drug delivery aspects. *J. Pharm. Pharmacol., 60,* 671–688.

Antonietti, M., Forster, S., Hartmann, J., & Oestreich, S., (1996). *Macromolecules, 29,* 3800–3806.

Asthana, A., Chauhan, A. S., Diwan, P. V., & Jain, N. K., (2005). Poly(amidoamine) (pamam) dendritic nanostructures for controlled site specific delivery of acidic anti-inflammatory active ingredient. *AAPS Pharm. Sci. Tech., 6,* Article 67.

Bai, S., Thomas, C., & Ahsan, F., (2007). Dendrimers as a carrier for pulmonary delivery of enoxaparin, a low molecular weight heparin. *J. Pharm. Sci., 96,* 2090–2096.

Barbara, K., & Maria, B., (2001). Dendrimers: properties and application. *Acta. Biochimica. Polonica., 48*(1), 199–208.

Battah, S. H., Chee, C. E., Nakanishi, H., Gerscher, S., MacRobert, A. J., & Edwards, C., (2001). Synthesis and biological studies of 5-aminolevulinic acid containing dendrimers for photodynamic therapy. *Bioconjug. Chem., 12,* 980–988.

Beezer, A. E., King, A. S. H., Martin, I. K., Mitchel, J. C., Twyman, L. J., & Wain, C. F., (2003). Dendrimers as potential drug carriers, encapsulation of acidic hydrophobes within water soluble PAMAM dendrimers. *Terahedron., 59,* 3873–3880.

Bergbreiter, D. E., & Liu, Y. S., (1997). Water-soluble polymer-bound, recoverable palladium (0)-phosphine catalysts. *Tetrahedron Lett., 38,* 7843–7846.

Bhuleier, E., Wehner, W., & Vogtle, F., (1978). "Cascade"- and "nonskid-chain-like" syntheses of molecular cavity topologies. *Synthesis,* 155–158.

Blum, R. H., & Carter, S. K., (1974). Adriamycin: a new anticancer drug with significant clinical activity. *Ann. Intern. Med., 80*(2), 249–259.

Boas, U., & Heegaard, P. M. H., (2004). Dendrimers in drug research. *Chem. Soc. Rev., 33,* 43–63.

Bonnemann, H., & Richards, R. M., (2001). Nanoscopic metal particles-synthetic methods and potential applications, *Eur. J. Inorg. Chem.,* 2455–2480.

Bosman, A. W., Janssen, H. M., & Meijer, E. W., (1999). About dendrimers: Structure, physical properties, and applications. *Chem. Rev., 99,* 1665–1688.

Buhleier, E., Wehner, W., & Vogtle, F., (1978). Cascade- and nonskid-chain-like syntheses of molecular cavity topologies. *Synthesis, 2,* 155–158.

Buri, P. F., et al., (1995). Les formes ophtalmiques a` action prolongée, in: M. Van Ooteghem (Ed.), *Preparations Ophtalmiques,* Lavoisier, Paris, 127–215.

Carnahan, M. A., Middleton, C., Kim, J., Kim, T., & Grinstaff, M. W., (2002). Hybrid dendritic-linear polyester-ethers for in situ photopolymerization. *J. Am. Chem. Soc., 124,* 5291–5293.

Ceroni, P., & Juris, A., (2003). Photochemical and photophysical properties of dendrimers. In *Handbook of Photochemistry and Photobiology,* Nalwa, H. S., (ed.), Volume *3*: Supramolecular photochemistry, American Scientific Publishers: Stevenson Ranch, CA, USA, Chapter 4, pp. 157.

Chatterjee, D., & Mahata, A., (2002). Visible light induced photodegradation of organic pollutants on the dye adsorbed TiO_2 surface. *J. Photochem. Photobiol., A: Chem., 153,* 199–204.

Chauhan, A. S., Sridevi, S., Chalasani, K. B., Jain, A. K., Jain, S. K., Jain, N. K., & Diwan, P. V., (2003). Dendrimer-mediated transdermal delivery: enhanced bioavailability of indomethacin. *J. Control Release, 90,* 335–343.

Cheng, Y., (2008). Dendrimers as drug carrier: applications in different routes of drug administration. *J. Pharm. Sci., 97,* 33–36.

Cheng, Y., Man, N., Xu, T., Fu, R., Wang, X., Wang, X., & Wen, L., (2007). Transdermal delivery of nonsteroidal anti-inflammatory drugs mediated by polyamidoamine (PAMAM) dendrimers. *J. Pharm. Sci.*, *96*, 595–602.

Choi, Y., Thomas, T., Kotlyar, A., Islam, M. T., Baker Jr., J. R., (2005). Synthesis and functional evaluation of DNA-assembled polyamidoamine dendrimer clusters for cancer cell-specific targeting. *Chem. Biol.*, *12*, 35–43.

Christine, D., Ijeoma, F. U., & Andreas, G. S., (2005). Dendrimers in gene delivery. *Adv. Drug Deliv. Rev.*, *57*, 2177–2202.

Committee on Drugs (COD), (1997). Alternative routes of drug administration advantages and disadvantages, Pediatrics, *100*, 143–152(and references cited therein).

Crooks, R. M., Lemon, B. I., Yeung, L. K., & Zhao, M., (2000). Dendrimer- encapsulated metals and semiconductors: Synthesis, characterization, and applications. *Top. Curr. Chem.*, *212*, 81–135.

Daniels, M. C., & Astruc, D., (2004). Gold nanoparticles: assembly, supramolecular chemistry, quantum-size-related properties, and applications toward biology, catalysis, and nanotechnology *Chem. Rev.*, *104*, 293–346.

Decher, G., (1997). Fuzzy Nanoassemblies: Toward layered polymeric multicomposites, *Science*, *277*(5330), 1232–1237.

Devarakonda, B., Hill, R. A., & De Villiers, M. M., (2004). The effect of PAMAM dendrimer generation size and surface functional group on the aqueous solubility of nifedipine, *Int. J. Pharm.*, *284*, 133–140.

Doubrovin, M., Serganova, I., Mayer-Kuckuk, P., Ponomarev, V., & Blasberg, R. G., (2004). Multimodality *in vivo* moleculargenetic imaging, *Bioconjug. Chem.*, *15*, 1376–1388.

Duncan, R., (2006). Polymer conjugates as anticancer nanomedicines, *Nat. Rev. Cancer*, *6*, 688–701.

D'Emanuele, A., Jevprasesphant, R., Penny, J., & Attwood, D., (2004). The use of a dendrimer-propranolol prodrug to bypass efflux transporters and enhance oral bioavailability. *J. Control. Release*, *95*, 447–53.

El-Khouly, M. E., Chen, Y., Zhuang, X., & Fukuzumi, S., (2009). Long-lived charge-separated configuration of a push-pull archetype of disperse red 1 end-capped Poly[9, 9-Bis(4-diphenylaminophenyl) fluorene]. *J. Am. Chem. Soc.: Part A: Polymer Chem.*, *131*(18), 6370–6371.

El-Khouly, M. E., Kang, E. S., Kay, K. Y., Choi, C. S., Araki, Y., & Ito, O. A., (2008). New blue-light emitting polymer: Synthesis and photoinduced electron transfer process. *J. Polymer Sci.: Part A: Polymer Chem.*, *46*(12), 4249–4253.

Frechet, J. M. J., & Tomalia, D. A., (2001). *Dendrimers and Other Dendritic Polymers*. New York, Chichester, Wiley.

Fu, H. L., Cheng, S. X., & Zhang, X. Z., (2007). Dendrimer/DNA complexes encapsulated in a water-soluble polymer and supported on fast degrading star poly (DL-lactide) for localized gene delivery. *J. Gene. Med.*, *124*(3), 181–188.

Fu, H. L., Cheng, S. X., Zhang, X. Z., & Zhuo, R. X., (2008). Dendrimer/DNA complexes encapsulated functional biodegradable polymer for substrate-mediated gene delivery. *J. Gene. Med.*, *10*(12), 1334–1342.

Godbey, W. T., Wu, K. K., & Mikos, A. G., (2001). Recent progress in gene delivery using non-viral transfer complexes. *J. Control. Release*, *72*, 115–125.

Gorman, C. B., & Smith, J. C., (2001). Structure–property relationships in dendritic encapsulation. *Acc. Chem. Res.*, *34*, 60–71.

Greaves, J. L., & Wilson, C. G., (1993). Treatment of diseases of the eye with mucoadhesive delivery systems, *Adv. Drug Deliv. Rev.*, *11*, 349–383.

Grinstaff, M. W., (2002). Biodendrimers: new polymeric materials for tissue engineering. *Chem. Eur. J.*, *8*, 2838–2846.

Grinstaff, M. W., (2007). Designing hydrogel adhesives for corneal wound repair. *Biomaterials.*, *28*(35), 5205–5214.

Hawker, C., (2006). Dendrimers: Novel polymeric nanoarchitectures for solubility enhancement. *Biomacromolecules*, *7*(3), 649–58.

Hawker, C. J., & Frechet, J. M. J., (1990). Preparation of polymers with controlled molecular architecture. A new convergent approach to dendritic macromolecules, *J. Am. Chem. Soc.*, *112*(21), 7638–7647.

Hay, B. P., Werner, E. J., & Raymond, K. N., (2004). Estimating the number of bound waters in Gd(III) complexes revisited. Improved methods for the prediction of q-values, *Bioconjug. Chem.*, *15*, 1496–1502.

Holana, V., Chudickova, M., Trosan, P., Svobodova, E., Krulova, M., Kubinova, S., Sykova, E., Sirc, J., Michalek, J., Juklickova, M., Munzarova, M., & Zajicova, A., (2011). Cyclosporine A-loaded and stem cell-seeded electrospun nanofibers for cell-based therapy and local immunosuppression, *J. Control Release*, *156*, 406–412.

Hou, Y., Kondoh, H., Ohta, T., & Gao, S., (2005). "Size-controlled synthesis of nickel nanoparticles," *Appl. Surf. Sci.*, *241*(1–2), 218–222.

Hulteen, J. C., & Martin, C. R., (1998). Template synthesis of nanoparticles in nanoporous membranes. In: *Nanoparticles and Nanostructured Films*, Fendler, J. H., (ed.), Wiley-VCH: Weinheim, 235.

Hutnick, M., Ahsanuddin, S., Guan, L., Lam, M., Baron, E., & Pokorski, J., (2017). Pegylated dendrimers as drug delivery vehicles for the photosensitizer silicon phthalocyanine Pc 4 for candidal infections. *Biomacromolecules*, *18*, 379–385.

Jansen, J. F. G. A., Meijer, E. W., De Brabandervan, & Den Berg, E. M. M., (1995). "The dendritic box: shape-selective liberation of encapsulated guests," *J. Amer. Chem. Soci.*, *117*, 15, 4417–4418.

Jevon, P. P., (2010). *Miscellaneous Routes of Medication Administration, in Medicines Management*: A guide for nurses (eds Jevon, P., Payne, E. Higgins, D., & Endacott, R.), Wiley-Blackwell, Oxford, Miscellaneous routes of medication administration, medicines management, chapter 12, Wiley-HC, p. 239.

Jevprasesphant, R., Penny, J., Jalal, R., Attwood, D., McKeown, N. B. D., & Emanuele, A., (2003). Engineering of dendrimer surfaces to enhance transepithelial transport and reduce cytotoxicity. *Pharm. Res.*, *20*, 1543–1550.

Kang, S. J., Durairaj, C., Kompella, U. B., O'Brien, J. M., & Grossniklaus, H. E., (2009). Subconjunctival nanoparticle carboplatin in the treatment of murine retinoblastoma. *Arch. Ophthalmol.*, *127*(8), 1043–1047.

Katz, E., & Willner, I., (2003). Probing biomolecular interactions at conductive and semiconductive surfaces by impedance spectroscopy: Routes to impedimetric immunosensors, DNA-Sensors, and enzyme biosensors. *Electroanalysis*, *15*, 913.

Klajnert, B., & Bryszewska, M., (2001). Dendrimers: Properties and application. *Acta. Biochim. Pol.*, *48*, 199–208.

Kochanek, K. D., Murphy, S. L., Xu, J., & Arias, E., (2014). Mortality in the United States, 2013, *Centers for Disease Control and Prevention*, *8*.

Kojima, C., Kono, K., Maruyama, K., & Takagishi, T., (2000). Synthesis of polyamidoamine dendrimers having poly (ethylene glycol) grafts and their ability to encapsulate anticancer drugs, *Bioconjugate*, *11*, 910–917.

Kolhe, P., Misra, E., Kannan, R. M. Kannan, S., & Lai, M. L., (2003). Drug complexation, *in vitro* release and cellular entry of Dendrimers and hyperbranched polymers. *Int. J. Pharm.*, *259*, 143–160.

Koukourakis, M., Koukouraki, S., Giatromanolaki, A., Archimandritis, S., Skarlatos, J., Beroukas, K., Bizakis, J., Retalis, G., Karkavitsas, N., & Helidonis, E., (1999). Liposomal doxorubicin and conventionally fractionated radiotherapy in the treatment of locally advanced non–small-cell lung cancer and head and neck cancer. *J. Clin. Oncol.*, *17*(11), 3512–3521.

Kumara, A., Yellepeddia, V. K., Daviesb, G. E., Strycharc, K. B., & Palakurthia, S., (2010). Enhanced gene transfection efficiency by polyamidoamine (PAMAM) dendrimers modified with ornithine residues, *Int. J. Pharm.*, *392*, 294–303.

Lee, J., & Nan, A., (2012). Combination drug delivery approaches in metastatic breast cancer. *J. Drug Deliv.* vol., Article ID 915375, pp. 17, 2012. doi:10. 1155/2012/915375.

Lezzi, R., Guru, B. R., Glybina, I. V., Mishra, M. K., Kennedy, A., & Kannan, R. M., (2012). Dendrimer-based targeted intravitreal therapy for sustained attenuation of neuroinflammation in retinal degeneration, *Biomaterials*, *33*, 979–988.

Liu, M., & Frechet, J. M., (1999). Designing dendrimers for drug delivery. *Pharm. Sci. Technol. Today*, *2*, 393–401.

Liu, S., Leech, D., & Ju, H., (2003). Application of colloidal gold in protein immobilization, electron transfer, and biosensing. *Anal. Lett.*, *36*, 1–19.

Lucarelli, F., Marrazza, G., & Mascini, M., (2005). Enzyme-based impedimetric detection of PCR products using oligonucleotide-modified screen-printed gold electrodes *Biosens. Bioelectron.*, *20*, 2001–2009.

Madaan, K., Kumar, S., Poonia, N., Lather, V., & Pandita, D., (2014). Dendrimers in drug delivery and targeting: drug-dendrimer interactions and toxicity issues. *J. Pharm. Bioallied. Sci.*, *6*(3), 139–150.

Majoral, J. P., & Caminade, A. M., (1999). Dendrimers containing heteroatoms (Si, P, B, Ge, or Bi). *Chem. Rev.,* *99*, 845–880.

Malik, N., Evagorou, E. G., & Duncan, R., (1999). Dendrimer–platinate: A novel approach to cancer chemotherapy, *Anticancer Drugs*, *10*, 767–776.

Malik, N., Wiwattanapatapee, R., Klopsch, R., Lorenz, K., Frey, H., Weener, J. W., Meijer, E. W., Paulus, W., & Duncan, R., (2000). Dendrimers: relationship between structure and biocompatibility *in vitro*, and preliminary studies on the biodistribution of 125I-labelled polyamidoamine dendrimers *in vivo, J. Control. Release*, *65*, 133–148.

Martin, C. R., & Mitchell, D. T., (1999). Template-synthesized nanomaterials in electrochemistry. In *Electroanalytical Chemistry*, Bard, A. J., Rubinstein, I., (eds.), Dekker, New York, *21*, 1–74.

Mellott, A. J., Forrest, M. L., & Detamore, M. S., (2013). Physical non-viral gene delivery methods for tissue engineering. *Ann. Biomed. Eng.*, *41*(3), 446–468.

Mignani, S., Kazzouli, E. L. S., Bousmina, M., & Majoral, J. P., (2013). Dendrimer space concept for innovative nanomedicine: a futuristic vision in medicinal chemistry, *Prog. Polym. Sci.*, *38*(7), 993–1008.

Mishra, V., & Jain, N. K., (2014). Acatazolamide encapsulated dendritic nano-architectures for effective glaucoma management in rabbits. *Int. J. Pharm.*, *461*, 380–390.

Mohammad, N., & Antony, D., (2006). Crossing cellular barriers using dendrimer nanotechnologies. *Curr. Opin. Pharmacol.*, *6*, 522–527.

Morones, J. R., Elechiguerra, J. L., Camacho A., et al., (2005). The bactericidal effect of silver nanoparticles, *Nanotechnology*, *16*(10), 2346–2353.

Naik, A., Kalia, Y. N., & Guy, R. H., (2000). Transdermal drug delivery: overcoming the skin-barrier function, *Pharm. Sci. Technol. Today*, *3*, 318326.

Newkome, G. R., Moorefield, C. N., & Vögtle, F., (2001). *Dendrimers and Dendrons: Concepts, Syntheses, Applications.* Wiley, Weinheim.

Newkome, G. R., Yao, Z., Baker, G. R., & Gupta, V. K., (1985). Micelles. Part 1. Cascade molecules: a new approach to micelles A [27]-arborol. *J. Org. Chem.*, *50*(11), 2003–2004.

Nishiyama, N., Stapert, H. R, Zhang, G. D, Takasu, D., Jiang, D. L., Nagano, T., Aida, T., & Ataoka, K., (2003). Light-harvesting ionic dendrimer porphyrins as new photosensitizers for photodynamic therapy. *Bioconjug. Chem.*, *14*, 58–66.

Ooteghem, M. V., (1995) .Les Collyres guttae ophthalmicae, In: M. Van Ooteghem (Ed.), *Preparations Ophtalmiques*, Lavoisier, Paris, Chapter 3, 58–111.

Otterson, G. A., Villalona-Calero, M. A., Hicks, W., Pan, X., Ellerton, J. A., Gettinger, S. N., & Murren, J. R., (2010). Phase I/II Study of inhaled doxorubicin combined with platinum-based therapy for advanced non−small cell lung cancer. *Clin. Cancer Res.*, *16*(8), 2466−2473.

Patel, H. N., & Patel, D. R. P. M., (2013). Dendrimer applications–A review. *Int. J. Pharm. Bio. Sci.*, *4*(2), 454–463.

Patri, A. K., Kukowska-Latallo, J. F., & Baker, Jr. J. R., (2005). Targeted drug delivery with dendrimers: comparison of the release kinetics of covalently conjugated drug and non-covalent drug inclusion complex. *Adv. Drug Deliv. Rev.*, *57*, 2203–2214.

Pentek, T., Newenhouse, E., O'Brien, B., & Chauhan, A., (2017). Development of a topical resveratrol formulation for commercial applications using dendrimer nanotechnology. *Molecules*, *22*, 137.

Pezzoli, D., Chiesa, R., De Nardo, L., & Candiani, G., (2012). We still have a long way to go to effectively deliver genes! *J. Appl. Biomater. Funct. Mater.*, *10*, 182–191.

Pileni, M. P., (1998). Size and morphology control of nanoparticle growth in organized surfactant assemblie. In *Nanoparticles and Nanostructured Films*, Fendler, J. H., (ed.), Wiley-VCH: Weinheim, 71.

Pingarron, J. M., Yánez-Sedeno, P., & Gonzalez-Cortes, A., (2008). Gold nanoparticle-based electrochemical biosensors. *Electrochim. Acta.*, *53*, 5848–5866.

Quintana, A., Raczka E., Piehler, L., Lee, I., Myc, A., Majoros, I., Patri, A. K., Thomas, T., Mulé, J., & Baker, Jr., (2002). Design and function of a dendrimer-based therapeutic nanodevice targeted to tumor cells through the folate receptor. *Pharm. Res.*, *19*, 1310–1316.

Rajakumar, P., Raja, S., Satheeshkumar, C., Ganesan, S., Maruthamuthu, P., & Suthanthiraraj, S. A., (2010). Synthesis of triazole Dendrimers with a dimethyl isophthalate surface group and their application to dye-sensitized solar cells. *New J. Chem.*, *34*, 2247–2253.

Ramachandran, C., & Fleisher, R. D., (2000). Transdermal delivery of drugs for the treatment of bone diseases, *Adv. Drug Deliv. Rev.*, *42*, 197–223.

Ravikumar, S., Gokulakrishnan, R., & Boomi, P., (2012). *In vitro* antibacterial activity of the metal oxide nanoparticles against urinary tract infectious bacterial pathogens, *Asian Pac. J. Trop. Dis.*, *2*(2), 85–89.

Ren, G., Hu, D., Cheng, E. W. C., Vargas-Reus, M. A., Reip, P., & Allaker, R. P., (2009). Characterisation of copper oxide nanoparticles or antimicrobial applications, *Int. J. Antimicr. Agent, 33*(6), 587–590.

Roucoux, A., Schulz, J., & Patin, H., (2002). Reduced transition metal colloids: A novel family of reusable catalysts. *Chem. Rev., 102*(10), 3757–3778.

Ruparelia J. P., Chatterjee, A. K., Duttagupta, S. P., & Mukherji, S., (2008). Strain specificity in antimicrobial activity of silver and copper nanoparticles, *Acta. Biomaterialia, 4*(3), 707–716.

Ruth, D., & Lorella, I., (2005). Dendrimer biocompatibility and toxicity. *Adv. Drug Deliv. Rev., 57*, 2215–2237.

Schaaff, T. G., & Whetten, R. L., (2000). Giant gold-glutathione cluster compounds: intense optical activity inn metal based transitions. *J. Phys. Chem. B., 104*(12), 2630–2641.

Schonenberger, C., Van der Zande, B. M. I., Fokkink, L. G. J., Henny, M., Schmid, C., Kruger, M., Bachtold, A., Huber, R., Birk, H., & Staufer, U., (1997). Template synthesis of nanowires in porous polycarbonate membranes: Electrochemistry and morphology. *J. Phys. Chem. B., 101*, 5497–5505.

Seabrook, T. J., Jiang, L., Thomas, K., & Lemere, C. A., (2006). Boosting with intranasal dendrimeric Aβ1–15 but not Aβ1–15 peptide leads to an effective immune response following a single injection of Aβ1–40/42 in APP-tg mice, *J. Neuroinflammation, 3*, 1–10.

Shelimov, K. B., & Moskovits, M., (1999). Composite nanostructures based on template-grown boron nitride nanotubules. *Chem. Mater., 12*, 250–254.

Shemer, G., Shemer, O., Krichevski, G., Molotsky, T., Lbitz, I., & Kotlyar, A. B., (2006). Chirality of silver nanoparticles synthesized on DNA. *J. Am. Chem. Soc., 128*, 11006–11007.

Shulga, O., & Kirchhoff, J. R., (2007). An acetylcholinesterase enzyme electrode stabilized by an electrodeposited gold nanoparticle layer. *Electrochem. Commun., 9*, 935–940.

Sigal, G. B., Mammen, M., Dahmann, G, Whitesides, G. M., (1996). Polyacrylamides bearing pendant-sialoside groups strongly inhibit agglutination of erythrocytes by influenza virus: The strong inhibition reflects enhanced binding through cooperative polyvalent interactions. *J. Am. Chem. Soc., 118*, 3789–3800.

Sikwal, D., Kalhapure, R., Jadhav, M., Rambharose, S., Mocktar, C., & Govender, T., (2017). Non-ionic Self-assembling amphiphilic polyester dendrimers as new drug delivery excipients. *RSC Adv., 7*, 14233–14246.

Sun, M., Fan, A., Wang, Z., & Zhao, Y., (2012). Dendrimer-mediated drug delivery to the skin, *Soft Matter, 8*, 4301–4305.

Svenson, S., & Tomalia, D. A., (2005). Dendrimers in biomedical applications – reflections on the field. *Adv. Drug Deliv. Rev., 57*, 2106–2129.

Tajarobi, F., El-Sayed, M., Rege, B. D., Polli, J. E., & Ghandehari, H., (2001). Transport of poly amidoamine dendrimers across Madin-Darby canine kidney cells, *Int. J. Pharm., 215*(1–2), 263–267.

Tathagata, D., Minakshi, G., & Jain, N. K., (2008). Poly (propyleneimine) dendrimer and dendrosome based genetic immunization against hepatitis B. *Vaccine, 26*(27–28), 3389–3394.

Thomas, T. P., Patri, A. K., Myc, A., Myaing, M. T., Ye, J. Y., Morris, T. B., & Baker, J. R., (2004). *In vitro* targeting of synthesized antibody-conjugated dendrimer nanoparticles. *Biomacromolecules, 5*, 2269–2274.

Thorn, C. F., Oshiro, C., Marsh, S., Hernandez-Boussard, T., McLeod, H., Klein, T. E., & Altman, R. B., (2011). Doxorubicin pathways: pharmacodynamics and adverse effects. Pharmacogenet. *Genomics.*, *21*(7), 440.

Tomalia, D. A., (1995). Dendrimer molecules, *Sci. Am.*, *272*, 62–66.

Tomalia, D. A., (2004). Birth of a new macromolecular architecture: dendrimers as quantized building blocks for nanoscale synthetic organic chemistry, *Aldrichimica. Acta.*, *37*, 39–57.

Tomalia, D. A., Baker, H., Dewald, J., Hall, M., Kallos, G., Martin, S., Roeck, J., Ryder, J., & Smith, P., (1985). A new class of polymers: starburst-dendritic macromolecules. *Polym. J.*, *17*, 117–132.

Tomalia, D. A., Baker, H., Dewald, J., Hall, M., Kallos, G., Martin, S., Roeck, J., Ryder, J., & Smith, P., (1986). Dendritic macromolecules: synthesis of starburst dendrimers, *Macromolecules.*, *19*, 2466–2468.

Tomalia, D. A., Hedstrand, D. M., & Ferritto, M. S., (1991). Comb-burst dendrimer topology: New macromolecular architecture derived from dendritic grafting. *Macromolecules*, *24*, 1435.

Tomalia, D. A., Naylor, A. M., & Goddard, W. A., (1990). III. Starburst dendrimers: Molecular-level control of size, shape, surface chemistry, topology, and flexibility from atoms to macroscopic Matter. *Angew. Chem., Int. Ed. Engl.*, *29*, 138–175.

Tomalia, D. A., Naylor, A. M., & Goddard, W. A., (1990). Starburst dendrimers: Molecular level control of size, shape, surface chemistry topology and flexibility from atoms to macroscopic matter. *Angew. Chem., Int. Ed. Engl.*, *29*, 138–175.

Triesscheijn, M., Baas, P., Schellens, J. H., & Stewart, F. A., (2006). Photodynamic therapy in oncology. *Oncologist*, *11*, 1034–1044.

Undre, S. B., Pandya, S. R., Kumar, V., & Singh, M., (2016). Dendrimers as smart materials for developing the various applications in the field of biomedical sciences, *Adv. Mater. Lett.*, *7*(7), 502–516.

Vandamme, T. H. F., & Vandamme, L. B., (2005). Poly(amidoamine) dendrimers as ophthalmic vehicles for ocular delivery of pilocarpine nitrate and tropicamide, *J. Control. Release*, *102*, 23–38.

Velazquez, A. J., Carnahan, M. A., Kristinsson, J., Stinnett, S., Grinstaff, M. W., & Kim, T., (2004). New dendritic adhesives for sutureless ophthalmic surgeries: *in vitro* studies of corneal laceration repair. *Arch. Ophthalmol.*, *122*, 867–870.

Wathier, M., Jung, P. J., Carnahan, M. A., Kim, T., & Grinstaff, M. W., (2004). Dendritic macromers as in situpolymerizing biomaterials for securing cataract incisions. *J. Am. Chem. Soc.*, *126*, 12744–12745.

Willner, I., Baron, R., & Willner, B., (2007). Integrated nanoparticle–biomolecule systems for biosensing and bioelectronics. *Biosens. Bioelectron.*, *22*, 1841–1852.

Wrobel, D., Marcinkowska, M., Janaszewska, A., Appelhans, D., Voit, B., Klajnert-Maculewicz, B., Bryszewska, M., Stofik, M., Herma, R., Duchnowicz, P., et al., (2017). Influence of core and maltose surface modification of peis on their interaction with plasma proteins—human serum albumin and lysozyme. *Colloids Surf. B. Biointerfaces.*, *152*, 18–28.

Wu, G., Barth, R. F., Yang, W., et al., (2004). Site-specific conjugation of boron-containing dendrimers to anti-EGF receptor monoclonal antibody cetuximab (IMC-C225) and its evaluation as a potential delivery agent for neutron capture therapy. *Bioconjug. Chem.*, *15*(1), 185–194.

Xiangyang, Shi., Kai, Sun., Lajos, P. B., & James, R. B. Jr, (2006). Synthesis, characterization, and manipulation of dendrimer-stabilized iron sulfide nanoparticles, *Nanotechnology, 17*, 4554–4560.

Yang, H., & Kao, W. J., (2006). Dendrimers for pharmaceutical and biomedical application. *J. Biomater Sci. Polym. Ed., 17*, 3–19.

Yang, T. Z., Hussain, A., Bai, S. H., Khalil, I. A., Harashima, H., & Ahsan, F., (2006). Positively charged polyethylenimines enhance nasal absorption of the negatively charge drug, low molecular weight heparin. *J. Control. Release, 115*, 289–297.

Yavuz, B., Bozdag, P. S., & Unlu, N., (2013). Dendrimeric systems and their applications in ocular drug delivery. *Sci. World J., 732340*, 1–13.

Zhang, G. D., Harada, A., Nishiyama, N., Jiang, D. L., Koyama, H., Aida, T., & Kataoka, K., (2003). Polyion complex micelles entrapping cationic dendrimer porphyrin: effective photosensitizer for photodynamic therapy of cancer. *J. Control Release, 93*, 141–150.

Zhang, H., Huang. X., Sun, Y., Lu, G., Wang, K., Wang, Z., Xing, J., & Gao, Y., (2015). Studied improvement of pulmonary absorption of poorly absorbable macromolecules by hydroxypropyl-b-cyclodextrin grafted polyethylenimine (HP-b-CD-PEI) in rats. *Int. J. Pharm., 489*, 294–303.

Zhang, Y., Peng, H., Huang, W., Zhou, Y., & Yan, D., (2008). Facile preparation and characterization of highly antimicrobial colloid Ag or Au nanoparticles, *J. Colloid and Interface Sci., 325*(2), 371–376.

Zhao, M., & Crooks, R. M., (1999). Homogeneous hydrogenation catalysis using monodisperse, dendrimer-encapsulated Pd and Pt nanoparticles. *Angew. Chem., Int. Ed., 38*, 364–366.

Zhong, Q., & Da Rocha, S. R. P., (2016). Poly(amidoamine) dendrimer doxorubicin conjugates: *in vitro* characteristics and pseudosolution formulation in pressurized metered-dose inhalers. *Mol. Pharm., 13*(3), 1058–1072.

Zimselmeyer, B. H., Mackay, S. P., Schatzlein, A. G., & Uchegbu, I. F., (2002). The lower-generation polypropylenimine dendrimers are effective gene-transfer agents. *Pharm. Res., 19*(7), 960–967.

CHAPTER 2

DENDRIMERS: A TOOL FOR ADVANCED DRUG DELIVERY

PATRÍCIA MAZUREKI CAMPOS,[1] JULIANA PALMA ABRIATA,[1]
ABAYOMI TOLULOPE OGUNJIMI,[1]
JULIANA MALDONADO MARCHETTI,[1] and ROBERT LEE[2]

[1]*School of Pharmaceutical Sciences of Ribeirão Preto, University of São Paulo, Ribeirão Preto, São Paulo, Brazil*

[2]*Ohio State University, Columbus, Ohio, USA,*
E-mail: lee.1339@osu.edu

CONTENTS

ABSTRACT

Dendrimers are nanosized polymeric molecules, symmetrically organized around the core, with hyperbranched arms as a tree. They vary in diameter and shape as a function of their generations as well as their surface end groups, which characterize their polarity and determine the linkage type and their biological behavior. Dendrimers have been applied in several biological areas including diagnostic uses, antineoplastic drug delivery, topical and transdermal application, and gene therapy. To improve efficacy, surface modifications have been tried and have shown better biological and physicochemical interactions; likewise, the toxicological aspects could be attenuated and/or tailored with the dendrimers, showing their versatility. Their uses as nanoplatforms to carry molecules and nanoparticles may be conjugated by covalent bonds or complexed to the carried materials in order to maintain the controlled release for passive or active targeting.

This chapter intends to highlight the main issues and major contributions of dendrimer nanocarriers in the biomedical area with a view to encourage researchers to tap into the versatility of this promising drug delivery system and further engender the entrance of more dendrimer-based biomedical products into the market.

2.1 INTRODUCTION

Dendrimers are a new class of synthetic polymers, arranged in a tree-like form, with ordered symmetric molecules around a nucleus; they consist of branches or tree-arms surrounding a core. They are macromolecules with outstanding features of modern drug delivery systems, in which their scaffolds can carry loadings of different molecular weights and physicochemical characteristics such as drugs; macromolecules like enzymes, proteins, antibodies; and nucleic acids like DNA and RNA (Jain et al., 2010; Abbasi et al., 2014). These tridimensional structures were independently developed by two groups: Newkome et al. (1985) and Tomalia et al. (1985). They are made up of several units holding a rich architecture that can assemble different surface functionalities (Jain et al., 2010). Also, they exist as globular systems with low polydispersity index, with polymeric radial arrangements of the repeated units arising from a central spot. The term Dendrimer comes from the Greek words *dendron* and *meros*, meaning tree/branch and part/portion, respectively (Frechet, 2002).

Due to their remarkable features, dendrimers have been applied to several areas such as spectroscopy and electronic and molecular devices by integration of strength and miscellaneous chemical synthesis (Moorefield et al., 2012). They have been used in the electrochemistry of metallic nanoparticles in films (Losada et al., 2017), mass biosensors for quantification of intracellular microRNA (Guo et al., 2016), and incorporated into organoclays for adsorption of NH_3 and CO_2 as green technology to prevent air pollution (Shah et al., 2017) among others. In fact, the massive exploration of dendrimers has been found in biomedical applications, for example, they have been used as carriers for neuroblastoma cells (Yesil-Celiktas et al., 2017), combined drug transport for antihypertensive therapy (Singh et al., 2017), by functionalization with adhesive peptides to enhance cell responses for tissue engineering (Maturavongsadit et al., 2016), and in pulmonary delivery of antibiotics (Rajabnezhad et al., 2016).

Dendrimer technology is highly expensive, being limited to drugs considered to possess demanding development or to critical health issues. Based on available literature, their applications have also been directed toward solving elementary goals such as solubility, tuning, and innovations in the therapeutic fields as delivery systems (Stieger et al., 2012). The important considerations and topics related to dendrimers as well as an overview of dendrimers are dealt with in this chapter.

2.2 STRUCTURAL ASPECTS AND PROPERTIES

2.2.1 STRUCTURAL FEATURES AND PHYSICOCHEMICAL PROPERTIES

Dendrimers are highly defined polymers that in low generations are in open display as a circle with greater asymmetry, but when the generations increase above four, they grow in diameter tridimensionally, forming a globular symmetry. This globular form is seen as "nanoscale atoms" with the ability to congregate and create molecules (Selin et al., 2016). The macromolecular structure of the dendrimers has a well-defined size, is monodisperse, is flexible, and possesses a multivalent molecular surface (Klajnert and Bryszewska, 2001; Rodriguez-Prieto et al., 2016; Szulc et al., 2016; Gorzkiewicz and Klajnert-Maculewicz, 2017). The monodispersibility property occurs due to a well-defined molecular structure, often absent in other polymers (Nanjwade et al., 2009).

Their structures are assembled layer-by-layer around a central core in consecutive steps, which repeat twice at a time, making one generation called "G" (Deloncle and Caminade, 2010). The dendrimer structure consists of three parts (Figure 2.1): central core, branches, and functional end groups on the surface. These three parts are important for dendrimer performance and can affect the size, shape, and its properties. Dendrimers begin with a central atom or atom groups named the core and the growth occurs through its "dendrons," which are the branches of other atoms, arising from the central atom by several chemical reactions. The core is usually named "generation 0," and the "dendrons" added are named "generation 1st, 2nd, 3th, 4th," respectively, as a new layer creates a new generation with double functional end groups (Nanjwade et al., 2009; Abbasi et al., 2014; Gorzkiewicz and Klajnert-Maculewicz, 2017). The hyperbranched structure of dendrimers ensures a polyvalence surface, which can be modified with different functional end groups (Lombardo et al., 2016).

When the building blocks are assembled around the dendrimer core structure, it creates several possibilities to accommodate guest small molecules,

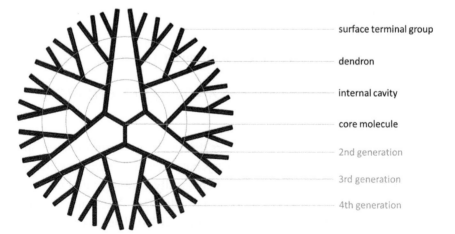

FIGURE 2.1   Representation of molecule structure of dendrimer (Reprinted with permission from Gorzkiewicz, M., & Klajnert-Maculewicz, B., (2017). Dendrimers as nanocarriers for nucleoside analogues. Eur. J. Pharm. Biopharm., 114, 43–56. © 2017 Elsevier.).

which could be within the interior protected by the extensive branches, near the core or in the voids of dendrimers, or on the multivalent surface, which allow interaction with molecules of several molecular weights. In addition, the dendrimer surface responds to the external environment through the end groups, which also determines the dendrimer characteristics (Noriega-Luna et al., 2014). Dendrimers provide multivalent interactions due to their peculiar structural organization, chemical composition, and monodispersibility, with reproducibility even at high scales (Kesharwani et al., 2014).

Dendrimers of high generations are larger in size and rich in peripheral groups compared with dendrimers of low generations (Table 2.1). They have sizes around 10 nm or less, with shapes similar to those of proteins and biomolecules, a unique property that allows them to fit and work easily within the biological environments (Lee et al., 2005). Each generation added to a dendrimer core increases its size, molecular weight, and surface functionalities. With continuous dendrimer growth, the outer shell becomes overwhelmed and dendrons tend to fold forward into the interior, limiting the generation number and consequently leaving the core with low molecular density. Thus, the ability of surface groups to make bonds such as hydrogen bonds and ion pairing will determine the properties of higher generations and a formation of a dense outer shell on the dendrimers ("dendritic box"),

TABLE 2.1 Physical Characteristics of PAMAM Dendrimers

Generation	Number of surface groups	Molecular weight (Da)[a]	Diameter (nm)[b]
0	4	517	1.5
1	8	1,430	2.2
2	16	3,256	2.9
3	32	6,909	3.6
4	64	14,215	4.5
5	128	28,826	5.4
6	256	58,048	6.7
7	512	116,493	8.1
8	1024	233,383	9.7
9	2048	467,162	11.4
10	4096	934,720	13.5

[a] Molecular weight is based on detect-free, ideal-structure, amine terminated dendrimers.

[b] Molecular dimensions determined by exclusion chromatography by size-exclusion chromatography.

preventing backfolding of end groups, a scenario that is fundamental to the structural chemistry of dendrimers. Furthermore, this density overlap illustrates the dendrimer flexibility, which is lower in low-generation dendrimers (with lesser amount of end groups).

The monomers used in building up dendrimers can be pure hydrocarbons like alkyl and aromatic moieties, synthetic monomers such as phosphorus, sulfur, and silicon or biomolecules such as carbohydrates, amino acids, and nucleotides, leading to the production of different designs and opportunities for chemical interactions on the dendrimer structure involving the three parts. For this reason, this tridimensional synthetic polymer with low molecular density and arrangement existing as an aggregated supramolecular was analyzed by X-ray and compared to medium-size proteins, particularly the higher generation dendrimers despite dendrimers being considered as less flexible than proteins. Furthermore, dendrimers respond to external stimuli such as solvent, pH, ionic strength, and temperature.

Dendrimers, covered with amine groups in the surface (primary amines) and in the interior (tertiary amines) such as PAMAM and PPI, have a basic characteristic in which they respond to low pH by amine protonation, thereby increasing electrostatic repulsion and extending their molecular conformation. At neutral pH, there is an increase in the backfolding of these dendrimers by the internal hydrogen bonds of amine groups. At high pH, there is a dendrimer contraction, consequently leading to a high degree of backfolding and weak repulsion forces among the dendrons. The pH effect is more pronounced for dendrimers with generations between five and seven as compared to the eighth-generation dendrimer.

The solvent effect describes the ability of solvating the dendrimer, and it is important to the conformational state of dendrimers. Different solvents exert varying effects on the molecular dynamics of dendrimers, being more pronounced with high-quality solvents. In addition, low-generation dendrimers are less solvated than high-generation ones. Other solvent features affecting the dendrimer is solvent polarity, for example, PPI dendrimer is little solvated by benzene (nonpolar solvent) and more solvated by chloroform (weakly acid solvent), because the latter acts as a hydrogen donor with the extension of dendrimer conformation and low backfolding. However, each dendrimer has specific or non-specific interactions, leading to higher or lower solvation, depending on the chemical features of the dendrimers.

In the same way, charged dendrimers are less affected by ionic strength if there is low salt condition; there is no increase in the backfolding of dendrons, thereby keeping its extended conformation. Based on the dendrimer concentration, flexible dendrimers are affected by larger and smaller structures, such as other dendrimers and solvents, respectively.

Dendrimers with amine and oxygen in their structures are endowed with fluorescent property; this occurs in dendritic polymers such as poly (amino ethers), polysiloxanes, poly (propyl ether imine), polyethyleneimines, and fullerodendrons. They have in common tertiary amine branching corresponding to oxygen molecules, which influences the emission of intrinsic fluorescence, that is, the excitation of the lone-pair electrons or formation of peroxyl radical. This specific feature is helpful in dendrimer characterization by image analysis (Boas et al., 2006; Imae, 2012).

2.2.2 PHARMACOKINETICS

The relationship between dendrimers and pharmacokinetics is dependent on physicochemical characteristics of the macromolecules and biological features where the macromolecules will be applied. For intravenous administration of dendrimers, they circulate in the blood stream until they are partitioned into organs and tissues; this transport is driven by the nature of macromolecules and their ability to transpose capillary vessels and reach the tissues to deliver the drugs without causing toxicity. For these steps to occur, dendrimers should have surface end groups and adequate size that improves their pharmacokinetic profile. In consequence, this will address specific dendrimer features pertaining to charge effect and low generations, which will enable to design them for targeting. For example, cationic PAMAM dendrimer (fifth generation) is easily biodistributed and excreted from the body as compared to uncharged PAMAM dendrimer (Kesharwani et al., 2014).

The other aspect is that molecules are transported in the dendrimers, into the voids, near the core, and/or on the surface through noncovalent interactions (van der Waals, hydrophobic, electrostatic interaction, and hydrogen bonding) via complexation or covalent bonds (prodrugs) via conjugation, which can be cleavable (esters, amides, and carbamates) or not. This will determine the release pattern of the loaded molecules from the dendrimers as it is one of the reasons of dendrimer use, that is, for improving drug pharmacokinetics profile and overcoming their physicochemical limitations by facilitating membrane crossing and consequent drug delivery to the target site. Besides the interaction of drugs with dendrimers, dendrimers can improve drug activity and decrease the toxicity. Cleavable linkages on the dendrimer surface are broken under conditions such as acid pH, redox potential, or enzyme-rich areas and are used for active targeting strategies in delivery systems for specific therapies such as chemotherapy in cancer. Covalent binding produces more stable association between drug and dendrimers, but the drug may lose its efficiency and efficacy. Thus, simple entrapment inside the dendrimers is proposed for small drugs through hydrophilic or hydrophobic interactions depending on the characteristics of drugs and chosen dendrimers, which is followed by passive diffusion. This kind of entrapment is associated with a fast release pattern of the guest molecule from the host-dendrimer, but it is slower than free drug (Najlah et al., 2012; Caminade and Turrin, 2014). Thus, for dendrimer development, a complete investigation of its pharmacokinetics is highly necessary so as to have a better choice among the diverse dendrimers in order to obtain satisfactory effects for *in vivo* applications and low toxicity (Abbasi et al., 2014; Caminade and Turrin, 2014).

2.2.3   SOLUBILITY ENHANCER

The solubilization power of dendrimers contributes to their use with poorly soluble drugs, thus widening drug applications to the treatment of cancer; parasitic, fungi, bacterial, and viral infections; and for anti-inflammatory and anti-hypertensive therapies among others (Kesharwani et al., 2014). The presence of functional end groups is responsible for high solubility, miscibility, and reactivity of dendrimers. Dendrimers' solubility is strongly influenced by the nature of these functional end groups. Dendrimers terminated in hydrophilic groups are soluble in polar solvents, while those with hydrophobic end groups are soluble in nonpolar solvents. A large number of functional end groups such as folate, antibodies, dextran, and other molecules are important to promote specific cell targeting, prolong blood half-life, and improve the bioavailability of substances (Klajnert and Bryszewska, 2001; Gorzkiewicz and Klajnert-Maculewicz, 2017). The increased solubility of dendrimers in organic solvents shows their rapid solubilization and easiness for physicochemical characterization purposes (Cheng et al., 2008).

This solubility enhancing property is also a function of the dendrimer structural properties such as size, compositions of core, and internal branching; dendrimer concentration; and environmental parameters such as pH and temperature. As all these features vary among dendrimer types, they influence their ability to form hydrogen bonds and ionic and hydrophobic interactions with loaded molecules, thereby fulfilling their solubility enhancing property (Cheng et al., 2008; Kesharwani et al., 2014).

2.3 SYNTHESIS

Dendrimers are synthesized by two usual methods which are the divergent and convergent growth methods. The dendrimers increase in size through repetitive additions of monomers on a central polyfunctional core and at same time dendrimer generations, until expressive end groups outward on the surface, reach high degree of package. The core reacts with one reactive or two dormant groups (monomers molecules), leading to the formation of the first generation. In the sequence, the peripheral groups are activated for subsequent reactions with more monomers (Abbasi et al., 2014; Kesharwani et al., 2014). The first production method was developed by Tomalia and involved the growth of dendrimers from the nucleus, while the second method was developed by Hawker and Fréchet (Nanjwade et al., 2009). However, other strategies of synthesizing dendrimers have been proffered which include hypercores and branched monomer growth, double exponential growth, lego chemistry, and click chemistry (Kesharwani et al., 2014).

In dendrimer synthesis, the use of intermediates is called half generation. A classic example is PAMAM, which is a half generation if it is complexed with carboxylic acid groups (Tripathy and Das, 2013).

2.3.1 DIVERGENT GROWTH METHOD

In the divergent growth method, dendrimer synthesis starts from a multifunctional core molecule and the arms are added step by step. In this way, the core molecule reacts with a molecule with one or two reactive domains proceeded by surface group activation and are called as first generation of dendrimers. In this method, there are inclusion and removal of protecting groups for the branching sites. Then, new layers are built with more domains and are called second, third and other generations respectively. This approach has been successful for the production of large quantities of dendrimers, but it has a disadvantage as it requires the use of excess reagents in order to prevent side reactions and/or incomplete reaction of the end groups, leading to structure defects. Due to excess reagents, the purification of the final product is always a difficult process. However, this method's advantage is that dendrimers are synthesized according to needs as well as configuration of the surface groups (Abbasi et al., 2014; Kesharwani et al., 2014; Klajnert and Bryszewska, 2001).

2.3.2   CONVERGENT GROWTH METHOD

The divergent method, on the other hand, involves the addition of one functional group type in a two-step process, which starts from the exterior to produce the molecular structure of the dendrimer before coupling with the protected/unprotected arm, which constitute the "dendrons." The second step involves anchoring of several "dendrons" on the divergent core to form the final structure. When the dendrimer grows large enough, it is linked to a core. The resultant dendrimer produced this way is easier to purify as it has few possibilities to have defects when compared to the first method. Furthermore, this method allows the control of dendrimer molecular weight and positioning of groups (Klajnert and Bryszewska, 2001; Abbasi et al., 2014).

2.4   TYPES OF DENDRIMERS

With advance in synthetic chemistry and the search for new smart drug delivery systems, dendrimers for different purposes and with diverse chemical compositions have been developed. Some of the most frequently used dendrimers are presented and discussed in this chapter; however, there exist a rich diversity of dendrimers such as polyether dendrimers, polyester dendrimers, or hybrid dendrimers whose cores and branches are made up of different chemical compositions and have been applied in the treatment of several diseases.

2.4.1   POLYAMIDOAMINE (PAMAM) DENDRIMERS

These are the most-studied type of dendrimers currently in science. Their synthesis occurs through the divergent method, and they possess ethylenediamine or ammonia as core. PAMAM is a non-immunogenic dendrimer, water soluble, and has polyamide branches with tertiary amines as focal points, which can attach to several molecules to address specific targets. They are commercially available up to the tenth generation with low polydispersibility; however, they possess concentration-dependent toxicity and can cause hemolysis (Najwade et al., 2009; Kersharwani et al., 2014).

Due to the diversity of PAMAM dendrimers, they can be linked to different surface groups like alkyl chains, oligomers, polymers, enzymes, and proteins, among others, which modify their physicochemical properties such as solubility and surface charge (Imae, 2012). They have been successfully linked to DNA molecules, followed by transfection due to their positively charged surface (Dufès et al., 2005).

Nowacka et al. (2015) studied three types of PAMAM dendrimers (at generations 2.5 and 3) having the surface groups of –COOH, –NH$_2$, and –OH. All these types have ethylenediamine core. They investigated the interaction between bovine insulin molecule and the three dendrimers. They showed that all interactions were electrostatic, with the strongest interaction occurring between positively charged dendrimer and insulin (negative charge at neutral pH). They also showed that further secondary interactions took place between the molecules beyond the positive portions of insulin interacting with anionic dendrimers.

2.4.2 POLY (PROPYLENEIMINE) (PPI) AND POLY (ETHYLENE IMINE) (PEI) DENDRIMERS

The PPI dendrimers are biocompatible amine-terminated hyperbranched macromolecules and are synthesized by the divergent method. They could have primary or secondary amine as core, and likewise, 1,4-diaminobutane. PPI dendrimer contains two types of nitrogen atoms from primary and tertiary amines. They are synthesized by the repetitive Michael addition reaction of acrylonitrile to primary amines followed by hydrogenation of nitriles (heterogeneously catalysis). This type of dendrimer has polyalkylamines as end groups and possesses, in their spaces, numerous tertiary tris-propylene amines. PPI dendrimers are commercially available up to fifth generation (Imae, 2012; Kesharwani et al., 2014).

The PEI dendrimers which are different from the PPIs have an ethylenediamine core, which is a homologue of the PPI. PEI dendrimers are synthesized by alkylation through Michael addition and Gabriel amine reaction, and they possess amine surface groups. PEI dendrimers have alkyl chains as spacer groups, thus making them more hydrophobic than PPI dendrimers (Imae, 2012).

2.4.3 PEPTIDE DENDRIMERS

Other biocompatible dendrimers are the peptide dendrimers that have amino acid sequences on the surface of usual dendrimers or incorporates amino acids into their branching or core during their synthesis, characterizing them as macromolecules with peptide bonds in their structure. Because of their composition and easiness to be internalized by the body, they have been used in different biological areas such as in the treatment of cancer, pain, allergy, and central nervous system disorders in addition to being antimicrobial and antiviral drugs. Furthermore, peptide dendrimers have been applied as a diagnostic tool when conjugated with contrast agents (Nanjwade et al., 2009). Poly (L-lysine) from hexamethylenediamine core and poly (L-ornithine) from an ethylenediamine core are examples of peptide dendrimers and can be synthesized by convergent and divergent methods (Imae, 2012).

2.4.4 TRIAZINE DENDRIMERS

Triazine dendrimers are very often applied to gene therapy to deliver RNA molecules and oligonucleotides. Their synthesis is based on sequential substitution of the trichlorotriazine moiety by amine nucleophiles. Triazine dendrimers reactivity is dependent on the mono, di, and/or trichlorotriazines moieties and possesses selective synthesis process with high yield due to reaction conditions such as temperature, time, and nucleophiles used (Lim and Simanek, 2012). Triazine dendrimer has its surface covered by 1,4-diazabicyclo[2.2.2]octane through hydrophilic linkers, with subsequent substitution by methyl, benzyl, and dodecyl groups. These dendrimers have been tested in antimicrobial assay, in which the dodecyl group exerted higher antimicrobial activity against *Staphylococcus aureus* and *Escherichia coli* (Sreeperumbuduru et al., 2016).

2.4.5 OTHER TYPES OF DENDRIMERS

There are other types of dendrimers like liquid crystalline, tecto, chiral, and PAMAMOS dendrimers. Liquid crystalline dendrimers consist of mesogenic monomers (36 units) functionalized with carbosilane dendrimers. The functionalization leads to the formation of liquid crystalline phases based on the added mesogenic monomers. Currently, they have been applied in gene and anti-HIV therapies (Nanjwade et al., 2009; Kesharwani et al., 2014).

Tecto dendrimers, also called core shell dendrimers, have a highly organized polymeric structure. They are synthesized by controlled covalent attachment of dendrimers with a dendrimer in the core. When a therapeutic molecule is vehiculated, it may or may not be in the structure of dendrimer, which is generally organized layer-by-layer. This organization can be designed by different types of dendrimers, with each one being added for a specific purpose or function (Nanjwade et al., 2009). They have been used with fifth generation PAMAM dendrimer as a core followed by the addition of 2.5th generation PAMAM dendrimers (carboxyl surface groups) in melanoma cell studies as nanoplatform for drug delivery (Kesharwani et al., 2014).

Chiral dendrimers are characterized by their chirality, which is present in the branches and core, constructed with chemically similar dendrimers in a defined stereochemistry. They have been applied in symmetric catalysis and chiral molecular recognition (Nanjwade et al., 2009).

Poly(amidoamine-organosilicon) dendrimers (PAMAMOS dendrimers) are inverted unimolecular micelles with the composition of hydrophilic and nucleophilic polyamidoamine (PAMAM) localized in their interior part, recovered by hydrophobic organosilicon (OS) and organized layer-by-layer around a core. In drug delivery applications, they possess interesting features for electronics and photonics purposes with their ability to precisely encapsulate several agents (Kesharwani et al., 2014).

2.5 CHARACTERIZATION OF DENDRIMERS

The nanoscaled dendrimers have well-defined organizational structure, morphology, and chemical composition; thus, their characterization is of great importance. The analytical methods applied in the characterization of these dendrimers are described above (Table 2.2). Besides the specific methods of

TABLE 2.2 Characterization Methods Used for Dendrimers

Analytical methods	Characterization parameter
Nuclear magnetic resonance (NMR)	It helps in determining chemical transformation undergone by end groups and hence applicable to structural analysis of dendrimers and step-by-step characterization of synthesis
Infrared spectroscopy and Raman spectroscopy	It ascertains the chemical transformation taking place during the synthesis or surface engineering of dendrimers
UV–visible spectroscopy	It helps in determining the change in chemical structure and synthesis method by detecting chromophores and auxochromes. Also used to test the purity of dendrimers
Fluorescence	It is used to characterize the structure and synthesis of dendrimers having photochemical groups and to quantify defects occurred during the synthesis
Circular dichroism	Characterization of structure of dendrimers having optical activity
Atomic force microscopy	Size, shape and structure
Transmission electron microscopy	
Electron paramagnetic resonance	Surface structure
X-ray diffraction	Chemical composition, size and shape
X-ray photoelectron spectroscopy	Chemical composition and size
Electrochemistry	It gives information about the structure of dendrimers
Electrophoresis	Purity and homogeneity of water-soluble dendrimers
Small-angle X-ray scattering (SAXS)	It gives average radius of gyration (Rg) in solution hence used for determination of average particle size, shape, distribution, and surface-to-volume ratio

TABLE 2.2 (Continued)

Analytical methods	Characterization parameter
Small-angle neutron scattering (SANS)	It gives average radius of gyration (Rg) in solution as well as detailed information about the internal structure of entire dendrimer
Laser light scattering (LLS)	Hydrodynamic radius of dendrimers
Mass spectrometry (FAB-MS, ESI-MS, FT-ICR MS, MALDI-TOF MS)	Determination of molecular mass and some structure information
Size exclusion (or Gel permeation) chromatography (SEC) (GPC)	Molecular weight and size
Intrinsic viscosity	Physical characterization and morphological structure
Differential scanning calorimetry (DSC)	Glass transition temperature (Tg), which is affected by the molecular weight, entanglement and chain-end composition of polymers
Dielectric spectroscopy	Study of molecular dynamics

(Reprinted with permission from Kesharwani, P., Jain, K., Jain, N. K. (2014). Dendrimer as nanocarrier for drug delivery. Progr Polym Sci, 39, 268–307. © 2014 Elsevier.)

dendrimers analysis, cytotoxicity and cytometry are performed to evaluate the response of cells and their cellular uptake. Cell apoptosis assay is a test that has been used to verify the cell membrane integrity after drug treatment and conducted using propidium iodide (PI) dye (Szulc et al., 2016).

Infrared spectroscopy is an analytical method important for dendrimer analysis as it identifies different functional groups present on the surface of dendrimers. It is important to determine the drug–dendrimer interactions through the comparison of the spectra of the dendrimers and drug (Gautam et al., 2012; Zain-ul-Abdin et al., 2017). Mainly, the chemical transformation on the dendrimer surface is manifested by absence of nitrile and aldehydes groups during PPI and phosphorous synthesis (Caminade et al., 2005).

Nuclear magnetic resonance spectroscopy analyzes and identifies the different groups present on the surface of dendrimers. It can determine the structure of molecules in solution and also confirm the different dendrimer generations (Zain-ul-Abdin et al., 2017). Several authors have used this method to detect degradation and to identify specific groups on dendrimer surface (Rodriguez-Pietro et al., 2016). Among the NMR analyzes for organic and heteroatom-containing dendrimers, the ^{1}H and ^{13}C types are very often used (Caminade et al., 2005).

Mass spectrometry is used to determine the mass of dendrimers and has also been employed to confirm the final product in the synthesis pathways of dendrimers. Among the mass spectrometry techniques available, the MALDI-TOF mass spectrometry (MALDI-TOF MS) is the most used method as it gives the average mass of the systems through the determination of the peaks of parent molecular ion. This method is applicable to high-molecular-weight dendrimers and has been used to determine the purity and polydispersibility of dendrimers (Kesharwani et al., 2014). As an application to define the fragmentation pattern of dendrimers, Gautam et al. (2012) studied the fragmentation pattern of different generations of PAMAM dendrimers. They discovered five common MALDI matrices: 2, 5-dihydroxybenzoic acid (DHB), 4- hydroxy-3-methoxycinnamic acid (FER), a-cyano-4- hydroxycinnamic acid (ACH), 2, 4, 6-trihydroxyacetophenone (THAP), and 3-hydroxypicolinic acid (HPA).

2.6   TOXICOLOGICAL ASPECTS

The unique characteristics of dendrimers are related to their ability to bind several types of molecules on the branched arms or surface groups making them a potential system, and, at the same time, with enhanced biological properties and strong influence on their toxicological aspects. The degree of biocompatibility depends on the compounds used in the synthesis of dendrimers, which generates the nontoxicity, non-immunogenicity, and biodegradability. For a static dendrimer, the dendrimetric structures are created using units of amide and alkyl or amino acids, sugar spacers, and amine branching. Dynamic dendrimer concept is how the structures with their adsorption and aggregation behave when used as drug delivery system (Imae, 2012). The other manner of defining biocompatibility is the ability of producing responses at the molecular, cellular, organ/tissue, and body levels and their correlation with DNA and proteins, cytotoxicity, organ damage, and systemic side effects. Furthermore, rapid clearance from the body and biodistribution can also confirm low biocompatibility (Li, 2012).

In particular, chemical composition exerts strong influence on the toxicity profile of dendrimers. PPI dendrimers are considered more cytotoxic than PAMAM dendrimers at the same generation and amount of surface groups, even as phosphorous and polyester dendrimers are less cytotoxic than PPI and PAMAM dendrimers (Caminade and Majoral, 2005; Lee et al., 2005; Morgan et al., 2006; Jain et al., 2010).

Moreover, toxicological issues are connected with physicochemical characteristics such as molecular size and surface charge. The amine groups on the surface confer positive charges, in which cationic dendrimers strongly interact with negative charges from cell surface and DNA molecules and chelate metal ions (Li, 2012). In this way, dendrimers toxicity is more attributed to cationic dendrimers than to anionic and neutral dendrimers, which are considered biocompatible as they do not interact with biological environment. Anionic and neutral dendrimers covering are, in general, of carboxylic acid and hydroxyl terminated PAMAM (Padilla de Jésus et al., 2002).

Regardless of candidates that reach clinical application, the cationic charge of dendrimers can limit their use. These include those that have amine groups or multiple cationic charge as PPI, PAMAM, and PLL (Jain et al., 2010). However, the toxicity elicited by cationic molecules is not only restricted to dendrimers (Hong et al., 2006).

One way to study the effect of peripheral charge of dendrimers with lipid bilayer of cells is through the affinity with red blood cells, as they demonstrate surface toxicity related to hemolytic activity due to cell membrane composition, which is rich in glycolipids and glycoproteins (negative charge). With higher dendrimer generation and consequent larger branches and terminal surface groups, more end groups will be available for interactions (Ziemba et al., 2012; Kesharwani et al., 2014). Studies with red blood cells were made with full (G1-G4) and half (G1.5-9.5) generations of PAMAM, with amino terminal and carboxylic terminal groups, respectively. All cationic dendrimers exerted hemolytic activity above 1 mg/mL of concentration and anionic dendrimers confirmed their less toxicity, even up to concentration of 2 mg/mL or after 24 hours of incubation (Malik et al., 2000).

In addition to molecular size, one study with different generations of PAMAM dendrimers (G3.0, 3.5, 4.0, 4.5, and 5.0) and PPI dendrimers (G3.0, 4.0, and 5.0) demonstrated the *in vivo* toxicity in zebrafish embryos and cytotoxicity in human cell lines. The 96-h treatment in zebrafish assay had toxicity over time for cationic PAMAM and PPI dendrimers, with the toxicity effects related to mortality and reduction of heartbeat and blood circulation. Apoptosis in zebrafish embryos was correlated in a concentration-dependent manner. Cytotoxicity in hepatocarcinoma and prostate carcinoma cells increased with increasing generations of cationic PAMAM and PPI dendrimers. These toxicity and cytotoxicity thresholds were not detected for anionic PAMAM (G3.5 and 4.5) (Bodewein et al., 2016). This fish embryo toxicity test is considered as an assay for acute results and offer possibilities of elucidating systemic and specific effects (Sträle et al., 2012).

Another comparative study considered PAMAM and thiophosphoryl dendrimers using the same acute model to evaluate the toxicity. The toxicity due to the increase in dendrimer generations, subsequent size increase, and number of functional groups at the surface was determined as these three features are interconnected. PAMAM dendrimer was relatively more toxic than thiophosphoryl dendrimer, showing evidence of surface covering linked to dendrimer generation. Furthermore, the toxicity increased as the generation of PAMAM increased, while the same effect did not occur with the thiophosphoryl dendrimer, which showed little toxicity for the tested concentrations (0.016–250 ppm) at all generations (Pryor et al., 2014). It has been described that higher generation of dendrimers can create nanoscale holes in the cell membrane, while on the other hand, it has also been found that higher dendrimer generations have different conformations, which could reduce toxicity (Jevprasesphant et al., 2003; Heiden et al., 2007; Naha et al., 2010; Naha and Byrne, 2013).

An acute oral toxicity study with PAMAM dendrimers evaluated the function of size and charge on immunocompetent CD-1 mice model. The maximum tolerated dose (MTD) was determined, and the animals were monitored for clinical signs of toxicity. Again, cationic dendrimers were more toxic than anionic counterparts, and the former had 10 times higher doses. The acute oral toxicity study of G3.5-COOH, G4.0-NH$_2$, G4.0-OH, and G6.5-COOH showed that they are safe with MTD up to 500 mg/kg compared with G7.0-NH$_2$, G7.0-OH, which displayed toxicity signs at low MTD (30–200 mg/kg) (Thiagarajan et al., 2013). These aspects of toxicity may be related to the ability of dendrimers in opening the tight junctions of intestinal epithelium or to be easily internalized by endocytic manner (Kitchens et al., 2008; Goldberg et al., 2010).

2.7 THERAPEUTIC APPLICATIONS OF DENDRIMERS

Through different chemical interactions between dendrimers and host molecules both in the peripheral and/or internal sites and diversity of dendrimer sizes, dendrimers have several applications in the biomedical field. Dendrimers can encapsulate proteins, enzymes, diagnostic agents, or therapeutic drugs through hydrophilic, hydrophobic, ionic, or nonionic interactions, and/or chemical bonding (prodrug) (Imae, 2012). In general, hydrophobic interactions occur in the interior, while hydrogen bonding and ionic interactions occur on the surface of dendrimers (Jain et al., 2010).

As delivery systems, they aid better solubility and bioavailability of drugs, favoring their delivery at specific sites in a controlled release pattern. In fact, they have been used as modifiers of physicochemical properties of drugs for various purposes (Stieger et al., 2012). The host molecules can be carried, conjugated, or complexed into dendrimers, further showing their versatility as nanocarriers (Figure 2.2).

Dendrimers are also interesting due to their high drug-loading capacity and precise physicochemical features such as size, shape, and defined surface groups (Najlah et al., 2012). Due to the versatility of their branched arms and high molecular weight, they can mimic the globular proteins of similar sizes. However, they are more robust, possess voids to encapsulate molecules, and are resistant to denaturing stress such as temperature (Noriega-Luna et al., 2014). In this context, they have been applied in several areas such as to mimic the surface of endostatin, a protein involved in

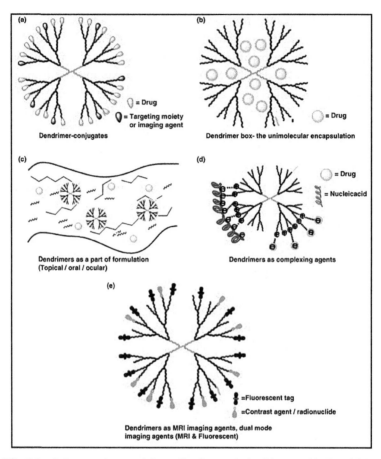

FIGURE 2.2 Scheme of potential applications of dendrimers. (a) Dendrimer drug conjugates, dendrimers linked to targeting moieties and imaging agents. (b) Encapsulation of the drugs in the dendritic interiors. (c) Dendrimers incorporated into various delivery systems for enhancing permeation, solubility, and so on. (d) Dendrimers as complexing agents. (e) Dendrimers as carriers for MRI and fluorescent imaging (Reprinted with permission from Menjoge, A. R., Kannan, R. M., & Tomalia, D. A., (2010). Dendrimer-based drug and imagingconjugates: Design considerations for nanomedical applications. Drug Discovery Today, 15, 171–185. © 2010 Elsevier.).

angiogenesis inhibition; to modulate the crystallization in dental enamel though biomineralization with dendrimers caped with carboxylic acids (Kasai et al., 2002; Chen et al., 2003; Noriega-Luna et al., 2014). Considering the wide diversity of therapeutic applications, more examples (Table 2.3) and specific uses of dendrimers are described in the following sections of this chapter.

TABLE 2.3 Therapeutic Applications of Dendrimers

Aim	Dendrimer Type	Carried Molecule	Finding	References
Enhance solubility	PAMAM dendrimers with amine terminal groups (G3), hydroxyl terminal groups (G3) and carboxylated terminal groups (G2.5, G3.5)	Albendazole	Improvement of aqueous solubility of anthelmintic drug by lipophilic and hydrogen bonds, contributed to the guest-host association with G3 dendrimers	Fernández et al, 2011
Increase bioavailability	PAMAM dendrimer with amine terminal groups (G3)	Doxorubicin	Transport efficiency of the drug-dendrimer complex across mucosal of rat intestine after oral administration with bioavailability 200-fold higher than free drug	Ke et al, 2008
Enhance cell uptake	Poly(glycerol-succinicacid) dendrimer with hydroxyl terminal groups (G4) and carboxylated terminal groups (G4.5)	Camptothecins	Increased cellular uptake for MCF-7 breast adeno-carcinoma cells in 16-fold and cell retention for G4.5 dendrimer	Morgan et al, 2006
Immuno-therapy	Poly(phosphorhydrazone) den-drimers capped with amino-bis(methylene phosphonate) end groups (G4, G5 e G6)	-	Dendrimer enabled the proliferation of human NK cells responsible for innate immunity	Poupot et al, 2016
Vaccine	Poly(propylene imine) dendrimer (G4) with surface covered by maltose	-	Production of immonomodulatory effect via stimulation of NF-κβ pathway	Jatczak-Pawlik et al, 2017
Complex-ation	PAMAM dendrimer with car-boxyl end groups (G4.5)	Risperidone	Dendrimer-risperidone complexation at 46 risperidone molecules per dendrimer and modification in the dopa-minergic neurons and motoneurons compared with free drug in the zebrafish model	Prieto et al, 2014
Drug carry-ing	PAMAM dendrimers (G4.5 to G7.5) coated magnetic nanoparticles	Gemcitabine	Highest gemcitabine loading in PAMAM dendrimer (G5.5) with high stability and toxicity for MCF-7 and SKBR-3 cell lines	Parsian et al, 2016

2.7.1 DIAGNOSTIC FIELD

The "dendritic box" property of dendrimers characterized by a surface of rigid shell full of hydrogen bonds with solid characteristic and an internal cavity able to entrap different molecules (Jansen et al., 1995) can be used in the diagnostic field. This encapsulation ability considers the shape and size of inner compartment of dendrimers. They have been applied to carry molecular probes for radiotherapy, X-ray and molecular resonance imaging (MRI) due their high molecular weight, which prevented the fast equilibrium with extracellular compartment and avoided spreading to other body tissues. They confer high degree of contrast, thereby improving the image quality and thus diagnosis of lesions and internal organs (Kesharwani et al., 2014).

For MRI, the contrast agents for clinical uses are gadolinium chelates and paramagnetic iron oxide particles, which possess the relaxation rate of surrounding water protons; however, both of them can accumulate due to long residence time, thereby increasing toxicity (Noriega-Luna et al., 2014). Therefore, the class of MRI agents may be conjugated with PAMAM dendrimers in order to form an ideal MRI agent with lower toxicity, side effects, higher relaxation, and specificity (Yan et al., 2010). A practical application of this concept has already been used for the functional imaging of kidney, a situation in which diaminobutane core with polyalkylenimine branches and PAMAM dendrimers were conjugated with gadolinium II and III chelates allowing for the visualization of the renal anatomy of mice (Brechbiel et al., 2005).

There are examples in the literature of dendrimers' use due to the fluorescence of amine branches at the dendrimer surface for confocal laser scanning microscopy and MRI. Furthermore, they have been used for complexation with inorganic materials such as metal nanoparticles for thermal therapy in cancers (Imae, 2012). As molecular probes, dendrimers can immobilize the sensor units on the surface of inorganic materials, for example, platinum coordination complexes are sensitive to sulfur dioxide, in which their crystalline structures have expansion-reduction behavior characterized with a display of orange color sign (Albrecht et al., 2000).

A new platform of dendrimers has been created involving linking magnetic iron oxide nanoparticles and dendrimers. The macromolecular characteristics and surface properties of dendrimers are adequate for magnetic resonance imaging. In general, the dendrimers are used as stabilizers of magnetic iron oxide nanoparticles or the magnetic nanoparticles can functionalize the dendrimer surface or can be assembled on the surface of magnetic nanoparticles. The presence of dendrimer prolongs the half-life of magnetic contrast agents which allows better image detection and diagnosis of solid tumors. However, its limited specificity can be overcome by targeting the system with folic acid, which binds with overexpressed receptors of cancer cells (Sun et al., 2016).

In totality, dendrimers can adapt to several uses in the diagnostic field, allowing the visualization of organs and specific sites by improving the physicochemical features of molecular probes and contrast agents with adequate loading and carrying, as well as to ameliorate biodistribution and excretion. Considering the foregoing, this opens opportunities for the synthesis of new complexes addressed to computer tomography, optical imaging, magnetic resonance, and radiotherapy (Longmire et al., 2008).

2.7.2 ANTINEOPLASTIC DRUGS

Cancer treatment is a challenge due to the intensive proliferation of cancer cells in the tumors and the local pathophysiology such as extensive angiogenesis, altered vascularization, and lymphatic drainage constituting the known enhanced permeation and retention (EPR) effect. For these reasons, the dendrimers can serve as a system to provide local delivery of antineoplastic drugs by improving drug solubility and chemotherapy outcomes using adequate doses and leading to less systemic side effects and further providing better biodistribution. All these could occur because dendrimers are small enough to interact and accumulate in the tumors (Sopczynski, 2008; Abbasi et al., 2014).

Dendrimers may entrap antineoplastic drugs in the void spaces or covalently bound to the surface terminal groups. When the drug is inside the dendrimers, a controlled release of such drug can be achieved; this is possible with low-molecular-weight anticancer molecules (Somani and Dufès, 2014). Examples of these applications have been described in the literature such as cisplatin with low aqueous solubility encapsulated into PAMAM dendrimers which accumulated in melanoma cells in murine model through the EPR effect with a 15-fold higher accumulation than in free cisplatin and decreases systemic toxicity after intravenous administration (Malik et al., 1999). Doxorubicin, a potent antineoplastic, also had its solubility increased after association with third and fourth generations of pegylated PAMAM dendrimers (Kojima et al., 2000). Pegylated PAMAM dendrimer (G4) improved 5-fluorouracil solubility with increased blood circulation time and less toxicity than the nonpegylated counterparts (Bhadra et al., 2003).

Due to the limited space inside the dendrimers, they can carry molecules conjugated to their surfaces (Somani and Dufès, 2014). Various anticancer drugs have been attached in this way, such as 2-methoxyestradiolagent complexed to PAMAM dendrimers (G5) with different end groups such as amine, hydroxyl, and acetamide, which showed positive, slightly positive, and close to neutral charges, respectively. The degree of complexation was dependent on the charge. The inhibition of cancer cell growth was very notable for the dendrimers described above; however, carboxylated PAMAM dendrimer did not show the same effect exhibited by 2-methoxyestradiol including the drug release pattern. Thus, an adequate surface charge is also an aspect to be considered for cancer treatment with antineoplastic drugs (Shi et al., 2010).

2.7.3 TOPICAL AND TRANSDERMAL DELIVERIES (CUTANEOUS, OCULAR, AND NASAL)

Topical and transdermal deliveries are adequate delivery routes because of lesser side effects, hepatic first-pass metabolism, and adequate local and plasma levels of drugs. They are applied for hydrophilic and lipophilic small molecules and macromolecules. Besides, they are recommended for medications with frequent administrations and drugs with short half-life (Paudel et al., 2010). However, to be able to deliver drugs through this route, it is necessary to overcome the skin and mucosal barriers through strategic drug-delivery tools, physical methods, and/or use of dermal absorption promoters (Prausnitz et al., 2004). The topical and transdermal pathways can contribute to local and systemic therapies and are considered noninvasive methods. On the other hand, these pathways are preferred for drugs with molecular weight less than 500 Da and/or octanol-water partition coefficient ($\log P$) between 1 and 3 (Bolzinger et al., 2012). But, the association of delivery systems with physical methods can expand these parameter's limits (Kalia et al., 2004).

Dendrimers, through their unique features, are adequate for several routes of administration including topical and transdermal. Their potential is related to their ability to cross cellular barriers through paracellular and transcellular pathways (Najlah et al., 2012; Mignani et al., 2013). Furthermore, they aim to incorporate drugs with poor aqueous solubility, increase their entry into the biological compartments, and elicit low toxic response (Kesharwani et al., 2014). The drug entrapped in the dendrimers can enter the skin through three different mechanisms (Figure 2.3). First, the entrapped drug is released from the drug–dendrimer complex, in which the high level of the dissociated drug solubilizes in the vehicle, forcing skin permeation. The second mechanism involves the function of dendrimers as penetration enhancers, modifying the lipid organization in the stratum corneum, thereby increasing drug diffusion into the skin. This may occur when a pre- or co-treatment with dendrimer is performed. The third mechanism is related to follicular delivery, where dendrimers accumulate in the follicle releasing the drug into the deep skin layers.

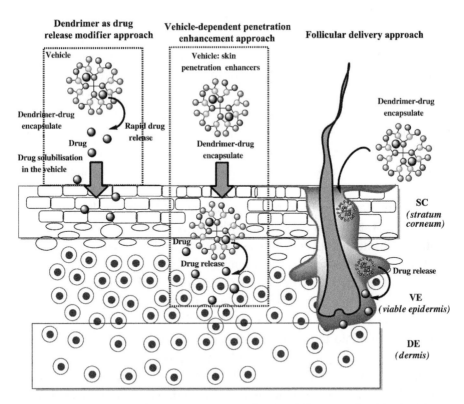

FIGURE 2.3 Three possible entry mechanisms for drug delivery mediated by dendrimers applied to the skin (Reprinted with permission from Mignani, S., Kazzouli, S. E., Bousmina, M., & Majoral, J. P., (2013). Expand classical drug administration ways by emerging routes using dendrimer drug delivery systems: a concise review. Adv. Drug Deliv. Rev., 65, 1316–1330. © 2013 Elsevier.).

There have been reports in literature that discussed the physicochemical properties of dendrimers and skin permeation. They studied the effects of size and surface groups of PAMAM dendrimers on the skin layers. It was observed that G2-dendrimer penetrated deeper than G4-dendrimer. Furthermore, the G2-dendrimer surface covered with acetyl or carboxyl groups were effective as penetration enhancer through extracellular pathway and amine end groups of G2-dendrimer promoted more cell internalization and enhanced retention with few skin permeation. Already, oleic acid surface groups of G2-dendrimer have been shown to have higher solubilizing effect over the skin layers (Yang et al., 2012). Similar results were reported by Venuganti et al. (2011) with PAMAM dendrimers having amine end groups of generations 2, 3, 4, 5, and 6; carboxyl end groups of generation 3.5; hydroxyl end group of fourth generation. They utilized the *in vitro* model of excised porcine skin and evaluated the skin penetration of FITC-labeled dendrimers by confocal laser scanning microscope. Amine terminal groups of dendrimers were more effective in promoting skin penetration than carboxyl and hydroxyl terminal groups. In addition, with increasing time, skin permeation increases in a linear manner. The association with iontophoresis induced skin penetration of cationic and neutral dendrimers. High-molecular-weight dendrimers had less penetration compared than low-molecular-weight ones for both passive diffusion and iontophoresis. They suggested the skin penetration of dendrimers occurs through intercellular and follicular paths.

The delivery of 8-methoxypsoralene through PAMAM dendrimers (G2.5 and G3.5) was studied in the skin of hairless mice model. The increasing permeation of 8-methoxypsoralene was quantified and its cutaneous distribution visualized by confocal laser scanning microscopy for both dendrimers. This opened an alternative delivery strategy for 8-methoxypsoralene in the treatment of skin diseases such as psoriasis and vitiligo using UVA therapy (Borowska et al., 2012). In spite of PAMAM dendrimers being mostly used to deliver drugs into the skin, it is important to still shed light on its toxicity. Thus, peptide dendrimers have been proposed due to its advantages related to low toxicity and biodegradability due to its amino acids composition. 5-Fluorouracil was conjugated with peptide dendrimer covered with arginine as end groups. This association improved the aqueous solubility and increased the deposition and permeation into the skin of 5-fluorouracil in function of peptide dendrimer concentration. Specially, peptide dendrimer with eight positive charge promoted the greatest flux and accumulation of

5-fluorouracil in the human skin after 48 hours, further showing the potential of peptide dendrimers as skin permeation enhancers (Mutalik et al., 2014).

In relation to ocular diseases, the main drawbacks are the little success in curing the posterior part of the eyes, including the vitreous humor, retina, choroid and optic nerve as compared with clinical results produced in diseases from the anterior part such as the cornea, iris, ciliary body, and lens. These aspects are related with drugs administered systemically and their difficulty to reach eye target tissues with therapeutic amounts of drugs due to the barriers encountered during blood circulation. Besides, the low residence time on the eyes after topical application of conventional formulations decreases the treatment potential (Amrite and Kompella, 2006). Considering dendrimers for this issue, they accomplish the task by their nanosize, permeability enhancer property across the cornea and sustained release of drugs, thereby making them reach the posterior part of the eyes (Vandamme and Brobeck, 2005). For ocular application, dendrimers should be in formulations with adequate pH, osmolality, and viscosity (Villanueva et al., 2016).

Important diseases of the eyes include diabetic retinopathy and glaucoma. PAMAM dendrimers (G3.5 and G4.5) conjugated with dexamethasone were proposed to enhance ocular penetration after subconjunctival and intravitreal injections. They observed drug levels above those reported in the literature are cleared after 3 h from the vitreous due to local clearance. Further, they found higher dendrimer levels in the retina-choroid than in the vitreous, indicating that dendrimers diffused throughout the eye after intravitreal injection; besides, dexamethasone levels were increased in the retina following subconjunctival injection (Yavuz et al., 2016).

Another study utilized PAMAM dendrimers (G3.5, G4, G4.5, and G5) conjugated with an isoflavone in different ratios for ocular drug delivery system and named puerarin. The complexation occurred by hydrogen bonds, and the puerarin release was lower from full generation dendrimer than from half generation one. However, both dendrimer generations increased the residence time in rabbit's eyes and no damage was observed in the corneal permeation study (Yao et al., 2010).

PAMAM dendrimers of generations of 3.5 (anionic) and 4.0 (cationic) were applied with iontophoresis on the cornea to deliver dexamethasone. Iontophoresis promoted a deep distribution of dexamethasone into the corneal tissue in the *ex vivo* studies, and the distribution was more pronounced using the G4-PAMAM dendrimer. Although in *in vivo* studies, the G3.5-PAMAM dendrimer increased dexamethasone concentration in the aqueous humor and prevented its rapid elimination and increased the drug solubility. This study represented a successful iontophoresis-dendrimer associated therapy to treat pathologies involving the eyes (Souza et al., 2015).

In relation to mucosal delivery, recently, a poly-L-lysine dendrimer (G4) with naphthalene disulfonate end groups was developed to treat and prevent bacterial vaginosis and sexually transmitted infections. It is a product formulated with a mucoadhesive gel named Vivagel® and which presents several clinical studies proving its efficacy and safety (Starpharma, 2016).

Intranasal delivery is a new approach to transport drugs to the brain. It is employed for drugs that are not appropriate for oral administration. It is applied in the nasal cavity and through extracellular and paracellular routes, following the olfactory nerve and respiratory region until it reaches the central nervous system. Drug delivery to the brain is a challenge, because the blood–brain barrier has several mechanisms to prevent the entry of substances. However, the advantages of this delivery route are direct transport avoiding systemic circulation, reduced side effect and metabolism, long-term administration, and immediate effect after administration (Illum, 2000; van Woensel et al., 2013). The use of dendrimers has been studied through an intranasal instillation (3 or 15 µg/animal) containing PAMAM dendrimers in male BALB/c mice. Gene expression and up-regulation of mRNA of brain derived-neurotrophic factor in the olfactory bulb, hippocampus and cerebral cortex were observed without changes in standard serum biomarkers from the blood. These profiles confirm that this pathway provokes neuronal effects in the brain without causing toxicological effects (Win-Shwe et al., 2014).

Haloperidol, an antipsychotic drug with low water solubility, was loaded into PAMAM dendrimer (G5) for intranasal delivery. This drug-dendrimer complex allowed more than 100-fold increase in haloperidol solubility. In addition, dendrimer-haloperidol formulation elicited expressive biodistribution producing behavioral responses comparable to the intraperitoneal administration (Katare et al., 2015).

A thermosensitive gel containing a complex compounded using radioactive ^{32}P-small interference RNA and PAMAM dendrimer (G7) was applied on the nasal cavity of rats and through the nose-to-brain route, delivery to the brain was achieved. The sequentially administered intranasal mucoadhesive gel did not damage the mucosa. Besides, it released the radiolabel material with silencing activity, proving that dendrimer association was decisive in mediating brain delivery (Perez et al., 2012).

2.7.4   GENE THERAPY

Dendrimers have also been applied for gene therapy in which the inner empty spaces and the reactive surface can host genetic material including nucleic acids, oligonucleotides, and antisense oligonucleotides (Eichman et al., 2000). Gene therapy is based on delivering the oligometric material to control or to modulate a specific target, for the production of the specific protein or for blocking it. When a gene therapy is designed, some points should be taken into consideration, such as those related to the genetic material to be transported like: high molecular weight and negative charge of the material, necessity of an appropriate packaging to compact it, protection from enzymatic degradation, and an efficient traffic until the target and release for eliciting an action (Dufès et al., 2005; Dehshahri and Sadeghpour, 2015).

Dendrimers are nonviral vectors and the positive charge of their surfaces helps the interaction with genetic materials and facilitates the cell adsorption followed by cell uptake. One example is PAMAM dendrimers with amine terminal groups, which adsorb the negative charge of nucleic acids, condensing it in a compact dendriplex. This complexation protects against enzymatic degradation and enables endocytosis. The high concentration of amine groups brings buffer effect, which is the driving force for endosomal escape and proper delivery of the genetic material to its destination which is the cytoplasm for oligonucleotides and siRNAs and to the nucleus for DNA. This allows nonspecificity; however, there are ways to improve and access targets in tumor or organs (Dufès et al., 2005; Dehshahri and Sadeghpour, 2015). The stability of dendriplexes can be characterized by different methods including laser light scattering, UV light absorption, and use of labeled genetic material (Eichman et al., 2000). The electrostatic interactions are responsible for the complexation between dendrimers and genetic materials, and they depend on dendrimer properties such as size, dendrimer generation, and number of end surface groups. PAMAM dendrimer has these interactions due to its amine groups, in which higher generations create more compact complexes compared to lower generations that generate looser complexes. Although this reasoning is not true for the cell uptake because the lower generation dendrimers tightens more efficiently and, consequently, the release of genetic material (Dehshahri and Sadeghpour, 2015)

Due to toxicity related with amine groups of PAMAM dendrimers, it was grafted with poly (L-glutamic acid) and polyethylenimine of low molecular weight to surround PAMAM dendrimer (G4.0) and to decrease its cytotoxicity. This system carried the pDNA of VEGF165 and was studied in an *in vivo* model of rabbit with injured carotid arteries, being able to inhibit significantly restenosis by an adequate delivery and action, what increased the VEGF165 expression in the vessels, besides the complexion promoted good stability, particle size around of 200 nm, and low polydispersity index (Zeng et al., 2011).

Especially for siRNA delivery, dendrimers are very useful tools by having controlled size and shape, besides their physicochemical features favors efficient transfection. Through siRNA technology, it is possible to congregate specific characteristics of dendrimers to improve structural aspects, which appears as potential delivery system (Tekade et al., 2015). With dendrimer carrying a non-viral vector, Waite and cols (2009) acetylated the primary amine of PAMAM dendrimer (G5) with acetic anhydride, complexed with siRNA and studied the siRNA delivery in U87 malignant glioma cells. They observed that acetylation up to 60% formed particles with size of 200 nm. In addition, higher rates of acetylation reduced the cytotoxicity in U87 cells and in parallel improved the dissociation of siRNA/PAMAM dendrimer complex. On the one hand, acetylation decreased the endosomal escape due to diminution of buffering effect from the amine groups of PAMAM dendrimers (G5).

In order to maintain the buffering capacity, PAMAM (G3 and G4) and PPI (G4 and G5) dendrimers had their amine surfaces partially modified with histidine, pyridine and piperazine groups, all with buffer effect. It was obtained that dendrimers with higher generation and higher substitution promoted better cell uptake in murine neuroblastoma cells for a plasmid DNA of *E. coli*. Indeed, the substituents pyridine and piperazine for PPI and PAMAM dendrimers, respectively, had best effect for the cell transfection, despite none have demonstrated expressive cytotoxicity profiles. The physicochemical profiles of modified dendrimers were having size in the range of 198–298 nm for PPI and 126–187 nm for PAMAM, both with positive zeta potential (Hashemi et al., 2016).

2.7.5 STIMULI RESPONSIVE

Stimuli-responsive dendrimers possess specific structural organization and chemical composition subject to change due to local alterations, in order to deliver the bioactive molecules. There are several pathologies that present these altered conditions, which can be addressed to stimulate the dendrimers. The specific features of these dendrimers are based on biodegradability and/ or chemical intermediates similar to sub-products of metabolism. Examples given as peptide-based dendrimers, rich in polylisine and polyester dendrimers based on 2,2-bis(hydroxymethyl) propionic acid. The environmental conditions can be redox potential, pH, overexpressed enzymes, host–guest interactions as well as external physical stimuli such as light, strength of electrical or magnetic field and temperature. The uses of these conditions ensure specificity to the dendrimer with release at limited sites, improving the therapeutic effect and decreasing the side effects (Calderón et al., 2010).

In general, there are two manners of molecules release from the stimuli-responsive dendrimers. One is the structural change of dendrimers with subsequent release and the other is disruption of chemical linkages of the dendrimer-drug complex (Calderón et al., 2010). Some examples of sensitive chemical linkers as follows: acetal bond and hydrazine bond are sensitive to pH; carbamate and ester linkage are sensitive to pH and proteases; o-nitro benzyl group is sensitive to light; and short peptide sequence of substrate of cathepsin B is sensitive to light (Gingras et al., 2007).

Alterations in the pH to lower than seven are related with inflammation, infection, and cancer, though, the cellular organelles as lysosomes are acid, compared with normal tissues, extracellular media, and blood (pH around 7.4). Acid-labile molecules can be inserted in the core or surface of the dendrimers (Calderón et al., 2010; Zhu et al., 2014). The alteration at tumor sites of acid pH can be taken as a stimulus to design dendrimers that are pH-dependent in cancer therapy, like small size at neutral pH (blood circulation) and large size for reaching tumor periphery, through the EPR effect, where the loaded drug is released. Besides, groups responsive to pH variation on the dendrimer surfaces may be added through chemical linkages mentioned before (Stieger et al., 2012). Acetylated PPI dendrimer (G4) increased the encapsulation of methotrexate and doxorubicin, being acetylated higher than 90% and exhibited pH-responsive behavior with release of entrapped drugs and toxicity reduction of PPI dendrimer. The dendrimer acetylation provided a new option to delivery anticancer drugs, being a biocompatible tool to design new platforms and to decline their cytotoxicity (Wang et al., 2012). Conformational dendrimer changes can also be related to polyethylene glycol (PEG)-dendrimer hybrids with polylysine or polyester dendrons linked through acetal bonds that destabilized the dendrimer in acidic pH, exposing the hydrophilic core with drug release (Gillies et al., 2004).

Damaged tissues are in a constant oxidative stress; consequently, there are unbalanced redox microenvironments and overexpressed oxidative enzymes. Stimuli-responsive dendrimers can have disulfide bond or be conjugated to glutathione enzyme, which involve them in the control of potential redox local, in which disulfide bond is broken by the high concentration of glutathione with followed release of entrapped molecule. There are results proving that a dendrimer-drug system bounded by disulfide linkage could deliver up to 60% of the drug payload after 1 hour at intracellular GSH concentration in tumor cells compared to zero release in incubation at plasma GSH concentration (Kurtoglu et al., 2009).

The strategy of using a redox responsive-dendrimer allowed specificity for pegylated dendrimers administered endovenously, as demonstrated in B16 tumor-bearing mice. They showed a precise and effective doxorubicin concentration around the tumor due the long circulation time and EPR effect, together with the ability to release the doxorubicin at the tumor periphery at acidic pH and altered potential redox by the tumor presence associated with high concentration of GSH, resulting in improved antitumor efficiency, without causing weight loss in the mice. Likewise, there were decreased side effects and high safety; this was possible by the redox/pH triggered release of doxorubicin from the dual responsive-dendrimer (Hu et al., 2016).

Ester-derived moieties and short peptide sequences can serve as substrate for overexpressed enzymes that occur in tumor sites (Calderón et al., 2010). The carriers are pro-drug systems that possess chemical bonds that will be cleaved by specific enzymes present in high concentration in tumor areas. For example, it was synthesized as PAMAM dendrimer with ester bond to be hydrolyzed by esterases in order to carry paclitaxel to the tumor area. There was an increased cytotoxicity due to high intracellular concentration of paclitaxel delivered into the tumor cells (Khandare et al., 2006).

Micelle-like structures of dendrimers based on amphiphilic biaryl molecules were designed to respond to an enzymatic trigger. It was installed with enzyme cleavable ester groups, as lipophilic units to be degraded and exposed the hydrophilic core with release of guest molecules. This dendrimer-based amphiphilic assembly is the protein-stimuli system directed to pathological sites with unbalanced enzymes through non-covalent binding (Azagarsamy et al., 2009).

Utilization of external physical stimulus can vary according the method to be applied in order to concentrate the dendrimers in certain areas and/or to release the encapsulated molecules. The thermo-responsiveness is reached with molecular structures that change when a thermal stimulus exists. PAMAM dendrimers (G4) were reacted with n-butyramide, isobutyramide, and cyclopropanecarboxylic acid amide groups via amine groups. These alkylamides groups are thermo-responsive polymers (Kojima et al., 2009). The light stimulus offers easiness and controllable time of application and relative safety, where the active units are influenced by specific wavelengths of radiation (Calderón et al., 2010). Azobenzene derivatives are examples of light-driven compounds. They undergo *iso/cis* isomerization, a clean pho-tochemical process, and can be located at the surface or among the core of dendrimers, upon light stimulus, they modify their organizational structures to liquid crystalline, vesicles, nanofibers, and organogels. The localization of azobenzene groups can be applied for drug delivery to encapsulate and release molecules or by N=N bond cleavage (Deloncle and Caminade, 2010).

2.8 SURFACE MODIFICATIONS

The surface modification of a dendrimer is a tool to diversify the interaction types and to increase the specificity of a drug delivery system, aiming to enlarge its functionality (Imae, 2012). With this in mind, Teow and cols (2013) investigated the ability of PAMAM dendrimer (G3) with surface modified by lauryl chains and conjugated with paclitaxel in reducing the cell viability of human colon carcinoma cell line and primary porcine brain endothelial cells. They observed that dendrimer covered with lauryl chains increased the cytotoxicity on these cell lines by the permeability increased in the apical and basolateral directions compared with non-modified dendrimer and free drug. These barriers, intestinal and blood-brain, are challenges for drug delivery systems. Therefore, surface chemical modification is a strategy to overcome the bypass efflux transporters and enhance the permeability to deliver molecules (Najlah and D'Emanuele, 2006).

Due to the potential of multifunctionalization, dendrimers have been widely applied in several areas. The end groups define the nature of bindings and solubility of dendrimers. They can be PEG molecules to prevent the elimination by the reticuloendothelial system and to increase the solubility;

hydrophobic moieties to favor cell interaction and to reduce the cytotoxicity; cyclodextrin to protect against degradation; and amino acid sequences (arginine and TAT peptides) and peptides (transferrin) to improve the linkage to target receptors and increase cell penetration and endosomal escape (Dehshahri and Sadeghpour, 2015).

The surface modifications to decrease the toxicity, mainly, due to the positive charged end groups, are means to widen the dendrimer use and bring more biocompatibility. The transformation of amine groups of PAMAM dendrimer (G4) into pyrrolidone derivatives was analyzed in *in vitro* assays of hemolytic activity, with human serum albumin through circular dichroism, in addition to determining cell viability with a mouse neuroblastoma cell line. Surface modification with the pyrrolidone groups considerably decreased the hemolytic activity compared to G4-PAMAM-NH$_2$, known to present toxic effect. In addition, the cell viability only showed diminution to the modified dendrimer for doses higher than 2 mg/mL. The assay with human serum albumin indicates about few interaction with G4-PAMAM-pyrrolidone dendrimer, which supports its biocompatibility (Ciolkowski et al., 2012). The toxicity of the same modified dendrimer was also tested in cell lines of hamster fibroblasts, embryonic mouse hippocampal cells, and rat liver cells, and minor levels of toxicity were noted, without any influence of the oxidative responses such as intracellular ROS level and mitochondrial membrane potential (Janaszewska et al., 2013).

An alternative way to overcome the toxicity and circulation clearance of PAMAM dendrimers through endovenous is to bind PEG molecules to the dendrimer surface; this prevents plasma protein and cell interactions, prolongs the circulation time, and improves the EPR effect due to the increased size of the complex. Three different degrees of pegylation were prepared and compared to non-pegylated PAMAM dendrimer (G3). They promoted better solubilization of the lipophilic drug with fast release and stability in human plasma, however, they had low activity in the human epidermoid carcinoma KB cells. Furthermore, the pharmacokinetic studies and *in vivo* evaluation of antitumor effect confirmed the pegylation of dendrimers, thereby showing the effectiveness of targeted drug delivery (Jiang et al., 2010).

A surface covering of maltose was formed on PPI dendrimer (G4) and showed biocompatibility in cellular models, due to decrease in the exposure of amine groups that cause hemolysis and cytotoxicity. Further, this glycodendrimer with molecular recognition potential was evaluated for its immunomodulatory activity on the THP-1 monocytic cell line, and the effect was determined through the expression levels of the NF-κβ pathway proteins by quantitative real-time PCR, since carbohydrates play an important role in immune recognition. The potential of glycodendrimer in mRNA was noted in the expression of two gene markers of the NF-κβ pathway in a dose-dependent manner, being expressive for immune system function and regulation. This led to pro-inflammatory cytokines secretion, which may be an adjuvant for vaccines (Jatczak-Pawlik et al., 2017).

Different chain lengths of tertiary amines were inserted on third-generation PAMAM dendrimer that constitutes quaternary ammonium salts. These salts possess antimicrobial activity and together with the polymers exhibit antimicrobial activity with high efficiency. Dendrimers functionalized with antimicrobial agents tune their action against Gram-positive and Gram-negative bacteria (*Staphylococcus* and *Escherichia coli*) because they improved the binding on the cell surface of bacteria, followed by membrane disruption and cell death. The adsorption was greater in Gram-negative bacteria compared with Gram-positive bacteria for ammonium salts (Charles et al., 2012)

2.8.1 TARGETED DRUG DELIVERY

The targeting modalities for dendrimers are reached by binding specific molecules on the top of dendrimers. These surface molecules are directed to restricted sites at the cells and tissues, for example, folic acid, carbohydrates, IgG/Fab-type antibodies, and overexpressed receptors as VEGF (vascular endothelial growth factor) present in tumors. The designing of these platforms assembles the inclusion of molecules of cell membranes as proteins, lipids, carbohydrates, and cell receptors, as well as to interior compartments, which are DNA and RNA. All is to amplify the interactions and to increase the therapeutic effect of drug delivery mediated by the dendrimers (Menjoge et al., 2010; Madaan et al., 2014).

The aggressiveness of tumors is related with the overexpression of receptors, which stimulate its growth in a rapid manner; this is observed in HER2-positive breast cancer, specifically, human epidermal growth factor receptor 2 (HER2). HER2 receptor can act as a target for breast cancer therapy with PAMAM dendrimers. The authors grafted the trastuzumab antibody on the dendrimer surface, which binds to HER2 receptor in order to block the downstream signaling. This selective approach delivered the docetaxel and exerted expressive antiproliferative effect by the increased cell internalization, in addition to the reduction of toxicity as shown in an *in vivo* study. Trastuzumab was linked to the G4-PAMAM dendrimer through the heterocrosslinker MAL-PEG-NHS (Kulhari et al., 2016). With a similar aim, a specific dendrimer directed to the epidermal growth factor receptor (EGFR) was designed. The development involved synthesis of the antibody named Erbitux, which inhibits the EGFR and prevents cell proliferation and tumor growth. Erbitux was conjugated on a PEGylated PAMAM dendrimer (G4) to improve the specificity and effect of oncolytic adenovirus in the treatment of lung cancer with overexpressed EGFR. Indeed, systemic administration of this specific complex had longer blood circulation and higher tumor retention. It also decreased the immunogenicity and the liver sequestration of the oncolytic adenovirus, which is a characteristic of oncolytic adenovirus (Yoon et al., 2016).

Another target to treat cancer is the folate receptors. They are present in many cancers because cells need more folic acid to maintain the tumor growth, which is not the case for normal tissues. Head and neck squamous cell carcinoma have high concentration level of folate receptors. For this reason, they can be targeted through dendrimers for developing active targeting cancer therapy, which was applied for this type of cancer through gene therapy with PAMAM dendrimers as the nonviral vectors. Because several cell lines have high pattern of folate receptor expression, the conjugate folate-PAMAM dendrimer was more internalized and consequently more DNA plasmids were intracellularly delivered compared with cell lines without overexpression of folate receptors (Xu et al., 2016). Folate-conjugated dendrimers were used to deliver methotrexate, which was covalently coupled and not coupled. The covalent folate-conjugated dendrimer is more effective for the intracellular delivery in KB cells. The influx of drug occurring via endocytosis due to the overexpressed folate receptors of these cells showing more cytotoxicity, in comparison with methotrexate as inclusion complex into folate-conjugated dendrimers (Patri et al., 2005).

Other bioactive ligands are the saccharides and glycopeptides because at the cellular level, they can interact with avidity and specificity, enhancing the dendrimer ability to deliver several therapeutic molecules. The conjugation reactions with dendrimer are based on pre-derived glycans associated with reactive groups such as isothiocyanate and carboxylic acid, which react directly with PAMAM amine groups. However, this linkage is difficult because of the necessity to protect the hydroxyl groups in the sugar ring. These types of ligands were explored in hepatic treatments as the hepatocytes possess asialoglycoprotein that binds galactose, lactose, and *N*-acetylgalactosamine. Thus, covering the PAMAM surface with galactose through the hydrazine bond increased the avidity to HepG2 cells, and, with *N*-acetylgalctosamine through the thiourea bond, there was more liver accumulation. The level of surface functionalization determined the multivalent effect, which means the association strength between the dendrimer and cells (Liu et al., 2012).

2.8.2   LINKAGE TO NANOCARRIERS

With the view to amplify and diversify the dendrimer applications, they were linked to several types of delivery systems such as liposomes, carbon nanotubes, and other nanoparticles. These associations intend to improve the physicochemical stability and effectiveness of the delivery systems, for example, better dispersibility and drug entrapment, enhanced cell internalization and therapeutic effects (Kesharwani et al., 2014). One example is the protection of photocatalytic activity of TiO_2 nanoparticles with PAMAM dendrimers against the chemical degradation agent as 2,4-dichlorophenoxy-acetic. The dendrimer allowed a reservoir of it, which restrained its oxidation mechanism, due to the PAMAM dendrimer had surrounded the TiO_2 nanoparticles preserving the photocatalytic effect (Nakanishi and Imae, 2005). The layer-by-layer technique was used to cover gold nanoparticles alternating with PAMAM dendrimer and TiO_2 nanoparticles to produce a pH biosensor membrane (Vieira et al., 2012).

For the photodynamic therapy, PAMAM dendrimer (G3) was grafted on the porous silica nanoparticles followed by surface charge tuning with gluconic acid. These three-part carriers effectively encapsulated the aluminum phthalocyanine tetrasulfonate, which maintained the nanometric size around 150 nm with 30 nm of shell thickness and released the drug at the target tumor site with generation of singlet oxygen for oxidative reaction to kill cancer cells. The photosensitive drug carrier was more effective upon irradiation in comparison with free aluminum phthalocyanine tetrasulfonate (Tao et al., 2013).

In order to detect cancers that overexpress HER2 receptor, a dendrimer containing gold nanoparticles and a gadolinium imaging agent for dual mode of imaging was synthesized. The surface of fifth generation PAMAM dendrimer was modified by PEG molecules and 1,4,7,10-tetraazacyclodo-decane-1,4,7,10-tetraacetic acid to guest the gold nanoparticles and the gadolinium, which were further conjugated with Herceptin antibody. This assembly allowed better details and imaging to detect the HER2-positive cancers in breast and lung cancer patents, which can be used in clinical approach for early diagnosis. The dendrimers' ability propitiated the comput-erized tomography and magnetic resonance imaging by adequate encapsula-tion and circulation time of gold nanoparticles and gadolinium associated with the specificity conferred by antibody conjugation to HER2 receptors Furthermore, it increased the stability of gold nanoparticles and relaxivity of gadolinium (Otis et al., 2016). Another nanoplatform with dendrimer asso-ciated with gold nanoparticles and elastin-like peptides was developed by Fukushima et al. (2015) for photothermogenic and thermosensitive stimuli, respectively. They used PAMAM dendrimer (G4) modified with acetylated surface for production-controlled cell association and photocytotoxicity by temperature. Besides, the phase transition changes at body temperature, a controllable system for biomedical applications is shown.

Through an external magnetic field, magnetic nanoparticles can accumu-late in the tumor areas for cancer therapy. For this reason, they were coated with half generations of PAMAM dendrimer (G4.5 to G7.5), followed by the gemcitabine encapsulation. The intrinsic feature of half generation dendrimer of negative surface charge (carboxylic groups) decreases the dendrimer toxic-ity and the gemcitabine had high drug loading with high stability by the use of dendrimer (G5.5). In addition, cell internalization and cell toxicity were increased with gemcitabine-loaded dendrimer under magnetic field, showing the targetability characteristic (Parsian et al., 2016). In the same token, two different length chains of PEG were grafted on PAMAM dendrimer to amelio-rate, the pharmacokinetics and targeting abilities and, further to stabilize the magnetite nanoparticles (Fe_3O_4) and encapsulate doxorubicin. *In vitro* studies achieved high doxorubicin loading with its controlled release in the vicinity of tumor cells, guided by pH and temperature changes, followed by enzymatic degradation with cathepsin B. A potential magnetic-hyperthermic drug deliv-ery system is used for clinical application (Chandra et al., 2015).

2.9 CONCLUSION AND FUTURE PERSPECTIVE

The large amount of studies in the development and efficacy evaluations of dendrimers is based on their features as well-defined structures and particle size with low polydispersibility, besides offering high versatility and functionality to accomplish several purposes in the biomedical area. Dendrimers can host molecules of different molecular weights and physicochemical characteristics by entrapping them inside and/or on their surface through covalent or noncovalent linkages with sustained and controlled release of the substances. Moreover, they have attracted interest as they increase drug solubility, modify drug pharmacokinetics, and improves bioavailability that allows delivery to specific sites to exert therapeutic activity. The dendrimer functionalization or complexations with other nanoparticles subjected to external stimuli are alternatives to improve the specificity and efficacy of drug molecules. In addition, there are miscellaneous kinds of dendrimers with intrinsic properties and toxicities that can be applied in function of the interest area such as anticancer, diagnostic tools, topical, or systemic delivery tools.

This chapter offered an overview and insights to encourage the research on the opportunities that abound with this type of delivery system and to show the dendrimer's tridimensional architecture and diverse possibilities of modification and uses. In the near future, dendrimers' scaffolds will be increasingly explored in the development of new products and technologies for biomedical and other uses.

KEYWORDS

- biocompatibility
- dendrimer
- nanoparticle
- PAMAM
- polymeric molecule
- surface modification

REFERENCES

Abbasi, E., Aval, S. F., Akbarzadeh, A., Milani, M., Nasrabadi, H. T., Joo, S. W., Hanifehpour, Y., Nejati-Koshki, K., & Pashaei-Asl, R., (2014). Dendrimers: synthesis, applications, and properties. *Nanoscale Res. Lett., 9*, 1–10.

Albrecht, M., Lutz, M., Spek, A. L., & Koten, G., (2000). Organoplatinum crystals for gas-triggered switches. *Nature, 406*, 970–974.

Amrite, A. C., & Kompella, U. B., (2000). Nanoparticles for ocular drug delivery. In: *Nanoparticle Technology for Drug Delivery*, Gupta, R. B., Kompella, U. B. (Eds.), Taylor & Francis, *2006*, 319–360.

Azagarsamy, M. A., Sokkalingam, P., & Thayumanavam, S., (2009). Enzyme triggered disassembly of dendrimer-based amphiphilic nanocontainers. *J. Am. Chem. Soc, 131*, 14184–14185.

Bhadra, D., Bhadra, S., Jain, S., & Jain, N. K., (2003). A PEGylated dendritic nanoparticulate carrier of fluorouracil. *Int. J. Pharm., 257*, 111–124.

Boas, U., Christensen, J. B., & Heegaard, P. M. H., (2006). Dendrimers: design, synthesis and chemical properties. *J. Mater Chem., 16*, 3785–3798.

Bodewein, L., Schmelter, F., Di Fiore, S., Hollert, H., Fischer, R., & Fenske, M., (2016). Differences in toxicity of anionic and cationic PAMAM and PPI dendrimers in zebrafish embryos and cancer cell lines. *Toxicol. Appl. Pharmacol., 305*, 83–92.

Bolzinger, M. A., Briançon, S., Pelletier, J., & Chevalier, Y., (2012). Penetration of drugs through the skin, a complex rate-controlling membrane. *Cur. Opin. Colloid Interface Sci., 17*, 156–165.

Borowska, K., Wolowiec, S., Glowniak, K., Sieniawska, E., & Radej, S., (2012). Transdermal delivery of 8-methoxypsoralene mediated by polyamidoamine dendrimer G2. 5 and G3. 5-*in vitro* and *in vivo* study. *Int. J. Pharm., 436*, 764–770.

Brechbiel, M. W., Star, R. A., & Kobayashi, H., (2005). Methods for functional kidney imaging using small dendrimer contrast agents. Patent no. US 6,852,842 B2. Date of patent.

Calderón, M., Quadir, M. A., Strumia, M., & Haag, R., (2010). Functional dendritic polymer architectures as stimuli-responsive nanocarriers. *Biochimie, 92*, 1242–1251.

Caminade, A. M., & Majoral, J. P., (2005). Water-soluble phosphorus-containing dendrimers. *Prog. Polym. Sci., 30*, 491–505.

Caminade, A. M., & Turrin, C. O., (2014). Dendrimers for drug delivery. *J. Mater Chem. B., 2*, 4055–4066.

Caminade, A. M., Laurent, R., & Majoral, J. P., (2005). Characterization of dendrimers. *Adv. Drug Deliv. Rev., 57*, 2130–2146.

Chandra, S., Noronha, G., Dietrich, S., Lang, H., & Bahadur, D., (2015). Dendrimer-magnetic nanoparticles as multiple stimuli responsive and enzymatic drug delivery vehicle. *J. Magn. Magn. Mater, 380*, 7–12.

Charles, S., Vasanthan, N., Kwon, D., Sekosan, G., & Ghosh, S., (2012). Surface modification of poly(amido amine) (PAMAM) dendrimer as antimicrobial agents. *Tetrahedron. Lett., 53*, 6670–6675.

Chen, H., BanaszakHoll, M., Orr, B. G., Majoros, I., & Clarkson, B. H., (2003). Interaction of dendrimers (artificial proteins) with biological hydroxyapatite crystals. *J. Dent. Res., 82*, 443–448.

Cheng, Y., Xu, Z., Ma, M., & Xu, T., (2008). Dendrimers as drug carriers: applications in different routes of drug administration. *J. Pharm. Sci., 97*, 123–143.

Ciolkowski, M., Petersen, J. F., Ficker, M., Janaszewska, A., Christensen, J. B., Klajnert, B., & Bryszewska, M., (2012). Surface modification of PAMAM dendrimer improves its biocompatibility. *Nanomedicine, 8*, 815–817.

Dehshahri, A., & Sadeghpour, H., (2015). Surface decorations of poly(amidoamine) dendrimer by various pendant moieties for improved delivery of nucleic acid materials. *Colloids Surf. B., 132*, 85–102.

Deloncle, R., & Caminade, A. M., (2010). Stimuli-responsive dendritic structures: the case of light-driven azobenzene-containing dendrimers and dendrons, *J. Photoch. Photobio. C.*, *11*, 25–45.

Dufès, C., Uchegbu, I. F., & Schätzlein, A. G., (2005). Dendrimers in gene delivery. *Adv. Drug Deliv. Rev.*, *57*, 2177–2202.

Eichman, J. D., Bielinska, A. U., Kukowska-Latallo, J. F., & Baker, Jr. J. R., (2000). The use of PAMAM dendrimers in the efficient transfer of genetic material into cells. *Pharm. Sci. Technolo. Today*, *3*, 232–245.

Fernández, L., Sigal, E., Otero, L., Silber, J. J., & Santo, M., (2011). Solubility improvement of an anthelmintic benzimidazole carbamate by association with dendrimers. *Braz. J. Chem. Eng.*, *28*, 679–689.

Frechet, J. M. J., (2002). Dendrimers and supramolecular chemistry. *Proc. Natl. Acad. Sci.*, *99*, 4782–4787.

Fukushima, D., Sk, U. H., Sakamoto, Y., Nakase, I., & Kojima, C., (2015). Dual stimuli-sensitive dendrimers: photothermogenic gold nanoparticle-loaded thermo-responsive elastin-mimetic dendrimers. *Colloids Surf. B.*, *132*, 155–160.

Gautam, S. P., Gupta, A. K., Agrawal, S., & Sureka, S., (2012). Spectroscopic characterzation of dendrimers. *Int. J. Pharm. Pharm. Sci.*, *4*, 77–80.

Gillies, E. R., Jonsson, T. B., & Frechet, J. M., (2004). Stimuli-responsive supramolecular assemblies of linear-dendritic copolymers. *J. Am. Chem. Soc.*, *126*, 11936–11943.

Gingras, M., Raimundo, J. M., & Chabre, Y. M., (2007). Cleavable dendrimers. *Angew. Chem. Int. Ed. Engl.*, *46*, 1010–1017.

Goldberg, D. S., Ghandehari, H., & Swaan, P. W., (2010). Cellular entry of G3. 5 poly (amido amine) dendrimers by clathrin- and dynamin-dependent endocytosis promotes tight junctional opening in intestinal epithelia. *Pharm. Res.*, *27*, 1547–1557.

Gorzkiewicz, M., & Klajnert-Maculewicz, B., (2017). Dendrimers as nanocarriers for nucleoside analogues. *Eur. J. Pharm. Biopharm.*, *114*, 43–56.

Guo, Y., Wang, Y., Yang, G., Xu, J. J., & Chen, H. Y., (2016). Micro-RNA mediated signal amplification coupled with GNP/dendrimers on a mass-sensitive biosensor and its applications in intracellular microRNA quantification. *Biosens. Bioelectron.*, *85*, 897–902.

Hashemi, M., Tabatabai, S. M., Parhiz, H., Milanizadeh, S., Farzad, S. A., Abnous, K., & Ramezani, M., (2016). Gene delivery efficiency and cytotoxicity of heterocyclic amine-modified PAMAM and PPI dendrimers. *Mater Sci. Eng. C. Mater. Biol. Appl.*, *61*, 791–800.

Heiden, T. C., Dengler, E., Kao, W. J., Heideman, W., & Peterson, R. E., (2007). Developmental toxicity of low generation PAMAM dendrimers in zebrafish. *Toxicol. Appl. Pharmacol.*, *225*, 70–79.

Hong, S., Hessler, J. A., Holl, M. M. B., Leroueil, P., Mecke, A., & Orr, B. G., (2006). Physical interaction of nanoparticles with biological membranes: the observation of nanoscale hole formation. *J. Chem. Health. Saf.*, *13*, 16–20.

Hu, W., Qiu, L., Cheng, L., Hu, Q., Liu, Y., Hu, Z., Chen, D., & Cheng, L., (2016). Redox and pH dual responsive poly(amidoamine) dendrimer-poly(ethyleneglycol) conjugates for intracellular delivery of doxorubicin. *Acta. Biomater.*, *36*, 241–253.

Illum, L., (2000). Transport of drugs from the nasal cavity to the central nervous system. *Eur. J. Pharm. Sci.*, *11*, 1–18.

Imae, T., (2012). Physicochemical properties of dendrimers and dendrimer complexes. In: *Dendrimer-Based Drug Delivery Systems-From Theory to Practice*, Cheng, Y., (ed.), Wiley, Hoboken, 55–92.

Jain, K., Kesharwani, P., Gupta, U., & Jain, N. K., (2010). Dendrimer toxicity: let's meet the challenge. *Int. J. Pharm.*, *394*, 122–142.

Janaszewska, A., Ciolkowski, M., Wróbel, D., Petersen, J. F., Ficker, M., Christensen, J. B., Bryszewska, M., & Klajnert, B., (2013). Modified PAMAM dendrimer with 4-carbomethoxypyrrolidone surface groups reveals negligible toxicity against three rodent cell lines. *Nanomedicine: NBM*, *9*, 461–464.

Jansen, J. F. G. A., Meijer, E. W., & Brabander-van der B. E. M. M., (1995). The dendritic box: shape-selective liberation of encapsulated drugs. *J. Am. Chem. Soc.*, *117*, 4417–4418.

Jatczak-Pawlik, I., Gorzkiewicz, M., Studzian, M., Appelhans, D., Voit, B., Pulaski, L., & Klajnert-Maculewicz, B., (2017). Sugar-modified poly(propylene imine) dendrimers stimulate the NF-κβ pathway in an myeloid cell line. *Pharm. Res.*, *34*, 136–147.

Jevprasesphant, R., Penny, J., Attwood, D., McKeown, N., & D'Emanuele, A., (2003). Engineering of dendrimer surfaces to enhance transepithelial transport and reduce cytotoxicity. *Pharm. Res.*, *20*, 1543–1550.

Jiang, Y. Y., Tang, G. T., Zhang, L. H., Kong, S. Y., Zhu, S. J., & Pei, Y. Y., (2010). PEGylated PAMAM dendrimers as a potential drug delivery carrier: *In vitro* and *in vivo* comparative evaluation of covalently conjugated drug and non-covalent drug inclusion complex. *J. Drug Targeting, 18*, 389–403.

Kalia, Y. N., Naik, A., Garrison, J., & Guy, R. H., (2004). Iontophoretic drug delivery. *Adv. Drug Deliv. Rev.*, *56*, 619–658.

Kasai, S., Nahasawa, H., Shinamura, M., Uto, Y., & Hori, H., (2002). Design and synthesis of antiangiogenic/heparin-binding arginine dendrimer mimicking the surface of endostatin. *Bioorg. Med. Chem. Lett.*, *12*, 951–954.

Katare, Y. K., Daya, R. P., Gray, C. S., Luckham, R. E., Bhandari, J., Chauhan, A. S., & Mishra, R. K., (2015). Brain targeting of a water insoluble antipsycotic drug haloperidol via the intranasal route using PAMAM dendrimer. *Mol. Pharm.*, *12*, 3380–3388.

Ke, W., Zhao, Y., Huang, R., Jiang, C., & Pei, Y., (2008). Enhanced oral bioavailability of doxorubicin in a dendrimer drug delivery system. *J. Pharm. Sci.*, *97*, 2208–2216.

Kesharwani, P., Jain, K., & Jain, N. K., (2014). Dendrimer as nanocarrier for drug delivery. *Progr. Polym. Sci.*, *39*, 268–307.

Khandare, J. J., Jayant, S., Singh, A., Chandna, P., Wang, Y., Vorsa, N., & Minko, T., (2006). Dendrimer versus linear conjugate: influence of polymeric architecture on the delivery and anticancer effect of paclitaxel. *Bioconjug. Chem.*, *17*, 1464–1472.

Kitchens, K. M., Kolhatkar, R. B., Swaan, P. W., & Ghandehari, H., (2008). Endocytosis inhibitors prevent poly(amidoamine) dendrimer internalization and permeability across Caco-2 cells. *Mol. Pharm.*, *5*, 364–369.

Klajnert, B., & Bryszewska, M., (2001). Dendrimers: Properties and applications. *Acta. Biochim. Pol., 48*, 199–208.

Kojima, C., Kono, K., Maruyama, K., & Takagishi, T., (2000). Synthesis of polyamidoamine dendrimers having poly(ethylene glycol) grafts and their ability to encapsulate anticancer drugs. *Bioconjug. Chem., 11*, 910–917.

Kojima, C., Yoshimura, K., Harada, A., Sakanishi, Y., & Kono, K., (2009). Synthesis and characterization of hyperbranched poly(glycidol) modified with pH- and temperature sensitive groups. *Bioconjug. Chem.*, *20*, 1054–1057.

Kulhari, H., Pooja, D., Shrisvastava, S., Kuncha, M., Naidu, V. G. M., Bansal, V., Sistla, R., & Adams, D. J., (2016). Trastuzumab-grafted PAMAM dendrimers for the selective delivery of anticancer drugs to HER2-positive breast cancer. *Sci. Rep.*, *6*, 1–13.

Kurtoglu, Y. E., Navath, R. S., Wang, B., Kannan, S., Romero, R., & Kannan, R. M., (2009). Poly(amidomine) dendrimer-drug conjugates with disulfide linkages for intracellular delivery. *Biomaterials, 30*, 2112–2121.

Lee, C. C., MacKay, J. A., Fréchet, J. M. J., & Szoka, F. C., (2005). Designing dendrimers for biological applications. *Nat. Biotechnol., 23*, 1517–1526.

Li, Y., Zhao, L., & Cheng, Y., (2012). *Improving* the biocompatibility of dendrimers in drug delivery. In: *Dendrimer-Based Drug Delivery Systems–From Theory to Practice,* Cheng, Y., (ed.), Wiley, Hoboken, 207–237.

Lim, J., & Simanek, E. E., (2012). Triazine dendrimers as drug delivery systems: From synthesis to therapy. *Adv. Drug Deliv. Rev., 64*, 9, 826–835.

Liu, J., Gray, W. D., Davis, M. E., & Luo, Y., (2012). Peptide- and saccharide-conjugated dendrimers for targeted drug delivery: a concise review. *Interface Focus, 2*, 307–324.

Lombardo, D., Calandra, P., Bellocco, E., Laganà, G., Barreca, D., Magazù, S., Wanderlingh, U., & Kiselev, M. A., (2016). Effect of anionic and cationic polyamidoamine (PAMAM) dendrimers on a model lipid membrane. *Biochim. Biophys. Acta., Biomembr, 1858*(11), 2769–2777.

Longmire, M., Choyke, P. L., & Kobayashi, H., (2008). Dendrimer-based contrast agent for molecular imaging. *Curr. Top. Med. Chem., 8*, 1180–1186.

Losada, J., Armada, M. P. G., García, E., Casado, C. M., & Alonso, B., (2017). Electrochemical preparation of gold nanoparticles on ferrocenyl-dendrimer film modified electrodes and their application for the electrocatalytic oxidation and amperometric detection of nitrite. *J. Electroanal. Chem., 788*, 14–22.

Madaan, K., Kumar, S., Poonia, N., Lather, V., & Pandita, D., (2014). Dendrimers in drug delivery and targeting: drug-dendrimer interactions and toxicity issues. *J. Pharm. Bio. Allied. Sci., 6*, 139–150.

Malik, N., Evagorou, E. G., & Duncan, R., (1999). Dendrimer–platinate: a novel approach to cancer chemotherapy. *Anticancer Drugs, 10*, 767–776.

Malik, N., Wiwattanapatapee, R., Klopsch, R., Lorenz, K., Frey, H., Weener, J. W., Meijer, E. W., Paulus, W., & Duncan, R., (2000). Dendrimers: relationship between structure and biocompatibility *in vitro*, and preliminary studies on the biodistribution of ^{125}I-labelled polyamidoamine dendrimers *in vivo*. *J. Control Release, 65*, 133–148.

Maturavongsadit, P., Bi, X., Gado, T. A., Nie, Y. Z., & Wang, Q., (2016). Adhesive peptides conjugated PAMAM dendrimers as a coating polymeric material enhancing cell responses. *Chin. Chem. Lett., 27*, 1473–1478.

Menjoge, A. R., Kannan, R. M., & Tomalia, D. A., (2010). Dendrimer-based drug and imagingconjugates: Design considerations for nanomedical applications. *Drug Discov. Today, 15*, 171–185.

Mignani, S., Kazzouli, S. E., Bousmina, M., & Majoral, J. P., (2013). Expand classical drug administration ways by emerging routes using dendrimer drug delivery systems: a concise review. *Adv. Drug Deliv. Rev., 65*, 1316–1330.

Moorefield, C. N., Perera, S., & Newkome, G. R., (2012). Dendrimer chemistry: Supramolecular perpspectives and applications. In: *Dendrimer-Based drug Delivery Systems–From Theory to Practice,* Cheng, Y., (ed.), Wiley: Hoboken, 1–54.

Morgan, M. T., Nakanishi, Y., Kroll, D. J., Griset, A. P., Carnahan, M. A., Wathier, M., Oberlies, N. H., Manikumar, G., Wani, M. C., & Grinstaff, M. W., (2006). Dendrimer-encapsulated camptothecins: increased solubility, cellular uptake, and cellular retention affords enhanced anticancer activity *in vitro*. *Cancer Res., 66*, 11913–11921.

Mutalik, S., Shetty, P. K., Kumar, A., Kalra, R., & Parekh, H. S., (2014). Enhancement in deposition and permeation of 5-fluorouracil through human epidermis assisted by peptide dendrimers. *Drug Deliv., 21,* 44–54.

Naha, P. C., & Byrne, H. J., (2013). Generation of intracellular reactive oxygen species and genotoxicity effect to exposure of nanosized polyamidoamine (PAMAM) dendrimers in PLHC-1 cells *in vitro. Aquat. Toxicol., 132–133,* 61–72.

Naha, P. C., Davoren, M., Lyng, F. M., & Byrne, H. J., (2010). Reactive oxygen species (ROS) induced cytokine production and cytotoxicity of PAMAM dendrimers in J774A. 1 cells. *Toxicol. Appl. Pharmacol., 246,* 91–99.

Najlah, M., D'Emanuele, A., (2006). Crossing cellular barriers using dendrimer nanotechnologies. *Curr. Opin. Pharmacol., 6,* 522–527.

Najlah, M., Zhou, Z., D'Emanuele, A., (2012). *D*endrimer-based prodrugs: synthesis and biological evaluation. In: *Dendrimer-Based Drug Delivery Systems–From Theory to Practice,* Cheng, Y., (ed.), Wiley, Hoboken, 157–206.

Nakanishi, Y., & Imae, T., (2005). Synthesis of dendrimer-protected TiO_2 nanoparticles and photodegradation of organic molecules in an aqueous nanoparticle suspension. *J. Colloid. Interface Sci., 285,* 158–162.

Nanjwade, B. K. Bechraa, H. M., Derkara, G. K., Manvia, F. M., & Nanjwade, V. K., (2009). Dendrimers: Emerging polymers for drug-delivery systems. *Eur. J. Pharm. Sci., 38,* 185–196.

Newkome, G. R., Yao, Z. Q., Baker, G. R., & Gupta, V. K., (1985). Micelles. Part 1. Cascade molecules. A new approach to micelles. *J. Org. Chem., 50,* 2003–2004.

Noriega-Luna, B., Godínez, L. A., Rodríguez, F. J., Rodríguez, A., De Larrea, G. Z. L., Sosa-Ferreyra, C. F., Mercado-Curiel, R. F., Manríquez, J., & Bustos, E., (2014). Applications of dendrimers in drug delivery agents, diagnosis, therapy, anddetection. *J. Nanomater,* 1–19.

Nowacka, O., Milowska, K., & Bryszewska, M., (2015). Interaction of PAMAM dendrimers with bovine insulin depends on nanoparticle end-groups. *J. Lumin.,* 87–91.

Otis, J. B., Zong, H., Kotylar, A., Yin, A., Bhattacharjee, S., Wang, H., Baker, Jr. J. R., & Wang, S. H., (2016). Dendrimer antibody conjugate to target and image HER-2 overexpressing cancer cells. *Oncotarget, 7,* 36002–36013.

Padilla De Jésus, O. L., Ihre, H. R., Fréchet, J. M. J., & Szoka, F. C., (2002). Polyester dendritic systems for drug delivery applications: *in vitro* and *in vivo* evaluation. *Bioconjug. Chem., 13,* 453–461.

Parsian, M., Mutlu, P., Yalcin, S., Tezcaner, A., & Gunduz, U., (2016). Half generations magnetic PAMAM dendrimers as an effective system for targeted gemcitabine delivery. *Int. J. Pharm., 515,* 104–113.

Patri, A. K., Kukowska-Latallo, J. F., Baker, Jr. J. R., (2005). Targeted drug delivery with dendrimers: Comparison of the release kinetics of covalently conjugated drug and non-covalent drug inclusion complex. *Adv. Drug Deliv. Rev., 57,* 2203–2214.

Paudel, K. S., Milewski, M., Swadley, C. L., Brogden, N. K., Ghosh, P., & Stinchcomb, A. L., (2010). Challenges and opportunities in dermal/transdermal delivery. *Ther. Deliv., 1,* 109–131.

Perez, A. P., Mundiña-Weilenmann, C., Romero, E. L., & Morilla, M. J., (2012). Increased brain radioactivity bi intranasal ^{32}P-labeled siRNA dendriplexes within in situ-forming mucoadhesive gels. *Int. J. Nanomedicine, 7,* 1373–1385.

Poupot, M., Turrin, C. O., Caminade, A. M., Fournié, J. J., Attal, M., Poupot, R., & Fruchon, S., (2016). Poly(phosphorhydrazone) dendrimers: yin and yang of monocyte activation for human NK cell amplification applied to immunotherapy against multiple myeloma. *Nanomedicine, 12*, 2321–2330.

Prausnitz, M. R., Mitragotri, S., & Langer, R., (2004). Current status and future potential of transdermal drug delivery. *Nat. Rev. Drug Discov., 3*, 115–124.

Prieto, M. J., Zabala, N. E. R., Marotta, C. H., Gutierrez, H. C., Arévalo, R. A., Chiaramoni, N. S., & Alonso, S. V., (2014). Optimization and *in vivo* toxicity evaluation of G4. 5 pamam dendrimer-risperidone complexes. *Plos One, 9*, 1–10.

Pryor, J. B., Harper, B. J., & Harper, S. L., (2014). Comparative toxicological assessment of PAMAM and thiophosphoryl dendrimers using embryonic zebrafish. *Int. J. Nanomedicine, 9*, 1947–1956.

Rajabnezhad, S., Casettari, L., Lam, J. K. W., Nomani, A., Torkamani, M. R., Palmieri, G. F., Rajabnejad, M. R., & Darbandi, M. A., (2016). Pulmonary delivery of rifampicin microspheres using lower generation of dendrimers as a carrier. *Powder Technol., 291*, 366–374.

Rodriguez-Pietro, T., Barrios-Gumiel, A., De la Mataa, J., Sanchez-Nieves, J., & Gomeza, R., (2016). Synthesis of degradable cationic carbosilane dendrimers based on Si-O or ester bonds. *Tetrahedron, 72*, 5825–5830.

Selin, M., Peltronen, L., Hirvonen, J., & Bimbo, L. M., (2016). Dendrimers and their supramolecular nanostructures for biomedical applications. *J. Drug Delivery Sci. Technol., 34*, 10–20.

Shah, K. J., Imae, T., Ujihara, M., Huang, S. J., Wu, P. H., & Liu, S. B., (2017). Poly(amido amine) dendrimer-incorporated organoclays as efficient adsorbents for capture of NH_3 and CO_2. *Chem. Eng. J., 312*, 118–125.

Shi, X., Lee, I., Chen, X., Shen, M., Xiao, S., Zhu, M., Baker, Jr. J. R., & Wang, S. H., (2010). Influence of dendrimer surface charge on the bioactivity of 2-methoxyestradiol complexed with dendrimers. *Soft Matter, 6*, 2539–2545.

Singh, M. K., Pooja, D., Kulhari, H., Jain, S. K., Sistla, R., & Chauhan, A. S., (2017). Poly(amidoamine) dendrimer-mediated hybrid formulation for combination therapy of ramipril and hydrochlorotiazide, *Eur. J. Pharm. Sci., 96*, 84–92.

Somani, S., & Dufès, C., (2014). Applications of dendrimers for brain delivery and cancer therapy. *Nanomedicine (Lond), 9*, 2403–2414.

Sopczynski, B. P., (2008). A new anti-tumor drug delivery system: dendrimers. MMG445. *Basic Biotechnol. J., 2*, 87–92.

Souza, J. G., Dias, K., Silva, S. A. M., Rezende, L. C. D., Rocha, E. M., Emery, F. S., & Lopez, R. F. V., (2015). Transcorneal iontophoresis of dendrimers: PAMAM corneal penetration and dexamethasone delivery. *J. Control Release, 200*, 115–124.

Sreeperumbuduru, R. S., Abid, Z. M., Claunch, K. M., Chen, H. H., McGillivray, S. M., & Simanek, E. E., (2016). *RSC Adv., 6*, 8806–8810.

Stieger, N., Liebenberg, W., Aucamp, M. E., & De Villiers, M. M., (2012). The use of dendrimers to optimize the physicochemical and therapeutic properties of drugs. In: *Dendrimer-Based Drug Delivery Systems–From Theory to Practice,* Cheng, Y., (ed.), Wiley, Hoboken, 93–137.

Sträle, U., Scholz, S., Geisler, R., Greiner, P., Hollert, H., Rastegar, S., Schumacher, A., Selderslaghs, I., Weiss, C., Witters, H., & Braunbeck, T., (2012). Zebrafish embryos as an alternative to animal experiments–a commentary on the definition of the onset of protected life stages in animal welfare regulations. *Reprod. Toxicol., 33*, 128–132.

Sun, W., Mignani, S., Shen, M., & Shi, X., (2016). Dendrimer-based magnetic iron oxide nanoparticles: their synthesis and biomedical applications. *Drug Discov. Today*, *21*, 1873–1885.

Szulc, A., Pulaskib, L., Appelhansc, D., Voitc, B., & Klajnert-Maculewicz, B., (2016). Sugar-modified poly(propylene imine) dendrimers as drug delivery agents for cytarabine to overcome drug resistance. *Int. J. Pharm.*, *513*, 572–583.

Tao, X., Yang, Y. J., Liu, S., Zheng, Y. Z., Fu, J., & Chen, J. F., (2013). Poly(amidoamine) dendrimer-grafted porous hollow silica nanoparticles for enhanced intracellular photodynamic therapy. *Acta. Biomater.*, 9, 6431–6438.

Tekade, R. K., Maheshwari, R. G., Sharma, P. A., Tekade, M., & Chauhan, A. S., (2015). siRNA therapy, challenges and underlying perspectives of dendrimer as delivery vector. *Curr. Pharm. Des.*, *21*, 4614–4636.

Teow, H. M., Zhou, Z., Najlah, M., Yusof, S. R., Abbott, N. J., & D'Emanuele, A., (2013). Delivery of paclitaxel across cellular barriers using dendrimer-based nanocarrier. *Int. J. Pharm.*, *441*, 701–711.

Thiagarajan, G., Greish, K., & Ghandehari, H., (2013). Charge affects the oral toxicity of poly(amido amine) dendrimers. *Eur. J. Pharm. Biopharm.*, *84*, 1–12.

Tomalia, D. A., Baker, H., Dewald, J., Hall, M., Kallos, G., Martin, S., Roeck, J., Ryder, J., & Smith, P., (1985). A new class of polymers: starburst-dendritic macromolecules. *Polym. J.*, *17*, 117–132.

Tripathy, S., & Das, M. K., (2013). Dendrimers and their applications as novel drug delivery carriers, *J. Appl. Pharm. Sci.*, *3*, 142–149.

Van Woensel, M., Wauthoz, N., Rosière, R., Amighi, K., Mathieu, V., Lefranc, F., Van Gool, S. W., & Vleeschouwer, S., (2013). Formulation for intranasal delivery of pharmacological agents to combat brain disease: a new opportunity to tackle GBM? *Cancers*, *5*, 1020–1048.

Vandamme, T. F., & Brobeck, L., (2005). Poly(amidoamine) dendrimers as ophthalmic vehicles for ocular delivery of pilocarpine nitrate and tropicamide. *J. Control Release*, *102*, 23–38.

Venuganti, V. V. K., Sahdev, P., Hildreth, M., Guan, X., & Perumal, O., (2011). Structure-skin permeability relationship of dendrimers. *Pharm. Res.*, *28*, 2246–2260.

Vieira, N. S. C., Figueiredo, A., Faceto, A. D., Queiroz, A. A. A., Zucolotto, V., & Guimarães, F. E. G., (2012). Dendrimers/TiO_2 nanoparticles layer-by-layer films as extended gate FET for pH detection. *Sensors Actuators B. Chem.*, *169*, 397–400.

Villanueva, J. R., Navarro, M. G., & Villanueva, L. R., (2016). Dendrimers as a promising tool in ocular therapeutics: latest advances and perspectives. *Int. J. Pharm.*, *511*, 359–366.

Vivagel, Starpharma. Available at: http://www.starpharma.com/vivagel (Accessed on December 2016).

Waite, C. L., Sparks, S. M., Uhrich, K. E., & Roth, C. M., (2009). Acetylation of PAMAM dendrimers for cellular delivery of siRNA. *BMC Biotechnol.*, *9*, 1–10.

Wang, F., Cai, X., Su, Y., Hu, J., Wu, Q., Zhang, H., Xiao, J., & Cheng, Y., (2012). Reducing cytotoxicity while improving anti-cancer drug loading capacity of polypropylenimine dendrimers by surface acetylation. *Acta. Biomater.*, *8*, 4304–4313.

Win-Shwe, T. T., Sone, H., Kurokawa, Y., Zeng, Y., Zeng, Q., Nitta, H., & Hirano, S., (2014). Effects of PAMAM dendrimers in the mouse brain after a single nasal instillation. *Toxicol. Lett.*, *228*, 207–215.

Xu, L., Kittrell, S., Yeudall, W. A., & Yang, H., (2016). Folic acid-decorated polyamidoamine dendrimer mediates selective uptake and high expression of genes in head and neck cancer cells. *Nanomedicine (Lond).*, *11*, 2959–2973.

Yan, G. P., Ai, C. W., Li, L., Zong, R. F., & Liu, F., (2010). Dendrimers as carriers for contrast agents in magnetic resonance imaging. *Chin. Sci. Bull.*, *55*, 3085–3093.

Yang, Y., Sunoqrot, S., Stowell, C., Ji, J., Lee, C. W., Kim, J. W., Khan, S. A., & Hong, S., (2012). The effect of size, surface charge and hydrophobicity of poly (amidoamine) dendrimers on their skin penetration. *Biomacromolecules, 13*, 2154–2162.

Yao, W., Sun, K., Mu, H., Liang, N., Liu, Y., Yao, C., Liang, R., & Wang, A., (2010). Preparation and characterization of puerarin-dendrimer complexes as an ocular drug delivery. *Drug Dev. Ind. Pharm., 36*, 1027–1035.

Yavuz, B., Bozdaq, P. S., Sumer, B. B., Nomak, S. R., Vural, I., & Unlu, N., (2016). Dexamethasone–PAMAM dendrimer conjugates for retinal delivery: preparation, characterization and *in vivo* evaluation. *J. Pharm. Pharmacol., 68*, 1010–1020.

Yesil-Celiktas, O., Pala, C., Cetin-Uyanikgil, E. O., & Sevimli-Gur, C., (2017). Synthesis of silica-PAMAM dendrimer nanoparticles as promising carrier in Neuro blastoma cells. *Anal. Biochem., 519*, 1–7.

Yoon, A. R., Kasala, D., Li, Y., Hong, J., Lee, W., Jung, S. J., & Yun, C. O., (2016). Antitumor effect and safety profile of systemically delivered oncolytic adenovirus complexed with EGFR-targeted PAMAM-based dendrimer in orthotopic lung tumor model. *J. Control Release, 231*, 2–16.

Zain-ul-Abdin, Li Wang, Haojie Yu, Muhammad, S., Muhammad, A., Hamad, K., Nasir, M., & Abbasi, X. Y., (2017). Synthesis of ethylene diamine-based ferrocene terminated dendrimers and their application as burning rate catalysts. *J. Colloid. Interface. Sci., 487*, 38 51.

Zeng, X., Pan, S., Li, J., Wang, C., Wen, Y., Wu, H., Wang, C., Wu, C., & Feng, M., (2011). A novel dendrimer based on poly (L-glutamic acid) derivatives as an efficient and biocompatible gene delivery vector. *Nanotechnology, 22*, 375102.

Zhu, C. L., Wang, X. W., Lin, Z. Z., Xie, Z. H., & Wang, X. R., (2014). Cell microenvironment stimuli-responsive controlled-release delivery systems based on mesoporous silica nanoparticles. *J. Food Drug Anal., 22*, 18–28.

Ziemba, B., Matuszko, G., Bryszewska, M., & Klajnert, B., (2012). Influence of dendrimers on red blood cells. *Cell. Mol. Biol. Lett., 17*, 21–35.

CHAPTER 3

DENDRIMERS: GENERAL FEATURES AND APPLICATIONS

DURGAVATI YADAV,[1] KUMAR SANDEEP,[2] SHIVANI SRIVASTAVA,[1] and YAMINI BHUSAN TRIPATHI[1]

[1]*Department of Medicinal Chemistry, IMS, BHU, Varanasi-221005, India, E-mail: durgavati45yadav@gmail.com*

[2]*Department of Preventive Oncology, AIIMS, New Delhi, 110029, India*

CONTENTS

ABSTRACT

Dendrimers are the synthetic polymeric systems with improved physical and chemical properties due to their unique three-dimensional architecture. They have well-defined shape, size, molecular weight, and monodispersity. These structures are compatible with bioactive molecules like DNA, heparin, and other polyanions. These are also easily compatible with drug moieties. Their recognition abilities with nanoscopic size make them as most suitable building blocks for their self-assemblage and self-organization systems. The cavities formed inside the dendritic structure can be modified to incorporate hydrophobic or hydrophilic drugs. The dendritic terminal groups can be modified and transformed to attach various antibodies and bioactive substances for targeting purposes along with providing miscibility, reactivity, and solubility. Recently, dendrimers have aroused great interest for targeted delivery of drug molecules via different routes as a nanocarrier. Toxicity problems related to positive ion carrying dendrimers are overcome by PEGylation, which neutralizes the charge on them, making suitable for use. Dendrimers possess suitable properties to establish themselves as a potential carrier for delivery of therapeutic agents irrespective of certain synthetic and regulatory constraints. This chapter contains structural aspects and properties of dendrimers along with their pharmaceutical application as a potential novel drug delivery carrier.

3.1 INTRODUCTION

Dendrimers are a new class of polymeric materials. They are typically symmetrical around the core and generally adopts a spherical three-dimensional (3-D) architecture, which provides a high degree of surface functionality and versatility (Nanjwade et al., 2009). The term "cascade molecule" is also used for them. The dendritic structure (called as Dendron) is found in the central nervous system and the brain with large amount of cells growing to gain exchange of material and information with the surrounding tissue (Xu et al., 2014). A striking example of natural dendritic structures is discovered on gecko's feet with the maximum number of foot hair (Sitti and Fearing, 2003).

Comparison of size of dendrimers with biological actives are shown in Table 3.1. The dendrimer characterization using spectroscopic techniques is very useful and has wider application in the field of dendrimer chemistry. The characterization technique includes the study to decipher chemical composition, morphology, shape, homogeneity, molecular weight, synthesis, reaction rate, purity, conjugation, and dendrimer polydispersity (Crooks et al., 2001).

Various techniques used in dendrimer characterization are as follows:

- Ultra violet-visible spectroscopy
- Infrared spectroscopy

TABLE 3.1 Comparison of Size of Dendrimers with Biological Molecules

S.No	Dendrimers	Biological molecule
1	G3 dendrimer	Insulin
2	G4 dendrimer	Cytochrome C
3	G5 dendrimer	Haemoglobin
4	G6 dendrimer	Transthyretin
5	G7 dendrimer	Histone

- NMR
- Mass spectroscopy
- Raman spectroscopy
- Fluorescence spectroscopy
- Atomic force microscopy
- Electron microscopy (TEM & SEM)
- Chromatography
- Electrophoresis

Dendrimer characterization involves techniques of spectrophotometric analysis, electrophoresis, and microscopy as depicted in Figure 3.1.

3.2 TYPES OF DENDRIMERS

3.2.1 PAMAM DENDRIMER

Poly (amidoamine) dendrimers (PAMAM) are synthesized by the divergent method starting from ammonia or ethylenediamine initiator core reagents (Newkome and Shreiner, 2008). These are commercially available as methanol solutions. These dendrimers have unique 3-D architecture, which enabled them for use in siRNA condensing agents. Dendrimer-mediated RNA delivery has shown great potential for the management of several diseases. A type of dendrimer called "Starburst dendrimers" is applied as a trademark name for a sub-class of PAMAM dendrimers, which is based on the core of trisaminoethylene-imine (Dobrovolskaia et al., 2012). It is named after its star-like appearance when looking at the structure of the high-generation dendrimers of this type in two dimensions (Shcharbin et al., 2014). Further studies on high-generation dendrimers are being evaluated for their intrinsic toxicity as lower generation dendrimers are not showing intrinsic toxicity.

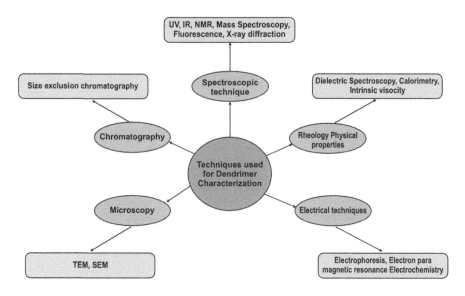

FIGURE 3.1   Techniques used for dendrimer characterization.

3.2.2   PAMAMOS DENDRIMER

Radially layered poly-(amidoamine-organosilicon) dendrimers (PAMA-MOS) are inverted mono-molecular micelles constituted of hydrophilic, nucleophilic polyamidoamine (PAMAM) interiors, and hydrophobic organo-silicon (OS) exteriors (Kao and Yang, 2006). These dendrimers are exceptionally useful precursors for the preparation of honeycomb-like networks, sheets, and coatings with nanoscopic PAMAM and OS domains. These dendrimers are first silicone-containing dendrimers and the best-characterized and utilized dendrimers of all families. These dendrimers have good use in purification and decontamination. These dendrimers have strong affinity toward electrophiles, making them suitable for the absorption and elimination of chemical compounds from air and aquatic streams.

3.2.3   PPI DENDRIMER

Poly (propylene imine) (PPI) dendrimers describes the propylamine spacer moieties. It is the oldest known dendrimer type developed initially by Vogtle. These dendrimers are generally polyalkyl amines; with primary amines as an end group. Dendrimer interior consists of numerous tertiary tris-propylene amines (Garg et al., 2011) and commercially available up to G5 with widespread applications in the field of material science and biology. The POPAM is sometimes used to describe this class of dendrimers. POPAM (stands for polypropylene amine), which closely resembles the PPI abbreviation (Krishna et al., 2005). In addition, these dendrimers are also sometimes denoted "DAB-dendrimers" where DAB refers to the core structure containing diamino butane. Dendrimers of this family of higher generations have porous structures with rough surfaces. There were several modifications that have been used for reducing the toxicity of these dendrimers for various applications, like maltotriose modifications significantly reduced the toxicity in series of G-4 dendrimers.

3.2.4 TECTO DENDRIMER

Tecto dendrimers are composed of a core dendrimer, surrounded by dendrimers of several steps in its periphery to perform necessary function for therapeutics. Different compounds perform varied functions ranging from diseased cell recognition, diagnosis of diseased state, drug delivery, reporting outcomes therapy, etc. These dendrimers hold a promise for the multidrug delivery and environmental remediation applications. They have a large number of potential reactive units. Their bioavailability is tuned by varying the number of tecto units present in the dendrimer.

3.2.5 MULTILINGUAL DENDRIMERS

In multilingual type of dendrimers, the surface contains numerous copies of a particular functional group.

3.2.6 CHIRAL DENDRIMERS

The chirality in these dendrimers is based upon the construction of a constitutionally different but chemically similar branch to the chiral core. It is generally used for enantiomeric resolutions and as catalysts for asymmetric synthesis. They are synthesized using click chemistry (CuAAC) and can be further modified by thiol-ene and thiol-yne click reactions with alcohols to form dendritic polyol sod dense structure.

3.2.7   HYBRID DENDRIMERS LINEAR POLYMERS

These are the hybrids (block or graft polymers) of dendritic and linear polymers with properties of both the types. They are utilized for developing sophisticated nanostructures with excellent performances for delivering versatile nanoplatform.

3.2.8   AMPHIPHILIC DENDRIMERS

They are built with two segregated sites of the chain end: one half is electron donating, while the other half is electron withdrawing. Amphipathic dendrimers are generally based on polyamidoamine (PAMAM) dendrimers, and these molecules display cell membrane affinity and hence helpful in reducing the cytotoxic effect. They are able to self-assemble into an aggregate, resulting in high multivalent structure. They are used for delivering small interference RNA (siRNA) and produce notable gene silencing effect.

3.2.9   MICELLAR DENDRIMERS

These are unimolecular micelles of water-soluble hyperbranched polyphenylenes. Micellization is a common process, and it is widely utilized in drug delivery to increase the bioavailability of nutrients and lipophilic drugs. Macromolecules are better suited for this due to their lower critical micelle concentration and high stability. Polymeric micelles can easily entrap hydrophobic drugs at the core that can be transported at concentrations higher than their intrinsic water solubility. They are very attractive in targeted drug delivery to cancer cells. Here, the drug can be encapsulated into micelles by two processes: by covalent attachment or by physical encapsulation.

3.2.10   MULTIPLE ANTIGEN PEPTIDE DENDRIMERS

It is a dendron-like molecular construct based upon a polylysine skeleton. Lysine with its alkyl group amino side chain serves as a good monomer for the introduction of numerous branching points. This type of dendrimer was first introduced by J.P. Tam in 1988 and has predominantly found its use in biological applications, for example, vaccine and diagnostic research (Patel and Patel, 2013). Multiple antigen peptide (MAP) dendrimer-based enzyme immunoassays have been developed for the detection of simian immunodeficiency virus (SIV) in non-human primates. Recently, MAPs were used for detecting rat and mouse growth hormone proteins for scientific studies.

3.2.11 FRECHET-TYPE DENDRIMERS

It is one of the recent types of dendrimer developed by Hawker and Frechet based on polybenzyl ether with various branched skeleton. This type of dendrimers have carboxylic acid groups as surface groups serving as good anchoring point for further surface functionalization and as polar surface groups to increase the solubility in polar solvents or aqueous media (Cheng et al., 2008). Generally, both symmetrical and unsymmetrical dendrimers are synthesized by click chemistry with a tripodal core.

3.3 PHARMACEUTICAL APPLICATIONS OF DENDRIMERS

Dendrimers have wide applications in pharmaceutical chemistry as shown in Figure 3.2 in different fields.

3.3.1 DENDRIMERS IN PULMONARY DRUG DELIVERY

Dendrimers have been reported to be used for pulmonary drug delivery of Enoxaparin. It is found that G2 and G3 generation of positively charged PAMAM dendrimers increase the relative bioavailability of Enoxaparin by 40% (Nanjwade et al., 2009). The positively charged dendrimer forms a complex with Enoxaparin, which is effective in deep vein thrombosis after its pulmonary administration (Lee and Larson, 2008).

3.3.2 DENDRIMERS IN TRANSDERMAL DRUG DELIVERY

Dendrimers improve the solubility and plasma circulation time via transdermal formulations and deliver drugs efficiently. PAMAM dendrimer complex

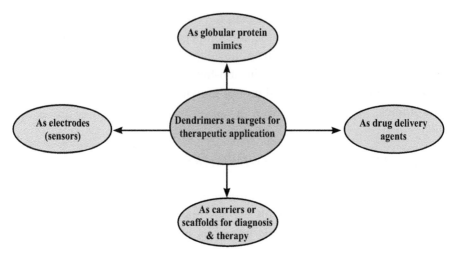

FIGURE 3.2 Application of dendrimers in drug delivery technology.

with NSAIDs (e.g., Ketoprofen, Diflunisal) have been reported to improve the drug permeation through the skin as it enhances the penetration. Ketoprofen and Diflunisal when conjugated with G5 PAMAM dendrimer showed 3.2 to 3.4 times higher permeation rate. Enhanced bioavailability of PAMAM dendrimers using indomethacin as the model drug in transdermal drug application was reported to be effective (Jevprasesphant, 2003; Jevprasesphant et al., 2003; Ahmed et al., 2016).

3.3.3 DENDRIMERS IN ORAL DRUG DELIVERY

Oral drug delivery studies using human colon adenocarcinoma cell line ($CaCO_2$) have indicated that low-generation PAMAM dendrimers crosses cell membranes through the combination of two processes, that is, paracellular transport and adsorptive endocytosis. The P-glycoprotein efflux transporters do not affect dendrimers; therefore, drug-dendrimer complexes are able to bypass the efflux transporters. PAMAM dendrimers conjugated with folic acid and fluorescein isothiocyanate used for targeting the tumor cells and imaging, respectively. DNA-assembled dendrimer conjugates allow the combination of different drugs with different targeting and imaging agents, as shown in Figure 3.2 (Tam and Spetzler, 2001; Pedziwiatr-Werbicka et al., 2011; Ziemba et al., 2011).

3.3.4 DENDRIMERS IN NASAL DRUG DELIVERY

The transmucosal routes via mucosal linings of the nasal, rectal, vagina, ocular, or oral cavity offer distinct advantages over peroral administration for systemic drug delivery. These advantages include possible bypass of the first pass effect and avoidance of pre-systemic elimination of gastro-intestinal tract depending on the particular drug (Svenson and Tomalia, 2012; Caminade and Turrin, 2014). For achieving systemic drug effects, administration through the nose offers an interesting alternative to the parenteral route, which can be inconvenient, or to oral administration that results in low bioavailabilities. PAMAM (polyamidoamine) dendrimer has brought the attention for nose-to-brain targeting and are repetitive branches that grow from a core. Many versatile molecules can be attached to their surface (Al-Jamal et al., 2005), like linking an arginine onto the surface of dendrimer, resulting in nanoparticles with a size of 188.7 ± 1.9 nm and a charge of +22.3 mV (He et al., 2010). siRNA targeting against the high mobility group box 1 (HMGB1) was electrostatically attached onto the nanoparticles (Hu et al., 2011). HMGB1 released by dying cells acting as a danger signal, thereby aggravating the damage of a stroke or other neurotoxic insults. One report says that upon intranasal administration, there is a wide distribution of the construct into the brain, including the hypothalamus, amygdala, cerebral cortex, and striatum. It is found that the localization of the PAMAM dendrimer and the siRNA is associated with an efficient knockdown of the protein HMGB-1. When a stroke was induced into experimental animals, the group receiving the intranasal administration of the construct had a significantly decreased infarction volume (Pavan et al., 2010; Biswas et al., 2013).

The potential of mucoadhesive gel of dendrimer formulations for nose to brain delivery was also observed (Shadab et al., 2013; Rassu et al., 2016). siRNA coupled to PAMAM dendrimers form dendriplexes, which are then formulated into muco-adhesive gels containing either 1% (w/w) chitosan or 0.25% (w/w) carbopol 974P NFTM as an effective target. These gels were prepared by blending the chitosan/carbopol with 23% (w/w) of thermosensible poloxamer to obtain *in situ* gelation. The resulting thermosetting gel has a phase transition below the temperature in the nasal cavity (32°C to 35°C) and above room temperature and can be given as a liquid. Different concentrations of the different gels were tested, and no toxicity was observed. Two intranasal doses were necessary to achieve high brain concentrations of radioactivity than achieved by intravenous administration of dendriplexes or intranasal administration of naked siRNA.

3.3.5 DENDRIMERS IN CELLULAR DELIVERY

PAMAM dendrimers with lauryl chains were used to reduce toxicity and enhance cellular uptake of the drug. For example, dendrimer–ibuprofen complexes entered the cells rapidly compared with pure drug (1 h versus >3 h), suggesting that dendrimers can efficiently carry drug complexes inside cells (Waite et al., 2009).

3.3.6 DENDRIMER HYDROGEL FOR OCULAR DRUG DELIVERY

Dendrimers are ideal for synthesizing hydrogels; hydrogels are cross-linked networks that increase in volume in aqueous solution and are more similar to living tissue than any other synthetic compound. Addition of PEG groups to the dendrimers found applications in cartilage tissue production and for sealing ophthalmic injuries. Hydrogel composed of PEGylated dendrimers containing ocular drug molecules attached to the dendrimers efficiently delivers the drug to the eye (Ludwig, 2005; Hamidi et al., 2008; Holden et al., 2012; Kambhampati and Kannan, 2013).

3.3.7 DENDRIMERS FOR CONTROLLED AND TARGETED DRUG DELIVERY

One of the most effective cell-specific targeting agents delivered by dendrimers is folic acid PAMAM dendrimers modified with carboxymethyl PEG5000 surface chains; this possessed reasonable drug loading, a reduced release rate, and reduced hemolytic toxicity compared with the non-PEGylated dendrimer (Cloninger, 2002). The star polymers were reported to give the most promising results regarding cytotoxicity and systemic circulatory half-life (72 h).

Anticancer drugs like adriamycin and methotrexate when encapsulated into PAMAM dendrimers (i.e., G3 and G4) with modified PEG monomethyl ether chains (i.e., 550 and 2000 Da, respectively) attached to their surfaces are better for effective release. A similar construct involving PEG chains and PAMAM dendrimers was used to deliver the anticancer drug 5-fluo-rouracil. Encapsulation of 5-fluorouracil into G4 increases the cytotoxicity and permeation of dendrimers. Controlled release of the flurbiprofen can be achieved by the formation of a complex with amine-terminated generation 4 (G4) PAMAM dendrimers (Gomez et al., 2009; Souza et al., 2015).

Polymeric dendrimer carriers facilitate the passive targeting of drugs to solid tumors, leading to the selective accumulation of macromolecules in tumor tissue, a phenomenon termed as "Enhanced Permeation and Retention" (EPR) effect. Therefore, the anticancer drug doxorubicin was reported to be covalently bound to this carrier via an acid-labile hydrazone linkage. The cytotoxicity of doxorubicin was significantly reduced (80–98%), and the drug was successfully taken up by several cancer cell lines (Greish, 2007; Torchilin, 2010; Miyata et al., 2011).

3.3.8   DENDRIMERS AS NANO-DRUGS

Poly(lysine) dendrimers modified with sulfonated naphthyl groups have been found to be useful as antiviral drugs against herpes simplex virus, potentially reducing the transmission of HIV and sexually transmitted diseases (STDs). This dendrimer-based nanodrug inhibits early stage virus/cell adsorption and later stage viral replication by interfering with reverse transcriptase or integrate enzyme activities (Tomalia and Fréchet, 2002; Boas and Heegaard, 2004; Svenson, 2009; Twibanire and Grindley, 2014). PPI dendrimers with tertiary alkyl ammonium groups attached to the surface have been shown to be potent antibacterial biocides against gram-positive and gram-negative bacteria. Poly(lysine) dendrimers with mannosyl surface groups are effective inhibitors of E. coli to infect horse blood cells in a hemagglutination assay. Chitosan dendrimer hybrids were also found to be useful as antibacterial agents (Sashiwa et al., 2003; Jayakumar et al., 2005).

3.3.9   DENDRIMERS IN GENE TRANSFECTION

Dendrimers can act as vectors in genetic treatment or gene-based therapy, for example, PAMAM dendrimers have been tested as genetic material carriers. Activated dendrimers carry a larger amount of genetic material than viruses. Amino-terminated PAMAM or PPI dendrimers have been reported as nonviral gene transfer agents, enhancing the transfection of

DNA by endocytosis. A transfection reagent called SuperFect™ consisting of activated dendrimers is commercially available (Dufès et al., 2005; Shcharbin et al., 2009; Liu et al., 2014). SuperFect-DNA complexes are characterized by high stability and provide more efficient transport of DNA into the nucleus than liposomes. The transfection efficiency of dendrimers is attributed to its well-defined shape and by the low pK of the amines (3.9 and 6.9) attached. The low pK allows the dendrimer to buffer the pH changes in the endosome. PAMAM dendrimers functionalized with cyclodextrin showed luciferase gene expression 100 times higher than non-functionalized PAMAM or for non-covalent mixtures of PAMAM and cyclodextrin. Dendrimers of high structural flexibility and partially degraded high-generation or hyperbranched are better suited for certain gene delivery operations (Paleos et al., 2006, 2007; Perumal et al., 2008; Santos et al., 2010) as shown in Table 3.2.

Dendrimers have proved to be very useful in health industry with many biological advances (Figure 3.3) in improving the diseased condition.

TABLE 3.2   Application of Dendrimers in the Treatment of Various Diseases

S No.	Uses of Dendrimers	Types used	Effective Treatment
1	Pulmonary drug delivery	PAMAM dendrimers complexed with Enoxaparin	Useful in deep vein thrombosis.
2	Transdermal drug delivery	PAMAM dendrimers complexed with NSAIDS	It improves the solubility and plasma circulation time.
3	Oral drug delivery	PAMAM dendrimers conjugate with folic acid	They are able to bypass the efflux transporters.
4	Nasal drug delivery	PAMAM dendrimers, Chitosans	By their drugs there is avoidance of pre-systemic elimination of gastro intestinal tract.
5	Cellular delivery	PAMAM dendrimers	Rapid entry of drugs into the cells.
6	Ocular drug delivery	Hydrogels	They are effective in cartilage tissue production & for sealing ophthalmic injuries.
7	Nano-drug	PDI, Chitosan dendrimers	They inhibit early stage cell adsorption & later stage viral replication.
8	Gene Transfection	PAMAM, mixer of PAMAM & cyclodextrin	They carry larger amount of genetic material.

3.4 TOXICITY AND PEGYLATION

It is believed that the dendrimers have toxicity mainly due to the interaction of the cationic surface with negative biological load membranes, thereby damaging cellular membranes and causing hemolytic toxicity and cytotoxicity. Therefore, PAMAM dendrimers are more cationic than anionic and hence decreases cytotoxicity. An example of interaction with lipid bilayers of cells occurs with the cationic dendrimer-G7 PAMAM, which form holes of 15–40 nm in diameter and disturbs the flow of electrolytes, causing cell death. Many toxic effects of dendrimers are attenuated at their surfaces with hydrophilic molecules, while PEG masks the surface charge (cationic dendrimers), thus improving biocompatibility and increasing the solubility of the polymers. The PEGylated dendrimers have lower cytotoxicity

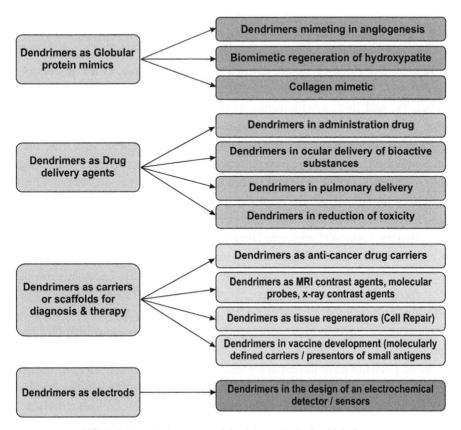

FIGURE 3.3 Various uses of dendrimers in the health industry.

and longer stay in the blood than non-PEGylated dendrimers (Casettari et al., 2010; Jevševar et al., 2010; Jokerst et al., 2011; Milla et al., 2012). PEGylation increases the dendrimers size, which in turn reduces renal clearance (Okuda et al., 2006; Veronese and Mero, 2008; Milla et al., 2012).

3.5 PROPERTIES THAT MAKE DENDRIMERS SPECIAL FOR DRUG DELIVERY

Dendrimers are useful nanoscale carriers for drug and gene delivery as both hydrophilic and hydrophobic drug molecules can be formulated and can be applied in intravenous, oral, pulmonary, nasal, ocular, and transdermal drug delivery systems. They have shown enormous potential as nanocarrier delivery systems because they can cross cell barriers by paracellular and transcellular pathways. The ability to modify and optimize the number and/or ratio of surface groups influences biodistribution, receptor-mediated targeting, therapy dosage, or controlled release of drugs from the dendrimer interior.

The 3-D structure of dendrimers gives them unique properties such as nanoscaled globular shape, defined functional groups at the periphery, hydrophobic or hydrophilic cavities in the interior, and extremely low polydispersity; and thus, a wide range of potential applications. For example, most dendrimers have globular structures (diameters < 10 nm); this property gives them similar sizes and shapes as specific proteins and other biomolecules and thereby makes them as perfect biomimics. Their regular branching pattern of dendrimers confers them dendritic architectures with well-defined periphery functional groups numbers, providing sites for the adherence of drug molecules, targeting moieties and solubilizing groups on the surface in a multivalent fashion (Smith and Diederich, 1998; Lukin et al., 2006; Antoni et al., 2007). The hydrophobic and hydrophilic cavities in the interior make them useful candidates as unimolecular micelles for the encapsulation of drug molecules, especially drugs (Narang et al., 2007; Cao et al., 2011). The low polydispersity of dendrimers assures the reproducibility of biodistribution of polymeric pro-drugs, when using them as scaffolds (Najlah and D'Emanuele, 2006; Shcharbin et al., 2014). Nanoparticles having a size range of 1–10 nm have the capacity to diffuse into tumor cells, thereby helping to overcome the limitations in chemotherapy by free drug, such as resistance exhibited by tumors. Dendrimers are useful in non-viral gene delivery systems. Their ability to transfect cells without inducing toxicity and to be tuned for stimuli-induced gene delivery confers a great advantage over other gene delivery vectors for use *in vivo* (Davis, 2002; Lungwitz et al., 2005; Liu et al., 2012). The well-defined hyperbranched structure of dendrimers has opened the gates for chemists to explore the possibility for mimicking protein functions with dendritic macromolecules, such as O_2-carrying hemoproteins and Coenzyme B_{12}. Thus, dendrimers can be used for: (i) stimuli-responsive nanocarriers, (ii) molecular tags, (iii) high payload efficiency, (iv) decrease in dosage requirements and re-dosage frequency, and (v) target delivery and minimization of drug migration, thus suppressing secondary effects during drug treatment. They possess their utility due to their multi-factorial properties (Figure 3.4).

Therefore, detailed *in vivo* toxicity examination of dendrimers is an important point and can facilitate the design of tailored dendrimer-mediated CNS drug-delivery systems. They can be synthesized and designed for specific applications. Due to their feasible topology, functionality, and dimensions, they are ideal drug-delivery systems, and their size also is very close to various important biological polymers and assemblies such as DNA and proteins, which is physiologically ideal.

The covalent attachment of drugs to the surface groups of dendrimers through hydrolysable or biodegradable linkages enhances the pharmacological properties of the drug and offers the opportunity for a greater control over drug release (Yang et al., 2008; Sato and Anzai, 2013).

3.6 MECHANISMS OF DRUG DELIVERY

Broadly, two mechanisms are involved for drug delivery of dendrimers:

(1) *In vivo* degradation of drug–dendrimer covalent bonding depending on the presence of suitable enzymes or factors responsible for cleaving the bonds.

(2) Release of drug due to changes in physical environment such as pH, temperature etc. This is independent of the external factors and takes place in cavities of the core (endoreceptor) or outer shell of the receptor (exoreceptor) (D'Emanuele and Attwood, 2005; Menjoge et al., 2010).

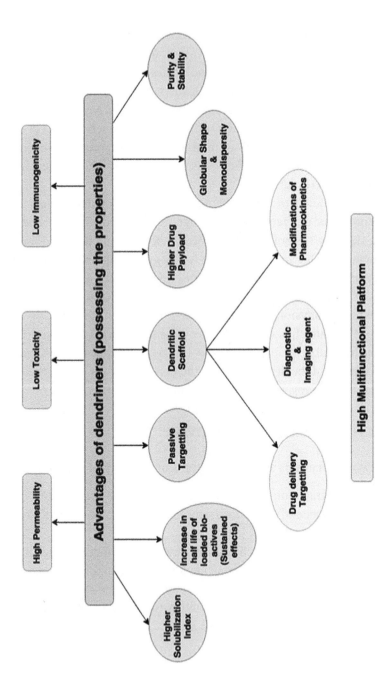

FIGURE 3.4 Properties of dendrimers that make them useful for various drug delivery technologies.

3.7 CONCLUSIONS

Dendrimer drug delivery technologies have an advantageous solution for delivering drugs and genes efficiently because of their nanosize, radially symmetric molecules with well-defined structure. They play a key role as enabling building blocks for nanotechnology. A variety of dendrimers exists, and each has biological properties such as polyvalency (responsible for multiple interactions with biological receptor sites, e.g., design of antiviral therapeutic agents), self-assembly, chemical stability, precise molecular weight, and low cytotoxicity. Liposomes and polymeric drug conjugates hold the maximum share of it, for example, doxorubicin HCl liposome injection. Development of synthesis techniques have led to the emergence of dendritic structures with a large number of surface groups that can be used for a wide range of display of biological molecules for targeting. Cascade reactions are the foundation of dendrimer synthesis, and generally, they follow either a divergent (synthesis starts from the core to which arms are attached by adding building blocks) or convergent approach (synthesis starts from the exterior starting with the molecular structure that finally becomes the outermost arm). Convergent approach is suitable for only lower generation of dendrimers. Hence, they possess huge diversity in their size and branching architectural structures. Further, they possess range of properties like high loading capacity, which makes them a good carrier molecule for the delivery of chemotherapeutic agents. PEGylated and non-PEGylated dendrimers encapsulate hydrophobic drug molecules into the hollow voids of their branching architecture enhancing the aqueous solubility and stability of the encapsulated drug molecules. Dendrimer-drug complexes successfully penetrate across the tumor's leaky vasculature and accumulate in the tissue. Targeted dendrimer drug complexes increase their residence time on the cell surface and enhance the internalization kinetics into the cell. However, there are limitations in the dendrimeric immunogen synthesis, which limits their use. Multiple antigen peptides are regarded as one of the most efficacious methods for antigen presentation; however, in the case of multiple antigen peptide preparation, the conjugation process in the solution is inconvenient.

Pharmacokinetic properties need to be considered for successful biomedical application of dendrimers, for instance, drug delivery, bioimaging, photodynamic therapy (treatment of tumor cells by photosensitizers), and boron neutron capture therapy (this method is also involved in cancer treatment).

Their targeted drug delivery reduces the side effects associated with conventional therapy. Their use is constrained due to their high manufacturing cost and some inherent toxicity issues. It still finds a good use in the therapeutic applications for a wide number of diseases, and it opens gates for further exploration in drug discovery and clinical applications.

KEYWORDS

- **dendritic structure**
- **drug delivery**
- **PEGylation**

REFERENCES

Ahmed, S., Vepuri, S. B., Kalhapure, R. S., et al., (2016). Interactions of dendrimers with biological drug targets: reality or mystery–A gap in drug delivery and development research. *Biomater. Sci., 4,* 1032–1050.

Al-Jamal, K. T., Ramaswamy, C., & Florence, A. T., (2005). Supramolecular structures from dendrons and dendrimers. *Adv. Drug Deliv. Rev., 57,* 2238–2270.

Antoni, P., Nystrom, D., Hawker, C. J., Hult, A., & Malkoch, M., (2007). A chemoselective approach for the accelerated synthesis of well-defined dendritic architectures. *Chem. Commun.,* 2249–2251.

Biswas, S., Deshpande, P. P., Navarro, G., Dodwadkar, N. S., & Torchilin, V. P., (2013). Lipid modified triblock PAMAM-based nanocarriers for siRNA drug co-delivery. *Biomaterials, 34,* 1289–1301.

Boas, U., & Heegaard, P. M. H., (2004). Dendrimers in drug research. *Chem. Soc. Rev., 33,* 43–63.

Caminade, A.-M., & Turrin, C. O., (2014). Dendrimers for drug delivery. *J. Mater. Chem. B., 2,* 4055.

Cao, W., Zhou, J., Mann, A., Wang, Y., & Zhu, L., (2011). Folate-functionalized unimolecular micelles based on a degradable amphiphilic dendrimer-like star polymer for cancer cell-targeted drug delivery. *Biomacromolecules, 12,* 2697–2707.

Casettari, L., Vllasaliu, D., Mantovani, G., Howdle, S. M., Stolnik, S., & Illum, L., (2010). Effect of PEGylation on the toxicity and permeability enhancement of chitosan. *Biomacromolecules, 11,* 2854–2865.

Cheng, Y., Xu, Z., Ma, M., Xu, T. 2008. Dendrimers as drug carriers: Applications in different routes of drug administration. *J. Pharm. Sci., 79.*

Cloninger, M. J., (2002). Biological applications of dendrimers. *Curr. Opin. Chem. Biol., 6,* 742–748.

Crooks, R. M., Zhao, M., Sun, L., Chechik, V., & Yeung, L. K., (2001). Dendrimer-encapsulated metal nanoparticles: Synthesis, characterization, and applications to catalysis. *Acc. Chem. Res., 34,* 181–190.

D'Emanuele, A., & Attwood, D., (2005). Dendrimer-drug interactions. *Adv. Drug Deliv. Rev., 57,* 2147–2162.

Davis, M. E., (2002). Non-viral gene delivery systems. *Curr. Opin. Biotechnol., 13,* 128–131.

Dobrovolskaia, M. A., Patri, A. K., Simak, J., Hall, J. B., Semberova, J., De Paoli Lacerda, S. H., & McNeil, S. E., (2012). Nanoparticle size and surface charge determine effects of PAMAM dendrimers on human platelets *in vitro. Mol. Pharm., 9,* 382–393.

Dufès, C., Uchegbu, I. F., & Schätzlein, A. G., (2005). Dendrimers in gene delivery. *Adv. Drug Deliv. Rev., 57,* 2177–2202.

Garg, T., Singh, O., Arora, S., & Murthy, R. S. R., (2011). Dendrimer- a novel scaffold for drug delivery. *Int. J. Pharm. Sci. Rev. Res., 7,* 211–220.

Gomez, M. V., Guerra, J., Velders, A. H., & Crooks, R. M., (2009). NMR characterization of fourth-generation PAMAM dendrimers in the presence and absence of palladium dendrimer-encapsulated nanoparticles. *J. Am. Chem. Soc., 131,* 341–350.

Greish, K., (2007). Enhanced permeability and retention of macromolecular drugs in solid tumors: a royal gate for targeted anticancer nanomedicines. *J. Drug Target., 15,* 457–464.

Hamidi, M., Azadi, A., & Rafiei, P., (2008). Hydrogel nanoparticles in drug delivery. *Adv. Drug Deliv. Rev., 60,* 1638–1649.

He, C., Hu, Y., Yin, L., Tang, C., & Yin, C., (2010). Effects of particle size and surface charge on cellular uptake and biodistribution of polymeric nanoparticles. *Biomaterials, 31,* 3657–3666.

Holden, C. A., Tyagi, P., Thakur, A., Kadam, R., Jadhav, G., Kompella, U. B., & Yang, H., (2012). Polyamidoamine dendrimer hydrogel for enhanced delivery of antiglaucoma drugs. *Nanomed. Nanotech. Biol. Med., 8,* 776–783.

Hu, H. C., Wang, T. Y., Chen, Y. C., Wang, C. C., & Lin, M. C., (2011). RNA interference inhibits high mobility group box 1 by lipopolysaccharide-activated murine macrophage RAW 264. 7 secretion. *J. Surg. Res., 168,* 181–7.

Jayakumar, R., Prabaharan, M., Reis, R. L., & Mano, J. F., (2005). Graft copolymerized chitosan - Present status and applications. *Carbohydr. Polym., 62,* 142–158.

Jevprasesphant, R., (2003). The influence of surface modification on the cytotoxicity of PAMAM dendrimers. *Int. J. Pharm., 252,* 263–266.

Jevprasesphant, R., Penny, J., Jalal, R., Attwood, D., McKeown, N. B., & D'Emanuele, A., (2003). The influence of surface modification on the cytotoxicity of PAMAM dendrimers. *Int. J. Pharm., 252,* 263–266.

Jevševar, S., Kunstelj, M., & Porekar, V. G., (2010). PEGylation of therapeutic proteins. *Biotechnol. J., 5,* 113–128.

Jokerst, J. V., Lobovkina, T., Zare, R. N., & Gambhir, S. S., (2011). Nanoparticle PEGylation for imaging and therapy. *Nanomedicine (Lond)., 6,* 715–728.

Kambhampati, S. P., & Kannan, R. M., (2013). Dendrimer nanoparticles for ocular drug delivery. *J. Ocul. Pharmacol. Ther., 29,* 151–65.

Kao, W. Y. J., & Yang, H., (2006). Dendrimers for pharmaceutical and biomedical applications. *J. Biomater. Sci. Ed., 17,* 3–19.

Krishna, T. R., Jain, S., Tatu, U. S., & Jayaraman, N., (2005). Synthesis and biological evaluation of 3-amino-propan-1-ol based poly(ether imine) dendrimers. *Tetrahedron, 61,* 4281–4288.

Lee, H., & Larson, R. G., (2008). Coarse-grained molecular dynamics studies of the concentration and size dependence of fifth-and seventh-generation PAMAM dendrimers on pore formation in DMPC bilayer. *J. Physcial Chem.*, *112*, 7778–7784.

Liu, H., Wang, Y., Wang, M., Xiao, J., & Cheng, Y., (2014). Fluorinated poly(propylenimine) dendrimers as gene vectors. *Biomaterials*, *35*, 5407–5413.

Liu, X., Rocchi, P., & Peng, L., (2012). Dendrimers as non-viral vectors for siRNA delivery. *New J. Chem.*, *36*, 256.

Ludwig, A., (2005). The use of mucoadhesive polymers in ocular drug delivery. *Adv. Drug Deliv. Rev.*, *57*, 1595–1639.

Lukin, O., Gramlich, V., Kandre, R., Zhun, I., Felder, T., Schalley, C. A., & Dolgonos, G., (2006). Designer dendrimers: Branched oligosulfonimides with controllable molecular architectures. *J. Am. Chem. Soc.*, *128*, 8964–8974.

Lungwitz, U., Breunig, M., Blunk, T., & Gopferich, A., (2005). Polyethylenimine-based non-viral gene delivery systems. *Eur. J. Pharm. Biopharm.*, 247–266.

Menjoge, A. R., Kannan, R. M., & Tomalia, D. A., (2010). Dendrimer-based drug and imaging conjugates: design considerations for nanomedical applications. *Drug Discov. Today,* *15*, 171–185.

Milla, P., Dosio, F., & Cattel, L., (2012). PEGylation of proteins and liposomes: A powerful and flexible strategy to improve the drug delivery. *Curr. Drug Metab.,* *13*, 105–119.

Miyata, K., Christie, R. J., & Kataoka, K., (2011). Polymeric micelles for nano-scale drug delivery. *React. Funct. Polym.,* *71*, 227–234.

Najlah, M., & D'Emanuele, A., (2006). Crossing cellular barriers using dendrimer nanotechnologies. *Curr. Opin. Pharmacol.*, *6*, 522–527.

Nanjwade, B. K., Bechra, H. M., Derkar, G. K., Manvi, F. V., & Nanjwade, V. K., (2009). Dendrimers: Emerging polymers for drug-delivery systems. *Eur. J. Pharm. Sci.,* *138*, 185–196.

Narang, A. S., Delmarre, D., & Gao, D., (2007). Stable drug encapsulation in micelles and microemulsions. *Int. J. Pharm.,* *345*(1–2), 9–25.

Newkome, G. R., & Shreiner, C. D., (2008). Poly(amidoamine), polypropylenimine, and related dendrimers and dendrons possessing different 1,2 branching motifs: An overview of the divergent procedures. *Polymer (Guildf).*, *49*, 1–173.

Okuda, T., Kawakami, S., Maeie, T., Niidome, T., Yamashita, F., & Hashida, M., (2006). Biodistribution characteristics of amino acid dendrimers and their PEGylated derivatives after intravenous administration. *J. Control. Release,* *114*, 69–77.

Paleos, C. M., Tsiourvas, D., & Sideratou, Z., (2006). Molecular engineering of dendritic polymers and their application as drug and gene delivery systems. *Mol. Pharm.*, *4*, 169–188.

Paleos, C. M., Tsiourvas, D., & Sideratou, Z., (2007). Molecular engineering of dendritic polymers and their application as drug and gene delivery systems. *Mol. Pharm.*, *4*, 169–188.

Patel, H. N., & Patel, P. M., (2013). Dendrimer applications - A review. *Int. J. Pharma Bio Sci.*, *4*, 2.

Pavan, G. M., Posocco, P., Tagliabue, A., Maly, M., Malek, A., Danani, A., Ragg, E., Catapano, C. V., & Pricl, S., (2010). PAMAM dendrimers for siRNA delivery: Computational and experimental Insights. *Chem. A. Eur. J.,* *16*, 7781–7795.

Pedziwiatr-Werbicka, E., Ferenc, M., Zaborski, M., Gabara, B., Klajnert, B., & Bryszewska, M., (2011). Characterization of complexes formed by polypropylene imine dendrimers and anti-HIV oligonucleotides. *Colloids Surfaces B. Biointerfaces, 83*, 360–366.

Perumal, O. P., Inapagolla, R., Kannan, S., & Kannan, R. M., (2008). The effect of surface functionality on cellular trafficking of dendrimers. *Biomaterials, 29*, 3469–3476.

Rassu, G., Soddu, E., Cossu, M., Gavini, E., Giunchedi, P., & Dalpiaz, A., (2016). Particulate formulations based on chitosan for nose-to-brain delivery of drugs. A review. *J. Drug Deliv. Sci. Technol., 32*, 77–87.

Santos, J. L., Pandita, D., Rodrigues, J., Pego, A. P., Granja, P. L., Balian, G., & Tomas, H., (2010). Receptor-mediated gene delivery using PAMAM dendrimers conjugated with peptides recognized by mesenchymal stem cells. *Mol. Pharm., 7*, 763–774.

Sashiwa, H., Yajima, H., & Aiba, S. I., (2003). Synthesis of a chitosan-dendrimer hybrid and its biodegradation. *Biomacromolecules, 4*, 1244–1249.

Sato, K., & Anzai, J. I., (2013). Dendrimers in layer-by-layer assemblies: Synthesis and applications. *Molecules, 18*, 8440–8460.

Shadab, M. D., Khan, R. A., Mustafa, G., Chuttani, K., Baboota, S., Sahni, J. K., & Ali, J., (2013). Bromocriptine loaded chitosan nanoparticles intended for direct nose to brain delivery: Pharmacodynamic, pharmacokinetic and scintigraphy study in mice model. *Eur. J. Pharm. Sci., 48*, 393–405.

Shcharbin, D. G., Klajnert, B., & Bryszewska, M., (2009). Dendrimers in gene transfection. *Biochemistry. (Mosc)., 74*, 1070–1079.

Shcharbin, D., Janaszewska, A., Klajnert-Maculewicz, B., Ziemba, B., Dzmitruk, V., Halets, I., Loznikova, S., Shcharbina, N., Milowska, K., Ionov, M., Shakhbazau, A., & Bryszewska, M., (2014). How to study dendrimers and dendriplexes III. Biodistribution, pharmacokinetics and toxicity *in vivo. J. Control. Release, 181*, 40–52.

Sitti, M., & Fearing, R. S., (2003). Synthetic gecko foot-hair micro/nano-structures as dry adhesives. *J. Adhes. Sci. Technol., 17*(19), 1055–1073.

Smith, D. K., & Diederich, F., (1998). Functional dendrimers: Unique biological mimics. *Chem. A. Eur. J., 4*, 1353–1361.

Souza, J. G., Dias, K., Silva, S. A. M., De Rezende, L. C. D., Rocha, E. M., Emery, F. S., & Lopez, R. F. V., (2015). Transcorneal iontophoresis of dendrimers: PAMAM corneal penetration and dexamethasone delivery. *J. Control. Release*, 115–124.

Svenson, S., (2009). Dendrimers as versatile platform in drug delivery applications. *Eur. J. Pharm. Biopharm., 71*, 445–462.

Svenson, S., & Tomalia, D. A., (2012). Dendrimers in biomedical applications-reflections on the field. *Adv. Drug Deliv. Rev., 64*, 102–115.

Tam, J. P., & Spetzler, J. C., (2001). Synthesis and application of peptide dendrimers as protein mimetics. *Curr. Protoc. Immunol.* Chapter 9, Unit 9. 6.

Tomalia, D. A., & Fréchet, J. M. J., (2002). Discovery of dendrimers and dendritic polymers: A brief historical perspective. *J. Polym. Sci. Part A. Polym. Chem., 40*, 2719–2728.

Torchilin, V. P., (2010). Passive and active drug targeting: Drug delivery to tumors as an example. *Handb. Exp. Pharmacol., 197*, 3–53.

Twibanire, J. D A. K., & Grindley, T. B., (2014). Polyester dendrimers: Smart carriers for drug delivery. *Polymers (Basel)., 6*, 179–213.

Veronese, F. M., & Mero, A., (2008). The impact of PEGylation on biological therapies. *Bio. Drugs. 22*(5), 315–329.

Waite, C. L., Sparks, S. M., Uhrich, K. E., & Roth, C. M., (2009). Acetylation of PAMAM dendrimers for cellular delivery of siRNA. *BMC Biotechnol., 9*, 38.

Xu, L., Zhang, H., & Wu, Y., (2014). Dendrimer advances for the central nervous system delivery of therapeutics. *ACS Chem. Neurosci., 5*, 2–13.

Yang, H., Lopina, S. T., DiPersio, L. P., & Schmidt, S. P., (2008). Stealth dendrimers for drug delivery: Correlation between PEGylation, cytocompatibility, and drug payload. *J. Mater. Sci. Mater. Med.*, *19*, 1991–1997.

Ziemba, B., Janaszewska, A., Ciepluch, K., Krotewicz, M., Fogel, W. A., Appelhans, D., Voit, B., Bryszewska, M., & Klajnert, B., (2011). *In vivo* toxicity of poly(propyleneimine) dendrimers. *J. Biomed. Mater. Res., Part A., 99, A.*, 261–268.

CHAPTER 4

COMPUTATIONAL APPROACH TO ELUCIDATE DENDRIMERS

GURPREET KAUR

Department of Biotechnology, CT Group of Institutions, Shahpur Campus, Jalandhar, Punjab, India – 144020, E-mail: kaur.gurpreet2809@gmail.com

CONTENTS

ABSTRACT

Dendrimers are polymeric monomers that have tree-like structure and are highly branched. The wide structure of dendrimer has great impact on the chemical and physical potential of working in a system. Due to these properties, dendrimers are suitable for a wide range of biomedical applications. Their unique characteristics can be used to design therapeutic and diagnostic agents. To synthesize a dendrimer, various methods have been used, and computational approach is one of the emerging techniques for dendrimer construction. Moreover, molecular modeling technique is a powerful tool for studying the interaction between the complexes and screening for suitable groups for binding. This chapter has tried to give an overview of the key factors to be considered during the computational synthesis of dendrimers. Computational models have provided the ability to predict and improve the dendrimer potential to work in the biological system. A particular emphasis has been made on the different factors influencing the molecular modeling of dendrimers.

4.1 INTRODUCTION

Monodisperse structure and homogeneity are the basic characteristics of radially symmetric nanosized dendrimers. Dendrimers have tree-like branches, which increase their area of work (Srinivasa-Gopalan et al., 2007). Since 1970s, researchers' interest in highly branched macromolecules have increased due to their characteristic of the wide surface area for use. The advancements in the dendrimer construction are highlighted in Table 4.1.

4.2 STRUCTURE OF DENDRIMERS

Dendrimers have three distinguished layers (as shown in Figure 4.1) which include:

TABLE 4.1   Discoveries in Dendrimer Construction

Type of dendrimer	Year	Reference
Branched macromolecule	1978	Vogtle et al., 1978
Three patents on polylysine dendrimer	early 1980s	Denkewalter et al., 1979; Denkewalter et al., 1981; Denkewalter et al., 1983
Published his work on Starburst dendrimer	1985	Tomalia et al., 1985
Synthesized Cascade molecules (laid the foundation for dendrimer synthesis)	1985	Newkome et al., 1985

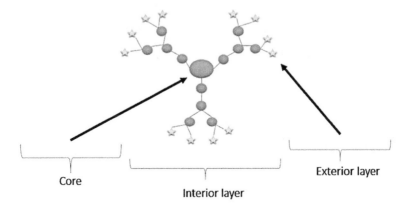

FIGURE 4.1   Architecture of a dendrimer.

1) An initiator core
2) An interior layer with repeating units outwardly attached to the initiator core
3) An exterior layer associated with the outermost portion

4.3   SYNTHESIS OF DENDRIMERS

The basic methods to synthesize dendrimers are either by the divergent method or by the convergent method.

1) **The divergent route for dendrimer synthesis**: In this method, the construction of dendrimer is in the outward direction starting from the core molecule. The fundamental procedure in the divergent method is to react the core molecule with a monomer molecule comprising one reactive group and two non-reactive groups. The new edge is activated for further reaction, and the reaction is repeated for several steps in order to create the layers of the dendrimer (Figure 4.2).

2) **The convergent route for dendrimers synthesis:** The drawback of the divergent method for dendrimer synthesis is side chain reactions or incomplete reactions. In order to prevent these complications, an excess amount of reaction mixture and reagent is needed for the desired reaction to occur, but this could further lead to problems in extracting and purifying the final product.

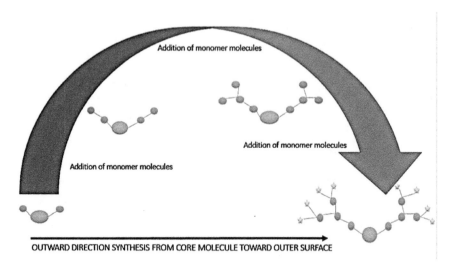

FIGURE 4.2   Divergent method of dendrimer synthesis.

Therefore, to overcome this limitation, the convergent method was developed in which the construction of dendrimer starts inwardly from end groups. When these constructions, namely, dendrons, are according to the desired size, they are attached to the multifunctional core molecule (Figure 4.3).

Although the limitations of the divergent method of dendrimer synthesis are covered in convergent approach, this method itself cannot be used to generate a high generation of dendrimers as the steric hindrance may occur between dendron and the core molecule (Gajjar et al., 2014).

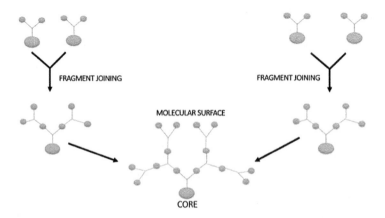

FIGURE 4.3   Convergent methods of dendrimer synthesis.

4.4   SYNTHESIS OF DENDRIMERS BY VARIOUS APPROACHES

4.4.1   *DOUBLE EXPONENTIAL GROWTH APPROACH*

This method is one of the type of convergent methods where the direction of dendron synthesis could be either inward toward the core by focal point activation or is headed toward surface-by-surface group activation (Kawaguchi et al., 1995; Klopsch et al., 1996; Vogtle et al., 2009)

4.4.2   *HYPERCORE OR HYPERMONOMER APPROACH*

New methods were developed that involved the pre-assemblage of oligomeric species, which could be linked together in order to form dendrimers (Wooley et al., 1991). This method has two requirements: block of dendrimer and hypermonomer/hypercore. The block of the dendrimer is constructed by reaction of a surface entity with the branched monomer in a single step followed by main point activation in the second step. The hypermonomer is built in a single step through the divergent method. The block of dendrimer and the hypercore can be reacted together to give a complete dendrimer.

4.4.3   *ORTHOGONAL COUPLING STRATEGY*

Orthogonal synthesis means the activities of the groups are inert toward the coupling reaction but can be activated in situ for the desired consecutive reaction for the formation of dendrimers. In this approach, two different branching entities with integral coupling functions are involved without the activation step (Spindler et al., 1993; Newkome et al., 2001) (Figure 4.4). Both the divergent and convergent methods of dendrimer synthesis could be

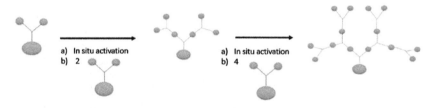

FIGURE 4.4   Orthogonal coupling strategy approach to synthesize dendrimers by computational approach.

used. However due to the limitation of meeting inflexible structural requirements of building entities, the orthogonal approach for dendrimer synthesis
is not widely followed (Greyson et al., 2001).

The modifications in the dendrimer's structure may lead to the alterations
in its biological activities. To predict the properties of dendrimers, computational technique is a valuable method to correlate the structure of dendrimer
with their property for the specific disease.

Several parameters such as generation number (size), type of monomer,
branching units, surface terminal groups, charges, and hydrophobicity, are
included to build a dendrimer. Computational techniques can explore the
properties of dendrimers, like molecular interactions, conformational analysis, and experimental validation, with less laborious efforts. Moreover,
molecular modeling allows controlling the different parameters of the experiment (Tian et al., 2009; Barata et al., 2011; Ouyang et al., 2011; Jain et al.,
2013). This approach has opened up a new pathway for interpreting and vali-
dating the experimental data and for designing and characterizing the biological interactions (Barnard et al., 2011). Knowledge of molecular interaction is
the basis to improve biological activities of dendrimers. As dendrimers are
a protein-like structure in three-dimensional way, molecular modeling techniques like docking studies, improving the design of dendrimer (Brocchini
et al., 2008; Uhlich et al., 2011) are better handled and performed through
the software. With computational facility, it is possible to perform complex
simulations such as temperature control, pressure, volume, and interaction
with drugs or proteins or nucleic acids altogether in one experiment.

4.5   STRUCTURAL GENERATION FOR DENDRIMERS

In order to define the structure and topology of the dendrimers, the understanding of the nomenclature is necessary. IUPAC nomenclature is mostly
acceptable for naming the dendrimers but with the increase in the size of
dendrimer molecules, clarity of naming becomes difficult as all the important structural features are not taken up (Lozac'h et al., 1979; Roberts et
al., 2008). Unfortunately, the manual assemblage of macromolecules gave
rise to tedious work and is likely to create mistakes with the increase in the
generations of dendrimers; hence, software packages have been proposed
for sequential assemblage of molecules. Specifically, Gromacs is proposed
for dynamic simulation of proteins (Hess et al., 2007), XPLOR generates

structure of dendrimers based on NMR and X-ray experimental data (Schwieters et al., 2003; Schwieters et al., 2006), and Starmaker (Charlmers and Roberts, 2014) and Dendrimer Building Toolkit (Maingi et al., 2012) contribute to dendrimer assembly. Other methods of naming the hyperbranched dendrimers are through Newkome-nomenclature (Florida et al., 1993) and cascadane (Friedhofen and Vogtle, 2006). These systems use repetitive units that establish dendrimers to a simple form. To further simplify the dendrimer nomenclature, dotted cap notation has been proposed. This method divides the dendrimer into a core unit that is attached to monomers for creating the framework of the dendrimer where the capping groups are attached (Roberts et al., 2008). However, this strategy does not provide much information about the core molecule or branching units.

The next step is to describe the topology and parameters of starting monomers for defining the force field. It is important to describe each individual atom information and the way they were brought together as a monomer.

4.6   SIMULATION OF DENDRIMER THROUGH A COMPUTATIONAL APPROACH

As dendrimers have large numbers of atoms, the creation becomes difficult due to steric hindrance. Molecular simulation is based on quantum mechanics, molecular mechanics, and molecular dynamics. Force field is applied to control the behavior of atoms within a dendrimer (Tian et al., 2009). Force field serves as the potential energy that includes the cumulative effect of forces as given in the following equation

Force field = Bonded forces (bond length, bond angle, torsion) + Non-bonded forces (electrostatic interaction, van der Waals interactions)

For parameterization of atom or molecule, force field use different methods that may later affect the outcomes. Some of the common force fields are AMBER (Ouyang et al., 2011; Lim et al., 2012), CHARMM (Kelly et al., 2008; Schneider et al., 2011; Mills et al., 2013), GROMOS (Javor and Reymond, 2009; Filipe et al., 2013), MARTINI (Zhong et al., 2011; Tian and Ma, 2012; Wang et al., 2012), CVFF (Quintana et al., 2002; Shi et al., 2010), OPLS (Tyssen et al., 2010; Stach et al., 2012; Al-Jamal et al., 2013), and

DREIDING FORCE (Miklis et al., 1997; Liu et al., 2009). The surrounding environment like the presence of water molecule or other solvent, ions, lipid molecules, and drugs should also be defined. Before proceeding for the actual measurements, energy minimization is recommended, which could be achieved by sort simulation with restrictions to the degree of freedom.

4.6.1 MOLECULAR DOCKING OF DENDRIMERS

In order to perform the docking of the drug into the structure, it is likely to know the dendrimer structure's capability to fit the drug into its cavity. The size of the cavity can also give an estimate of the maximum numbers of drugs that could be incorporated in the dendrimer (Tanis and Karatasos, 2009). Free energy between the dendrimer and drug is calculated in the form of docking score. Besides this, an alternative method of docking can be used where the drug can be placed randomly in the interior of the dendrimer with the assumption that the drug will find the docking site along the course of short span of time. But this approach is difficult for the large systems.

Dendrimers can also be treated as ligand instead of the receptor. The docking methods could be useful for estimating the free-energy binding, identifying the binding site, and exploring the binding potential of dendrimer as a therapeutic agent. Many docking experiments on dendrimers revealed that with the shape of cavity, electrostatic interactions also play an important role for the biological activity of dendrimers (Barata et al., 2011). Performing docking experiments could also be helpful in filtering the optimum dendrimer structure for binding the molecule. Moreover, the *in vivo* behavior and potential of the dendrimer and the drugs fitting in the cavity of dendrimer could also be evaluated by docking experiments. This method is similar to the high-throughput screening of the drugs performed on the drug targets.

4.6.2 IMPACT OF VARIOUS FACTORS ON MOLECULAR DOCKING

4.6.2.1 The Impact of Solvent

The structure and arrangements of atoms are dependent on the various internal factors like generation of dendrimers, monomer length, and their chemical properties and external factors like type of solvent, salt type, and ionic strength

(Evangelista-Lara and Guadarrama, 2005; Maiti et al., 2005). These factors are responsible for the receptor–ligand interactions as well as the addition of the functional moieties on the surface of the dendrimers. The representation of water and ions could be computationally demanding; therefore, the lower level of theory is considered, and the solvent is treated implicitly. Thus, the way to overcome this complication is to perform the all-atom simulation with explicit solvent on smaller generations and compare the radius of gyration (Rg) and atom distribution with models performed with different implicit parameters (Lee et al., 2002). It has been found that a good solvent system is essential for the dependable prediction of dendrimer size, conformation, and solvent presence in the dendrimer cavity (Gorman and Smith, 2000). The position of a water molecule (buried water, surface water at the dendrimer–solvent edge, and bulk water/solvent) in dendrimer decides its behavior. Buried water has lower entropy than the bulk water; therefore, the interaction of water molecule and dendrimer results in the release of different free energy (Blaak et al., 2008). It was also observed that the water molecule on the surface enters inside the dendrimer and tries to compete for the hydrogen bonds between dendrimer residue (Maiti et al., 2005; Lee et al., 2006).

4.6.2.2 Impact and Adaptability of the End Groups

The end groups of the dendrimers participate in many specific and non-specific interactions. Thus, the solution of the problem of multiple charges is to attach groups like acetyl, PEG (*polyethylene glycol*), or lipid molecules. Although the end groups can be modified, end group modification may alter the interactions between receptor and ligand which may, in turn, display the inefficiency of the moieties. The common example is the backfolding of the terminal end groups where two conditions arise: one is where labile molecule is protected, thus producing the effective structure, and in the second case, due to linked moiety, the structural arrangement makes the molecule ineffective in action (Maiti et al., 2005, 2006; Tian and Ma, 2013). Both these cases could be annotated with the help of molecular modeling. The molecular modeling approach can give an insight of the available location in the receptor for binding, and interactions between the complex and the possibility of the modified end groups to fit in the cavity of receptor, thus showing enhanced biological activity.

4.7 CONCLUSION

Dendrimers have proved to be an efficient molecule in many biological pathways. Dendrimers could be used in the diagnosis and therapeutic field. The conventional methods of dendrimer synthesis were divergent strategy and convergent strategy. But with time, the method of dendrimer synthesis has also become advanced. Computational approach has proposed various significant and efficient ways regardless of whether it includes designing of the drug or dendrimer. Several parameters like size of dendrimer, type of dendrimer, surface terminal group, charges etc., could be predicted through the use of various software. Molecular docking is an emerging technique for understanding the interaction study between dendrimer–ligand or protein–dendrimer. Impact of various internal and external factors are needed to be considered while performing docking experiments for the dendrimer construction. Although the molecular modeling techniques for the creation of dendrimers is a significantly helpful tool, there is still need of compelling software for these macromolecules.

KEYWORDS

- **docking**
- **drug design**
- **molecular modeling**
- **simulation dendrimer design**
- **structure activity relationship**
- **synthesis of dendrimers**

REFERENCES

Al-Jamal, K. T., Al-Jamal, W. T., Wang, J. T. W., Rubio, N., Buddle, J., Gathercole, D., Zloh, M., & Kostarelos, K., (2013). Cationic poly-L-lysine dendrimer complexes doxorubicin and delays tumor growth *in vitro* and *in vivo*. *ACS Nano, 7,* 1905–1917.
Barata, T. S., Teo, I., Brocchini, S., Zloh, M., & Shaunak, S., (2011). Partially glycosylated dendrimers block MD-2 and prevent TLR4-MD-2-LPS complex mediated cytokine responses. *PLoS Comput. Biol., 7,* e1002095.

Barata, T. S., Teo, I., Lalwani, S., Simanek, E. E., Zloh, M., & Shaunak, S., (2011). Computational design principles for bioactive dendrimer-based constructs as antagonists of the TLR4-MD-2-LPS complex. *Biomaterials, 32*(33), 8702–8711.

Barnard, A., Posocco, P., Pricl, S., Calderon, M., Haag, R., Hwang, M. E., Shum, V. W. T., Pack, D. W., & Smith, D. K., (2011). Degradable self-assembling dendrons for gene delivery: Experimental and theoretical insights into the barriers to cellular uptake. *J. Am. Chem. Soc., 133*, 20288–20300.

Blaak, R., Lehmann, S., & Likos, C. N., (2008). Charge-induced conformational changes of dendrimers. *Macromolecules, 41*, 4452–4458.

Brocchini, S., Godwin, A., Balan, S., Choi, J., Zloh, M., & Shaunak, S., (2008). Disulfide bridge based PEGylation of proteins. *Adv. Drug Deliv. Rev., 60*, 3–12.

Charlmers, D., & Roberts, B., (2014). Silico—A Perl Molecular Modeling Toolkit. Available online: http://silico. sourceforge. net/Silico/Home. html.

Denkewalter, R. G., Kolc, J., & Lukasavage, W. J., (1979). Preparation of lysine based macromolecular highly branched compound, *U. S. Patent No. 4*, 360, 646.

Denkewalter, R. G., Kolc, J., & Lukasavage, W. J., (1981). Macromolecular highly branched homogeneous compound based on lysine units, *U. S. Patent No. 4*, 289, 872.

Denkewalter, R. G., Kolc, J., & Lukasavage, W. J., (1983). Macromolecular highly branched homogeneous compound, *U. S. Patent No. 4*, 410, 688.

Evangelista-Lara, A., & Guadarrama, P., (2005). Theoretical evaluation of the nanocarrier properties of two families of functionalized dendrimers. *Int. J. Quantum Chem., 103*, 460–470.

Filipe, C. S., Machuqueiro, M., Darbre, T., & Baptista, M., (2013). Unraveling the conformational determinants of peptide dendrimers using molecular dynamics simulations. *Macromolecules, 46*, 9427–9436.

Florida, S., & Rouge, B., (1993). Systematic nomenclature for cascade polymers. *J. Polym. Sci. Part A Polym. Chem., 31*, 641–651.

Friedhofen, J. H., & Vogtle, F., (2006). Detailed nomenclature for dendritic molecules. *New J. Chem., 30*, 32–43.

Gajjar, D. G., Patel, R. M., & Patel, P. M., (2014). Dendrimers: Synthesis to applications: A review. *Macromolecules., 10*(1), 37–48.

Gorman, C. B., & Smith, J. C., (2000). Effect of repeat unit flexibility on dendrimer conformation as studied by atomistic molecular dynamics simulations. *Polymer, 41*, 675–683.

Greyson, S., & Fréchet, J. M. J., (2001). Convergent dendrons and dendrimers: From synthesis to applications *Chemical. Rev., 101*, 3819.

Hess, B., Kutzner, C., Van der Spoel, D., & Lindahl, E., (2007). GROMACS 4: Algorithms for highly efficient, load-balanced, and scalable molecular simulation. *J. Chem. Theory Comput., 4*, 436–447.

Jain, V., Maingi, V., Maiti, P. K., & Bharatam, P. V., (2013). Molecular dynamics simulations of PPI dendrimer–drug complexes. *Soft Matter, 9*, 6482–6496.

Javor, S., & Reymond, J. L., (2009). Molecular dynamics and docking studies of single site esterase peptide dendrimers. *J. Org. Chem.*, 74, 3665–3674.

Kawaguchi, T., Walker, K. L., Wilkins, C. L., & Moore, J. S., (1995). Double exponential dendrimer Growth. *J. Am. Chem. Soc., 117*, 2159.

Kelly, C. V., Leroueil, P. R., Orr, B. G., Banaszak, H. M. M., & Andricioaei, I., (2008). Poly(amidoamine) dendrimers on lipid bilayers II: Effects of bilayer phase and dendrimer termination. *J. Phys. Chem. B., 112*, 9346–9353.

Klopsch, R., Franke, P. A. D., & Schlüter, A. D., (1996). Repetitive strategy for exponential growth of hydroxy-functionalized dendrons. *Chem. A. Eur. J.*, *2*, 1330–1334.

Lee, H., Baker, J. R., & Larson, R. G., (2006). Molecular dynamics studies of the size, shape, and internal Structure of 0% and 90% acetylated fifth-generation polyamidoamine dendrimers in water and methanol. *J. Phys. Chem. B*, *110*, 4014–4019.

Lee, I., Athey, B. D., Wetzel, A. W., Meixner, W., & Baker, J. R., (2002). Structural molecular dynamics studies on polyamidoamine dendrimers for a therapeutic application: Effects of pH and generation. *Macromolecules*, *35*, 4510–4520.

Lim, J., Lo, S. T., Hill, S., Pavan, G. M., Sun, X., & Simanek, E. E., (2012). Antitumor activity and molecular dynamics simulations of paclitaxel-laden triazine dendrimers. *Mol. Pharm.*, *9*, 404–412.

Liu, Y., Bryantsev, V. S., Diallo, M. S., & Goddard, W., (2009). A PAMAM dendrimers undergo pH responsive conformational changes without swelling. *J. Am. Chem. Soc.*, *131*, 2798–2799.

Lozac'h, N., Goodson, A. L., & Powell, W. H., (1979). Nobel nomenclature—general principles. *Angew. Chem. Int. Ed. Engl.*, *18*, 887–899.

Maingi, V., Jain, V., Bharatam, P. V., & Maiti, P. K., (2012). Dendrimer building toolkit: Model building and characterization of various dendrimer architectures. *J. Comput. Chem.*, *33*, 1997–2011.

Maiti, P. K., & Goddard, W. A., (2006). Solvent quality changes the structure of G8 PAMAM dendrimer, a disagreement with some experimental interpretations. *J. Phys. Chem. B.*, *110*, 25628–25632.

Maiti, P. K., Çagın, T., Lin, S. T., & Goddard, W. A., (2005). Effect of Solvent and pH on the Structure of PAMAM Dendrimers. *Macromolecules*, *38*, 979–991.

Marta, S. M., & Urbanczyk-Lipkowska, Z., (2014). Advances in the chemistry of dendrimers. *New J. Chem.*, *38*, 2168–2203.

Marvinkumar, I., Patel, M. I., & Patel, R. R. (2016). Dendrimers: A novel therapy for cancer. http://www. pharmatutor.org/articles/denrimers-novel-therapy-for-cancer. Accessed on Aug. 12, 2016.

Miklis, P., Tahir, C., & Iii, W. A. G., (1997). Dynamics of Bengal rose encapsulated in the Meijer Dendrimer box. *J. Am. Chem. Soc.*, *7863*, 7458–7462.

Mills, M., Orr, B. G., Banaszak H. M. M., & Andricioaei, I., (2013). Attractive hydration forces in DNA-dendrimer interactions on the nanometer scale. *J. Phys. Chem. B.*, *117*, 973–981.

Newkome, G. R., Moorefield, C. N., & Vogtle, F., (2001). Dendrimers and Dendrons: Concepts, syntheses, applications, *Wiley-VCH, Weinheim*. Chapter 10.1–10.7.

Newkome, G. R., Yao, Z., Baker, G. R., & Gupta, V. K., (1985). Micelles. Part 1. Cascade molecules: a new approach to micelles. A [27]-arborol. *J. Org. Chem.*, *50*(11), 2003–2004.

Ouyang, D., Zhang, H., Parekh, H. S., & Smith, S. C., (2011). The effect of pH on PAMAM dendrimer-siRNA complexation: Endosomal considerations as determined by molecular dynamics simulation. *Biophys. Chem.*, *158*, 126–133.

Quintana, A., Raczka, E., Piehler, L., Lee, I., Myc, A., Majoros, I., Patri, A. K., Thomas, T., Mulé, J., & Baker, J. R., (2002). Design and function of a dendrimer-based therapeutic nanodevice targeted to tumor cells through the folate receptor. *Pharm. Res.*, *19*, 1310–1316.

Roberts, B. P., Scanlon, M. J., Krippner, G. Y., & Chalmers, D. K., (2008). The dotted cap notation: A concise notation for describing variegated dendrimers. *New J. Chem.*, *32*, 1543–1554.

Schneider, C. P., Shukla, D., & Trout, B. L., (2011). Effects of solute-solute interactions on protein stability studied using various counterions and dendrimers. *PLoS One*, *6*, e27665.

Schwieters, C. D., Kuszewski, J. J., Tjandra, N., & Clore, G. M., (2003). The Xplor-NIH NMR molecular structure determination package. *J. Magn. Reson.*, *160*, 65–73.

Schwieters, C., Kuszewski, J., & Mariusclore, G., (2006). Using Xplor–NIH for NMR molecular structure determination. *Prog. Nucl. Magn. Reson. Spectrosc.*, *48*, 47–62.

Shi, X., Lee, I., Chen, X., Shen, M., Xiao, S., Zhu, M., Baker, J. R., Jr., & Wang, S. H., (2010). Influence of dendrimer surface charge on the bioactivity of 2-methoxyestradiol complexed with dendrimers. *Soft Matter*, *6*, 20–27.

Spindler, R., & Frechet, J. M. J., (1993). Two-step approach towards the accelerated synthesis of dendritic macromolecules. *J. Chem. Soc. Perkins Trans.*, *1*, 913.

Srinivasa-Gopalan, S., & Yarema, K. J., (2007). Nanotechnologies for the life sciences: Dendrimers in cancer treatment and diagnosis, *7*, New York: Wiley.

Stach, M., Maillard, N., Kadam, R. U., Kalbermatter, D., Meury, M., Page, M. G. P., Fotiadis, D., Darbre, T., & Reymond, J. L., (2012). Membrane disrupting antimicrobial peptide dendrimers with multiple amino termini. *Med. Chem. Comm.*, *3*, 86–89.

Tanis, I., & Karatasos, K., (2009). Association of a weakly acidic anti-inflammatory drug (ibuprofen) with a poly(amidoamine) dendrimer as studied by molecular dynamics simulations. *J. Phys. Chem. B.*, *113*, 10984–10993.

Tian, W., & Ma, Y., (2009). Molecular dynamics simulations of a charged dendrimer in multivalent salt solution. *J. Phys. Chem. B*, *113*, 13161–13170.

Tian, W., & Ma, Y., (2012). Insights into the endosomal escape mechanism via investigation of dendrimer–membrane interactions. *Soft Matter*, *8*, 6378–6384.

Tian, W., & Ma, Y., (2013). Theoretical and computational studies of dendrimers as delivery vectors. *Chem. Soc. Rev.*, *42*, 705–727.

Tomalia, D. A., Baker, H., Dewald, J., Hall, M., Kallos, G., Martin, S., Roeck, J., Ryder, J., & Smith, P., (1985). A new class of polymers: Starburst-dendritic macromolecules. *Polymer Journal*, *17*, 117–132.

Tyssen, D., Henderson, S. A., Johnson, A., Sterjovski, J., Moore, K., La, J., Zanin, M., Sonza, S., Karellas, P., Giannis, M. P., et al., (2010). Structure activity relationship of dendrimer microbicides with dual action antiviral activity. *PLoS One*, *5*, e12309.

Uhlich, N. A, Darbre, T., & Reymond, J. L., (2011). Peptide dendrimer enzyme models for ester hydrolysis and aldolization prepared by convergent thioether ligation. *Org. Biomol. Chem.*, *9*, 7071–7084.

Vögtle, F., Buhleier, E., & Wehner, W., (1978). "Cascade"- and "Nonskid-Chain-like" syntheses of molecular cavity topologies. *Synthesis*, *2*, 155–158.

Vögtle, F., Richardt, G., & Verner, N., (2009). *Dendrimer Chemistry: Concepts, Syntheses and Applications.* Wiley-VCH, Chinchster. Chapter 2, pp. 25–48.

Wang, Y. L., Lu, Z. Y., & Laaksonen, A., (2012). Specific binding structures of dendrimers on lipid bilayer membranes. *Phys. Chem. Chem. Phys.*, *14*, 8348–8359.

Wooley, K. L., Hawker, C. J., & Frechet, J. M. J., (1991). Hyperbranched macromolecules via a novel double-stage convergent growth approach. *J. Am. Chem. Soc.*, *113*, 4252.

PART II

ADVANCEMENTS IN DENDRIMER APPLICATIONS

CHAPTER 5

AN OVERVIEW OF DENDRIMERS AND THEIR BIOMEDICAL APPLICATIONS

NIRUPAM DAS,[1, 2] PIYOOSH A. SHARMA,[2] ANKIT SETH,[2]
RAHUL MAHESHWARI,[3] MUKTIKA TEKADE,[4]
SUSHANT K. SHRIVASTAVA,[2] and RAKESH K. TEKADE[3]

[1]Pharmaceutical Chemistry Laboratory, Department of Pharmaceutical
Sciences, Assam University, Silchar–788011, India

[2]Department of Pharmaceutical Engineering & Technology,
Indian Institute of Technology (Banaras Hindu University),
Varanasi–221005, India

[3]National Institute of Pharmaceutical Education and Research
(NIPER) – Ahmedabad, Opposite Air Force Station Palaj,
Gandhinagar–382355, India

[4]TIT College of Pharmacy, Technocrats Institute of Technology,
Anand Nagar, Bhopal–462021, India

CONTENTS

ABSTRACT

The versatility of the terminal functionality and the flexibility toward structural modification allow a broad range of utilization of dendrimers. Since their discovery, the exploitation of dendrimers has touched many diversified areas and applications such as therapeutic, biomedical, and diagnostics, which are the most exploited avenues. Amongst the availability of various synthetic techniques, the widely adopted methodologies are the convergent and divergent approaches, and characterizations are carried out by a combination of spectroscopic, scattering, and electroanalytical methods. The very nanosized nature of the hyperbranched polymer efficiently functions as nanocarriers for the delivery of drugs, diagnostic agents, gene transfection agents, vaccines and immunomodulators, tissue regenerators, etc. They provide safe and efficient target delivery of therapeutic agents with considerable effectiveness because of their nanoparticulate nature. Numerous promising dendrimers and dendrimer-based delivery systems have been reported so far. The surface engineering of dendrimers allows modification of the surface with specific ligands for targeted drug delivery. Further, it also benefited in reducing the toxicity of dendritic carriers, particularly the cationic type of polymers. By considering their ability to traverse biological barriers, the dendritic nanocarriers served as a nontoxic and biocompatible polymer for targeting particular organ system for viable therapeutic outcomes. The chapter comprehensively emphasizes on the therapeutic, biomedical, and diagnostic applications of dendrimers.

5.1   HISTORY AND INTRODUCTION

The 21st century is the era of nanotechnology, and with its advent, various nanostructures and nanoconstructs have found myriad promising applications in the field of medicine. The size reduction of large materials to smallest possible structures and inherent nanosized molecules (1–100 nm) display properties different from those of conventional materials. Dendrimers are one such hyperbranched polymer whose size and surface functionality can be strategically manipulated for favorable pharmacodynamic and pharmacokinetic outcomes. The dendrimers are synthesized in the nanometric size range that can be smaller than 5 nm or size less than that of a virion. The fact is exemplified by a generation 13 (G13) stable dendrimer comprising triazines linked by diamines with dimensions that match the size of virus particles (~30 nm). The synthetic strategy itself became the subject of dendrimer-based conjugate drug-delivery systems (Lim et al., 2013; Kesharwani et al., 2015a).

The abiotic synthetically obtained novel polymeric macromolecules are typically symmetrical and repeatedly branched around the central core. The term "dendrimers" originated from the Greek word "dendron" connoting tree and "meros" implying part, although the molecules were referred initially as "cascade molecules" and "arborols." Ever since the dendrimers came into existence, the molecules have attracted the interest of organic, supramolecular, polymer, and coordination chemists. They also found broad application in drug development toward enhancing the pharmacokinetic profile and as diagnostic aids. The flexibility of synthesizing the polymers allows tailor-made approach for developing a growing number of new generation macromolecules with different structural framework and properties (Mekelburger et al., 1992; Fischer and Vögtle, 1999). The concept of synthesizing molecular-level multifunctional branching unit may be traced back to late 1941, when Paul John Florey, an American chemist, analyzed the molecular size distribution of various synthetic branched polymers (Flory, 1941). In 1974, he won the Nobel Prize in Chemistry for his work on theoretical and experimental physical chemistry of branched macromolecules.

Approximately four decades later, the first successful laboratory synthesis of cascade-like macromolecules with repetitive growth was reported by Buhleier et al. in 1978 (Buhleier et al., 1978). Subsequently, in 1981, Allied Corporation, America, patented (US4410688 and US4289872) two highly branched homogeneous macromolecular compounds based on successive layers of amino carboxylic acid units and lysine units. Accordingly, the patent claimed that the products could be utilized as surface modifying agents, metal chelating agents, and as substrates for preparing pharmaceutical formulations. However, the discovery of "true dendrimers" in 1985 is credited to Tomalia group of Dow Chemical Co., and the research revolutionized the whole concept of dendrimer designing. It emanates from the inspiration garnered from patterns of the dendritic root of trees, a lifetime hobby of Donald A. Tomalia who is a horticulturist and synthetic polymer chemist of the dendrimer research team. The polyamidoamine (PAMAM) dendrimers synthesized by the group were referred to as "starburst™ polymers" as the patterns of the new synthetic macromolecules resemble rays radiating from a central core. The polymers were unique and differed from the classical random coil polymers because of their extraordinary symmetry, high branching, and maximized functional terminal density. They possess the following fundamental architectural features: (a) an initial central core; (b) repeating units of interior layers (subsequent generations), linked to the core; and (c) peripheral functionality attached to the outermost generation (Figure 5.1).

The preparation of PAMAM dendrimer usually involves two-step strategies: (a) first, exhaustive Michael addition of ammonia with methyl acrylate to form the esters and (b) second, amidation of the esters with an excess of ethylenediamine (EDA) (Figure 5.2). The reason behind the use of EDA for the amidation step was that the excess of EDA could be removed under a condition that does not alter the functionality of dendrimer structure (Tomalia et al., 1985; Tomalia and Fréchet, 2002).

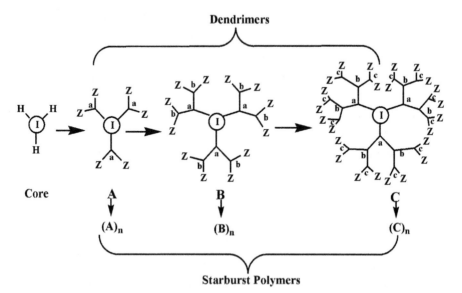

FIGURE 5.1 Flowchart for the construction of starburst polymers (initiator core) (I) to star-branched oligomers (generations $(A)_n$, $(B)_n$ and $(C)_n$} to starburst polymers (Z=terminal functionality) (Reprinted with permission from Tomalia, D. A., Baker, H., Dewald, J., Hall, M., Kallos, G., Martin, S., Roeck, J., Ryder, J., & Smith, P., (1985). A new class of polymers: starburst-dendritic macromolecules. Polym. J., 17, 117–132. © 1985 Nature Publishing Group.).

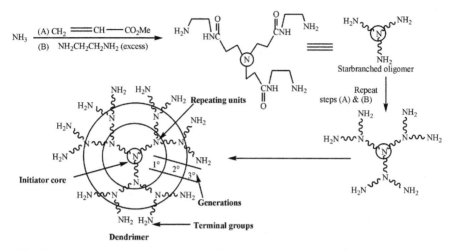

FIGURE 5.2 A divergent synthesis of dendrimers (ammonia as the interior core to starburst oligomers) (Reprinted with permission from Tomalia, D. A., Baker, H., Dewald, J., Hall, M., Kallos, G., Martin, S., Roeck, J., Ryder, J., & Smith, P., (1985). A new class of polymers: starburst-dendritic macromolecules. Polym. J., 17, 117–132. © 1985 Nature Publishing Group.).

Most of the dendrimers synthesized until 1985 followed the divergent synthesis approach, wherein the dendrimer is assembled from a central core and branched outward by the stepwise sequence of addition reactions (core to generations). The divergent approach presents the challenge of proper completion of reactions at the terminal groups and may lead to imperfections in successive generations and resulted in less monodisperse macromolecules. The approach, in turn, requires excess reagents for forceful completion of reactions, which is considered as one of the major factors hindering the purification of end products. Alternatively, the convergent approach takes advantage of the symmetrical nature of the dendrimers and the initiation point to build the macromolecules that become the part of the boundary. The subsequent growth utilizes the sole reactive group located at all branches of the polyfunctional core and leads to the generation of inner "wedge"-like cavities. The final step involves attachments of the wedge and gives rise to a monodisperse spheroidal macromolecule. In 1990, Hawker and Frechet first reported the concept of convergent synthesis of dendritic polyether macromolecules based on bromo derivative of 3,5-dihydroxybenzyl alcohol as the monomer unit (Figure 5.3). The convergent approach provides advantages over problems associated with divergent synthesis (Hawker and Frechet, 1990).

The initial progress of dendrimer research was slow because of the limited availability of the analytical methods to characterize the macromolecules.

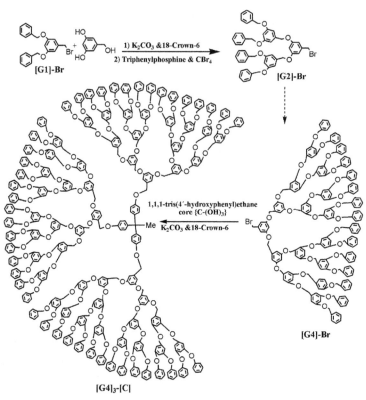

FIGURE 5.3 A convergent synthesis of dendrimers (bromo derivative 3,5-dihydroxybenzyl alcohol to penultimate dendritic wedge (periphery to the core to spheroid) (Reprinted with permission from Hawker, C., & Frechet, J. M. J., (1990). A new convergent approach to monodisperse dendritic macromolecules. J. Chem. Soc. Chem. Commun., 1010–1013. © 1969 Royal Society of Chemistry.).

Subsequently, the number of contemporary research, publications, and utilization of dendrimers and its congeners rapidly increased from the early 1990s that can be exemplified by applying the concept of click chemistry, hypercore, branched monomer growth, double exponential growth, and lego chemistry in combination with either divergent or convergent approach (Fischer and Vögtle, 1999; Madaan et al., 2014). Apart from PAMAMs, the various subunits employed in the design of dendrimers include PPI (polypropylenimine) (1) polyamide (2) polylysine (3) silicon (4) carbohydrate, and (5) metallodendrimer that were frequently reported among others (Boas and Heegaard, 2004) (Figure 5.4). Moreover, the terminal functionalities such as amino, carboxyl, hydroxyl, aryl, alkyl, alkenyl groups, etc., provide

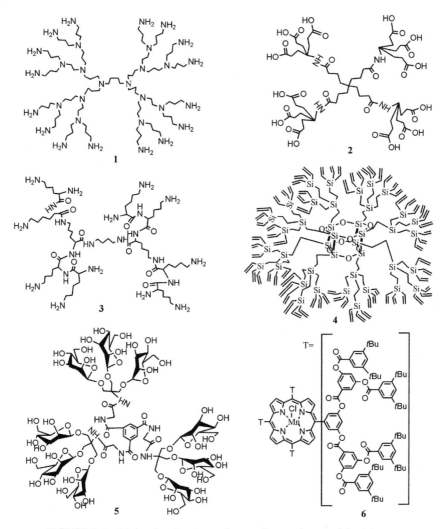

FIGURE 5.4 Molecular structures of some diverse classes of dendrimers.

the opportunity for surface modification with small molecules, fatty acids, and proteins and simultaneously enhancing biocompatibility and cellular uptake at the target site (Gajbhiye et al., 2009a; Saovapakhiran et al., 2009; Ciolkowski et al., 2012).

Dendrimers have been immensely exploited in the field of biomedical sciences and pharmaceutical industries in the last two decades. The chapter will provide a brief account of various fundamental aspects of dendrimers and an overview of their current and future applications and clinical qualifications.

5.2 NOMENCLATURE OF DENDRIMERS

In recent years, the design and synthesis of complex dendrimers necessitate universally accepted nomenclature for this novel class of macromolecules. The complexity of naming dendrimers increases with the increase in successive generations. Newkome et al. described the first systematic nomenclature of dendrimers, and 10 rules were setup for naming cascade polymers. The general line formula (**1**) of dendrimers is represented as:

$$C[R_1 (R_2 (\cdot \cdot \cdot R_1 (\cdot \cdot \cdot Rn (T)_{N_{b_n}} \cdot \cdot \cdot) \ X \ N_{b_i} \cdot \cdot \cdot)_{N_{b_2}})_{N_{b_1}}]N_C \ (1)$$

where C represents the initiator core moiety; R_i is the repeat unit; T: terminal moieties; N_{b_i}: multiplicity of the ith repeat unit of the branch; and N_c: branch multiplicity from the central core. For dendrimers having uniform branch unit, the above line formula (**1**) may be simplified as (**2**):

$$\left[\begin{array}{c} \textbf{Core} \\ \textbf{Unit} \end{array} \right] \left[\left(\begin{array}{c} \textbf{Repeat} \\ \textbf{Unit} \end{array} \right)^{G}_{Nb} \left(\begin{array}{c} \textbf{Terminal} \\ \textbf{Unit} \end{array} \right) \right] Nc \qquad (2)$$

The traditional system of naming dendrimers was found to be inappropriate as it does not indicate the number or type of terminal moieties and also fails to signify the hydrocarbon nature of the interior branches. The general form of the name is given by the formula (**3**).

$$\textbf{\textit{Z}} - \text{cascade:} \begin{array}{c} \textbf{Core} \\ \textbf{Unit} \end{array} [Nc] : \left(\begin{array}{c} \textbf{intermediate} \\ \textbf{repeat unit} \end{array} \right)^{l} : \begin{array}{c} \textbf{terminal} \\ \textbf{unit} \end{array} \qquad (3)$$

where Z: number of terminal moieties and l: layers of repeated units (i.e., the number of generations in this case). Now, applying formula (**3**) for methane nonylidynepropanol dendrimer (Figure 5.5), the nomenclature will be:

- First generation: 12-Cascade: methane [4]:(nonylidyne):propanol.
- Second generation: 36-Cascade: methane [4]: (nonylidyne)^2: propanol.
- Third generation: 108-Cascade: methane [4]: (nonylidyne)^3: propanol

A more detailed and interesting description of dendrimer nomenclature has been reported elsewhere (Newkome et al., 1993; Friedhofen and Vogtle, 2006). Dendrimers may also be represented as concentric circles depicting the interior core and various generations. A further simplified depiction of

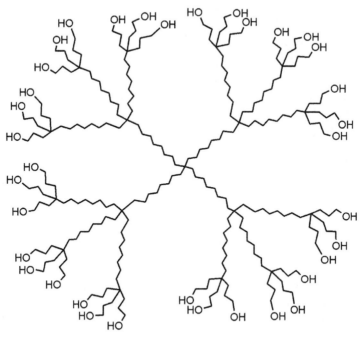

FIGURE 5.5 A 36-Cascade: methane [4]: (nonylidyne)^2: propanol dendrimer (Reprinted with permission from Newkome, G. R., Baker, G. R., Young, J. K., & Traynhama, J. G., (1993). Systematic nomenclature for cascade polymers. J. Polym. Sci. A. Polym. Chem., 31, 641–651. © 2003 John Wiley and Sons.).

surface-modified dendrimers is that the inner shells are shown as a black "ball." The number of the functional groups on the depicted surface is represented by an italic number beneath the ball. This circumvents the necessity of drawing the entire structure. The italic number for a G2.5-PAMAM dendrimer will be 16 when only the terminal carboxylic acids are depicted. The number of outer shell functionalities (pincers) in a G2.5-PAMAM dendrimer will be 8 (Figure 5.6) (Boas and Heegaard, 2004).

After the thorough examination of various nomenclatures of dendrimers, an editorial by Kesharwani et al. recently suggested that PPI dendrimer generations should be one less than their current nomenclature and must begin with core then G0, followed by G1, G2, G3, and so on (Kesharwani et al., 2015).

5.3 PROPERTIES OF DENDRIMERS

Unlike cross-linking or traditional linear polymers, the molecular architecture of dendrimers is more monodispersed and globular as a result of controlled

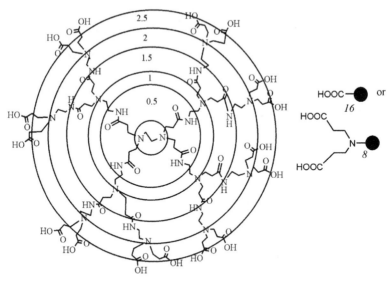

FIGURE 5.6 Concentric circle depiction of generations and black ball representation of a G 2.5 PAMAM dendrimer (Reprinted with permission from Boas, U., & Heegaard, P. M. H., (2004). Dendrimers in drug research. Chem. Soc. Rev., 33, 43–63. © 2003 Royal Society of Chemistry.).

synthesis. They are isotropically soluble functional polymers and in particular, the lower generation dendrimers with large surface areas to volume ratio (~1000 square meters per gram) and sufficient solubility may be exploited as carriers for catalyst (Alper, 1991). A supported PPI dendrimer on cross-linked polystyrene obtained *via* solid phase was utilized as an organocatalyst in Knoevenagel condensations. The closely packed amino groups (primary and tertiary) of the dendrimer were found to act as efficient organocatalysts. Many α,β-unsaturated nitriles were obtained in excellent yield (82–100%) in ethanol using 0.5 mol% of the catalyst (Panicker and Krishnapillai, 2014).

With the increase in generations, the molecular weights of dendrimers approximately double with simultaneous amplification of the terminal groups and branch shell. For a given molecular weight, a linear polymer is comparatively larger in size than dendrimers and also differs in term of physical properties such as diffusivity and ionic conductivity (Bhattacharya et al., 2013). Depending on the functionalities of the terminal groups, dendrimers may delineate an interface between a hydrophobic core and solvent accessible peripheral surface. Dendrimers with carboxylate chain end periphery retain the capability to solubilize water-insoluble hydrophobic building blocks such as pyrene by encapsulation (Fréchet, 1994).

Water-soluble dendrimers with hydrophobic interior have been designed to carry hydrophobic drugs in the systemic circulation (Kojima et al., 2000). The solubilization equals traditional micelles while not exhibiting critical micelle concentration. Similarly, the water solubility of hydrophobic drugs may also be enhanced by amphiphilic dendrimer synthesized by amide coupling of hydrophobic star-shaped poly(L-lactide) core and a benzyl ester-terminated PAMAM dendron shell. The polymer displayed a unique unimolecular micelle (14–28 nm) behavior in aqueous solution with lesser aggregation (205–344 nm) (Cao and Zhu, 2011). Alternatively, dendrimers having polar interior and a nonpolar periphery behave as unimolecular reverse micelles. They have the ability to extract and concentrate polar molecules from their solution in nonpolar solvents (Frechet, 2002). Therefore, derivatizing the terminal groups significantly increase the solubility that is further enhanced by its globular shapes (Grayson and Fréchet, 2001).

A recent molecular dynamic simulation of hybrid dendrimers utilizing PPI as core and PAMAM as the shell was studied. It was found that the branching chains of unhybridized PPI create a barrier against water penetration into the dendrimer core structure that encapsulates pyrene. After that, it was found that the addition of PAMAM to the surface of PPI removes the barrier and enhances the encapsulation capacity of the hybrid for hosting pyrene. The PAMAM-hybridized shell increases the size of the dendrimer cavities in number and volume. Kavyani et al. suggested that the shell may also open some pathways for the penetration of the water into the dendrimer interior structure (Kavyani et al., 2016). Additionally, modifying the architecture of dendrimers by attaching peptides to a template or core matrix has been exploited as immunogens and antigens, for de novo design of artificial proteins, as agonists and antagonists, and as new biopolymers and biomaterials (Tam and Spetzler, 2001).

The nanosize dendrimers have the potentiality to function as proteomimetic agents. In addition to the multivalent surface, dendrimers having a higher number of functional groups has similar molecular size (low molecular density) when compared to proteins. Peptide dendrimers are less compact than proteins but are more tightly packed than conventional linear polymers. X-ray analysis showed that dendrimers achieve an increase in globular shape with subsequent augmentation in generations. Depending on the solvent, pH, ionic strength and temperature, the nanosized dendrimers, like proteins, may respond to stimuli and attain either a tightly packed native conformation or an extended denatured conformation. Nevertheless, most dendrimers are less flexible than proteins and exhibit physicochemical properties different from that of proteins (Boas et al., 2006).

In contrast, few dendrimers do have innate ability to exhibit flexibility similar to certain enzymatic proteins. Wei et al. reported an interesting finding on PAMAM dendrimers constructed from a pyridoxamine core, wherein the terminal-free amines were coupled with *N,N*-dimethyl-L-phenylalanine. The specific dendrimers of various generations were allowed to undergo transamination reactions between dendrimers and phenylpyruvic acid or pyruvic acid in aqueous media. It was revealed that dendrimers having G4 PAMAM (shell) and pyridoxamine (core) were more efficient in mimicking some aspects of transaminase enzymes. The remote catalytic groups (*N,N*-dimethyl-L-phenylalanine) probably backfolded into the core by their flexibility, and the reaction was found to be enantioselective that favors the formation of L-amino acids (Wei et al., 2009).

5.4 SYNTHESIS OF DENDRIMERS

The synthesis of dendrimers involves the construction of inner and outer shell that revolves around a central core. The core molecule reacts with specific monomers that leads to the formation of first generation dendrimers. After that, the periphery of the first generation reacts further with other monomers to give subsequent generation and so on. The three entities have diverse functionalities with an aim to achieve the desired physicochemical properties by fine tuning the reactions. The two most widely adopted methods are divergent and convergent synthesis (Scheme 5.1), and they remain the cornerstone of dendrimer synthesis (Zhang and Simanek, 2000; Sharma et al., 2015).

5.4.1 DIVERGENT AND CONVERGENT APPROACH

In divergent synthesis, the core is the starting point of the growth and continues growing outward by a chain of repeated reactions involving coupling and activation steps. The repetition of the two steps depends on the multiplicity of the generations. The initial coupling step comprises reaction of peripheral functionalities of the core with the corresponding group of the first line of monomers at a particular focal point. The inactivated surface functionalities (protection) of the monomers attached to the core are then activated (deprotection) to further react with the second line of monomers

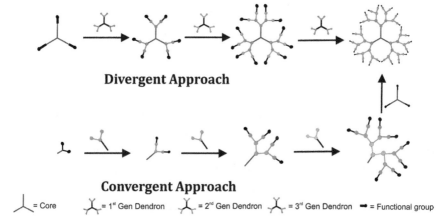

Divergent Approach

Convergent Approach

= Core   = 1ˢᵗ Gen Dendron   = 2ⁿᵈ Gen Dendron   = 3ʳᵈ Gen Dendron   ● = Functional group

SCHEME 5.1 Divergent and convergent synthetic approach (Reprinted with permission from Carlmark, A., Hawker, C., Hult, A., & Malkoch, M., (2009). New methodologies in the construction of dendritic materials. Chem. Soc. Rev., 38, 352–362. © 2008 Royal Chemical Society.).

for the increment of generations. The divergent approach is appropriate for large-scale preparation of dendrimers. Nevertheless, the approach is associated with inherent disadvantages with increment in generations such as identifying defects, separating the desired product from side products resulting from incomplete reactions, and obtaining the product in reasonable yields. The side reactions and imperfections may be avoided using the excess of reagents (Abbasi et al., 2014).

In contrary, convergent approach the growth starts from a unit that eventually becomes the part of the periphery of the dendrimer molecules. The reaction progresses from outward to inward and each branch of monomer are coupled at the end groups. As the reaction proceeds, it gradually linked the surface units of monomers together to give rise to a wedge-shaped dendritic fragment (higher-generation dendron). Ultimately, the activated dendritic fragment is attached to the polyfunctional core to produce globular dendrimers. However, the convergent approach suffers dendrimer size limitation as the dendrons are affected by steric hindrance during the final core attachments (Scheme 5.1) (Grayson and Fréchet, 2001).

5.4.2 OTHER APPROACHES

Dendrimers can also be synthesized applying the concept of click, lego chemistry, branched monomer growth, double exponential growth, etc. The concept has been applied in both convergent and divergent approaches for the viable synthetic outcome.

5.4.2.1 Click Chemistry

The goal of click chemistry is to generate substances by joining small units together, usually with carbon-heteroatom bond forming. The click reaction produces growing set of powerful, selective, and modular units. The formation of modular assembly and simultaneous joining occur in one pot. The word "click" signifies that the reaction starts with high energy (spring loaded) compounds that act as the driving force to click together to yield the final product (Kolb et al., 2001). Guerra et al. described the synthesis of various liquid crystalline dendrimers by copper-catalyzed azide alkyne cycloaddition (CuAAC) one pot reaction applying click chemistry.

The poly(arylester) click dendrimer (3) was prepared by stirring a mixture of azide (1) (diester of 1-azidoundecan-11-ol with terephthalic acid), an appropriate alkyne (2), and $[Cu(CH_3CN)_4][PF_6]$ in THF/H$_2$O (1:1) with considerable yield (78%) (Scheme 5.2) (Guerra et al., 2016).

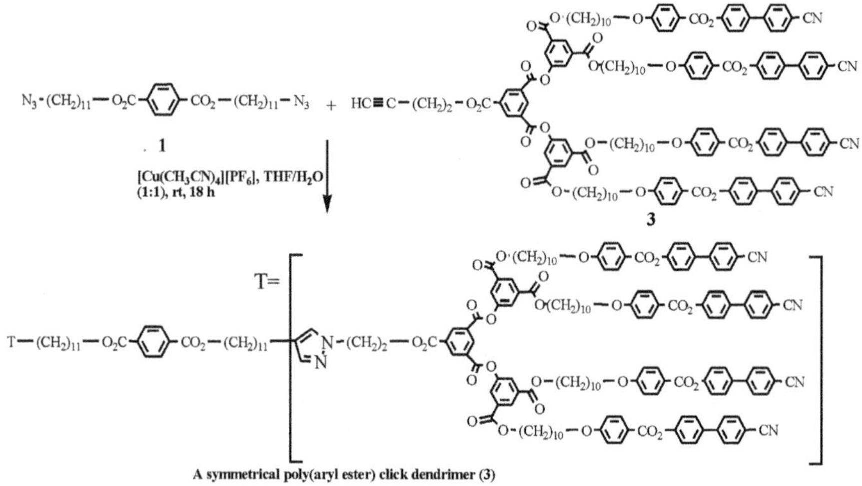

A symmetrical poly(aryl ester) click dendrimer (3)

SCHEME 5.2 Synthesis of symmetrical poly(aryl ester) dendrimer by click chemistry (Reprinted with permission from Guerra, S., Nguyen, T. L. A., Furrer, J., Nierengarten, J. F., Barbera, J., & Deschenaux, R.,(2016). Liquid-crystalline dendrimers designed by click chemistry. Macromolecules, 49, 3222–3231. © 2016 American Chemical Society.).

5.4.2.2 "Hypercores" and "Branched Monomer" Growth

The method accelerates the reactions by which dendrimers are synthesized *via* the convergent approach. The approach combines preassembly of oligomeric units' large dendritic core molecules (hypercores) and dendrons (wedge-branched monomer) to give dendrimers in higher yield and require fewer steps (Figure 5.7) (Nanjwade et al., 2009; Heise et al., 2015).

5.4.2.3 Double Exponential Growth

The double exponential growth method was initially developed for synthesizing linear oligomeric sequences. The process can be explained by taking a triprotected monomer (1) as the starting material. The monomer is then subjected to selective deprotection of A_p to A and B_p to B, and coupling of the deprotected focal point at A and B gives two products $A(B_p)_2$ and $A_p(B)_2$. The reaction between the latter gives rise to orthogonally protected trimer (2) that can be employed to repeat the growth by selective deprotection and coupling that result in the formation of 15-mer with 16 peripheral B_p groups (3) (Figure 5.8-Top) (Kawaguchi et al., 1995).

Balaji and Lewis successfully materialized the double exponential growth for the synthesis of aliphatic polyamide dendrimer scaffold. The first step involves alkylation of allylamine with ethyl bromohexanoate in the presence of K_2CO_3 and DMF to give a tertiary amine. After that, selective

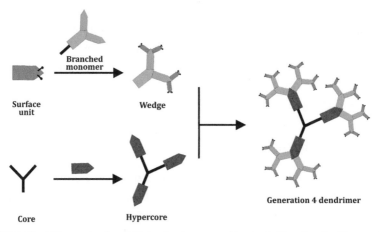

FIGURE 5.7 "Hypercores" and "branched monomer" growth (Reprinted with permission from Nanjwade, B. K., Bechra, H. M., Derkar, G. K., Manvi, F. V., & Nanjwade, V. K., (2009). Dendrimers: emerging polymers for drug-delivery systems. Eur. J. Pharm. Sci., 38, 185–196. © 2009 Elsevier.).

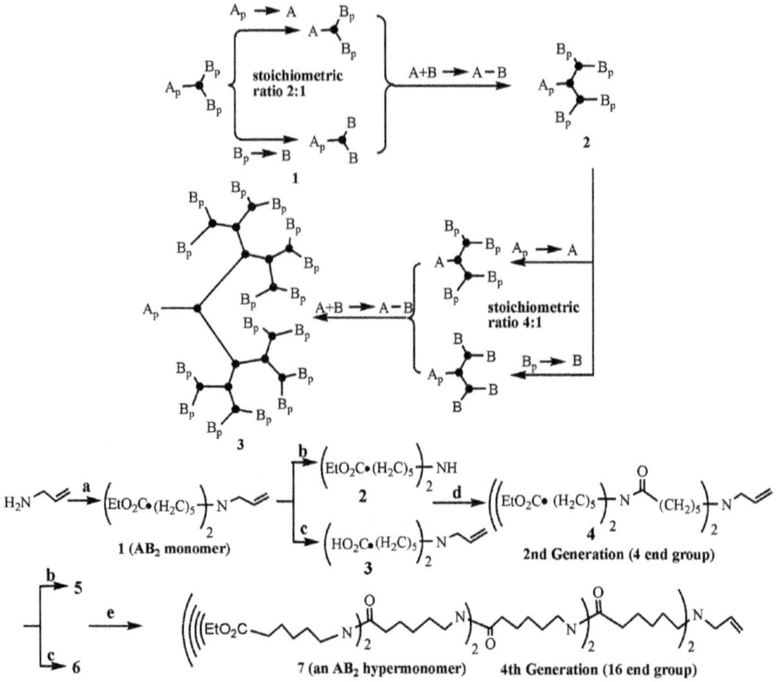

FIGURE 5.8 (Top)-double exponential dendrimer growth (Reprinted with permission from Kawaguchi, T., Walker, K. L., Wilkins, C. L., & Moore, J. S., (1995). Double exponential dendrimer growth. J. Am. Chem. Soc., 117, 2159–2165. © 1995 American Chemical Society.).

deprotection is carried out by removal of the allyl and ethyl groups by tetrakis(triphenylphosphine)palladium(0) and basic hydrolysis, respectively. The compounds deallylamino ester (**2**) and allylic diacid (**3**) were subjected to amide coupling to obtain the diamide having four end groups. Repeating the steps of selective deprotection and coupling yield the hexadecamer dendrimers (**7**) having 16 end groups (an AB₂ hypermonomer) (Scheme 5.3-Bottom) (Balaji and Lewis, 2009).

5.4.2.4 Lego Chemistry

Lego chemistry combines highly functionalized cores and branched monomers to generate phosphorus dendrimers. With rational modification within the synthetic route, lego chemistry facilitates one step terminal surface groups multiplicity from 48 to 250. Each generation of dendrimers can be synthesized per step and require least possible solvent volume, purification by simple washings, and produce nontoxic byproducts such as water and

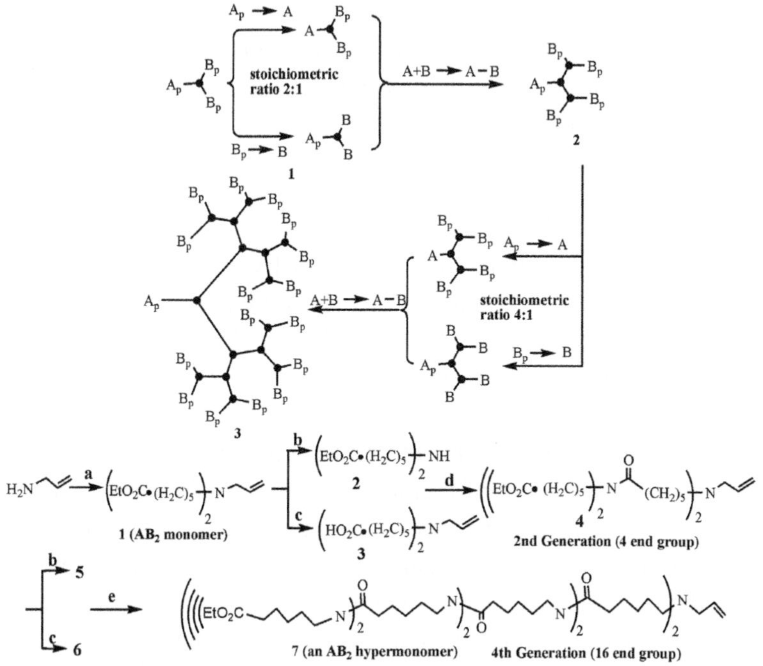

SCHEME 5.3 (Bottom)-a scheme for double exponential dendrimer growth–Reaction conditions (a) 2 eq. Br(CH$_2$)$_5$CO$_2$Et, K$_2$CO$_3$-DMF; (b) Pd(PPh$_3$)$_4$; (c) NaOH-McOH-H$_2$O; (d) 2 eq. (2) + (3), HBTU-DMF; (e) 2 eq. (5) + (6), HBTU-DMF (Reprinted with permission from Balaji, B. S., & Lewis, M. R., (2009). Double exponential growth of aliphatic polyamide dendrimers via AB2 hypermonomer strategy. Chem. Commun., 4593–4595. © 2009 Royal Society of Chemistry.).

nitrogen. The method has been applied for the synthesis of G5 dendrimers by employing three building blocks: two branched monomers (CA$_2$ and DB$_2$) and a triphosphine core (O,O,O-tris(4-(diphenylphosphino)phenyl) phosphorothioate). At the end of the fifth step, a G5 phosphate dendrimer with 250 aldehyde end groups was obtained (Maraval et al., 2003; Svenson and Tomalia, 2005).

5.5 TYPES OF DENDRIMERS

The past few years has seen a dramatic increase in dendrimer research, and information related to the polymers is humongous. A search on PubMed, the National Institute of Health's online research, shows 4,411 results for "dendrimer drug delivery" keyword as of October 2016. The inherent hyper-branched internal scaffold and existence of a large number of peripheral functionalities allow wide array of applications as dendritic nanodrugs, vaccines, biomimetics, diagnostic agents, and target drug-delivery systems. The types and utilization of various important classes of dendrimers are presented in tabular form (Table 5.1).

5.6   CHARACTERIZATION OF DENDRIMERS

With the advancement of dendrimer chemistry and associated intricate structures, the characterizations are also becoming complex and demand in depth and diligent analysis. However, the simultaneous development of analytical chemistry with the cargo of armamentarium makes the process less challenging. A single technique is rather inefficient in describing the structural features of dendrimers, and rather a heuristic approach is universally adopted. A wide array of methods has been exploited to characterize the structure of dendrimers in terms of chemical composition, morphology, shape, detection of imperfections, and homogeneity, etc. Many kinds of literature in the form of research and reviews dedicating to the characterization of dendrimers have been published, and among them, few are worth reading (Caminade et al., 2005; Biricova and Laznickova, 2009; Caminade, 2011; Kaur et al., 2016; Lizama et al., 2016). The important techniques exemplified by spectroscopic, spectrometric, scattering, microscopic, size exclusion chromatographic, electrometric, including the method to analyze the physical properties are presented in tabular form (Tables 5.2 and 5.3) citing brief description for each example.

5.7   APPLICATIONS

5.7.1   THERAPEUTIC APPLICATIONS

Over the years, dendrimers have demonstrated as potential, attractive, and efficient delivery system for various therapeutics including genes and vaccines (Gajbhiye et al., 2009b; Mody et al., 2014). The very nature of the highly branched polymers with adaptable peripheral functionalities and internal cavities makes dendrimers remarkable agents for targeted delivery of different molecules. Considerable advances have fructified toward the development of dendrimers itself as antimicrobials (Lazniewska et al., 2012; Lu et al., 2013; Abd-El-Aziz et al., 2015), vaccines, and immune stimulants (Heegaard et al., 2010); radiopharmaceuticals (Liko et al., 2016); and contrast agents among various other utilizations (Xiong et al., 2016). They are

TABLE 5.1 Various Types of Dendrimers and their Utilization

Sl no.	Dendrimers types	Examples	Core	Surface functional groups	Synthetic approach	Utilization	Ref.
1	PAMAM and PAMAM Starburst™	G1–G11	Ethylene diamine/ammonia	—NH_2/quarternary ammonium groups	Divergent and convergent, click chemistry	Nanocarriers, drug delivery, antimicrobial activity and solubility enhancer	Abid et al., 2016; Kesharwani et al., 2015b; Kalomiraki et al., 2016; Maiti et al., 2004; Milhem et al., 2000
2	PPI	G1–G5	1,4-diaminoethane	—NH_2/quarternary ammonium groups Surface engineered	Divergent and convergent	Nanocarriers for small interfering RNA, brain targeting anticancer drugs, antibacterial synergism	Patel et al., 2016; Taratula et al., 2009; Wrońska et al., 2015
3	PAMAMOS	G4 PAMAM core -an organosilicon (OS) layer and	Hydrophilic PAMAM interiors (core) and hydrophobic OS exteriors	Trimethylsilyl (inert), vinylsilyl (unsaturation),di methoxymethylsilyl (hydrolysable)	Hydrosilation and convergent attachment of OS to amine-terminated PAMAMs	Biological sensors for detection of chemical and biological threats	Dvornic, 2006
4	Tecto dendrimer	[[EDA]-dendri-PAMAM-$(NH_2)_{128}$]-amide-$\{[EDA]$-dendri-PAMAM-$(CO_2H)_{32}\}_{10}$	Ethylene diamine	Multifunctional surface	Self-assembly of the core-shell components followed by covalent bond formation	Identification of diseased cell and diagnosis of pathological state, drug delivery	http://www.nano.med.umich.edu; Uppuluri et al., 2000

TABLE 5.1 (Continued)

Sl no.	Dendrimers types	Examples	Core	Surface functional groups	Synthetic approach	Utilization	Ref.
5	Chiral dendrimer	Nonracemic dendrimer	Phenyltrisalanine	Methyl esters and tert-butyloxycarbonyl groups	Divergent and convergent	Asymmetric catalysis and chiral molecular recognition.	Ritzén and Frejd, 1999
6	Hybrid dendrimer	Guanidiniocarbonyl pyrrole conjugated G2 PAMAM	Ethylene diamine	Guanidiniocarbonyl pyrrole	Direct interaction of G2 PAMAM with guanidino pyrrole carboxylic acid in the presence of aminium-based coupling reagent	Vector for gene delivery, dendrimer hybrid nanocapsules for drug delivery	Jeong et al., 2016; Samanta et al., 2016
7	Amphiphilic micellar dendrimer	Hybrid hydrophilic PAMAM dendron bridged with two hydrophobic C_{18} alkyl chains	Hydrophobic micellar core loaded with doxorubicin	$-NH_2$ and C_{18} alkyl groups	Click chemistry	Nanomicelles for anticancer drug delivery	Wei et al., 2015
8	Multiple antigen peptide dendrimer	Polylysine MAPs	Polylysine core	$-NH_2$ groups modified for surface antigenicity	Divergent and convergent	Immunodiagnosis of viral diseases, biosensor and nanodiagnostics, synthetic vaccines, antivirals	Joshi et al, 2013; Sadler K., Tam, 2002

TABLE 5.1 (Continued)

Sl no.	Dendrimers types	Examples	Core	Surface functional groups	Synthetic approach	Utilization	Ref.
9	Liquid crystalline (LC) dendrimer	Ferrocenyl dendrimer	Tribenzoic acid core	A ferrocene and a cholesteryl mesogenic (induces the LC state) promoter	Convergent	Functional materials and molecular sensors molecular switching devices	Donnio et al., 2007
10	Peptide dendrimers	Polylysine dendrimers	PEG2000 core	$-NH_2$ groups	Divergent, convergent, click chemistry and solid-phase synthesis	Diagnostic (contrast agents) and therapeutic applications (gene and drug deliveries)	Wei et al., 2010
11	Glycodendrimers	G4 PPI-cytarabine complexed	1,4-diaminobutane	Maltose	Divergent and convergent	Functional antigens and antitumor vaccines, drug delivery	Shiao and Roy, 2012; Szulc et al., 2016
12	Metallodendrimer	Metalloporphyrins	Mn^{3+} ion core	Aryl (2,4 tert-butyl phenyl)	Convergent, divergent and click chemistry (CuAAC)	Biomimetic catalysts, imaging contrast agents, biomedical sensors, and therapeutic agents	Tang et al., 2011

TABLE 5.2 Spectroscopic, Spectrometric, and Scattering Method of Characterization

Sl no	Method	Examples	Brief description	Ref
1	Spectroscopic and spectrometric	Ultraviolet–visible (UV–Vis)	UV-Vis absorption spectrum is widely used to monitor the course of dendrimer synthesis. For example, the prominent absorption band at 280–285 nm has frequently been observed during the synthesis of metallodendrimer encapsulated nanoparticles. The further peak at 260 nm ensures the presence of DNA in the cluster of PAMAM DNA construct dendrimer for specific targeting of neoplastic cell-specific.	Choi et al., 2005; Pande and Crooks, 2011)
		FT-Infrared (IR) and Raman	IR spectroscopy is employed to analyze the microstructural changes of functionalized terminal groups. For example, stretching vibration of free OH groups for liquid crystal poly(amidoamine) dendrimer (PAMAM) functionalized with the terminal groups by one chain promesogenic calamitic units and two-chain promesogenic calamitic units at $3470\ cm^{-1}$ vanishes during sample heating and reappears on cooling but with a lower intensity phase transitions. Raman spectra also provide fingerprint *via* which dendrimer can be characterized. The microstructure of six generations of phosphorus dendrimers built from cyclotriphosphazene core with phenoxy and deuterophenoxy end groups were analyzed by Raman spectra. Upon comparison of line intensities in Raman spectra of generations of two molecules reveals their rather quick saturation and reflects strong homogeneity of dendrimer molecules.	(Furera et al., 2015; Popescu et al., 2006)
		Nuclear Magnetic Resonance (NMR)	NMR is one of the widely used instrumental techniques for analyzing dendrimers. Apart from ^{1}HNMR and ^{13}CNMR, correlation spectroscopy such as NOESY, TOCSY including heteronuclear correlation experiments, such as HSQC, HMQC, and HMBC are also used. A new side-reaction occurring during divergent synthesis of PAMAM dendrimers (generations G0–G2) was revealed using NMR analysis combined with mass spectroscopy.	(Tintaru et al., 2015)
		Electron Paramagnetic Resonance (EPR)	EPR spectroscopy is useful for studying metal complexes or radicals. Therefore, the capability to complex various external and internal sites of metallodendrimers and dendrimers having radicals either at the interface or within the interior branching unit can be characterized using EPR. A study complexation behavior of various PPI dendrimers with a maltose shell towards Cu(II) metal was carried out using EPR technique. Analysis of EPR spectra *via* computer indicates two pivotal complex geometries: axial (G1>>G2≥G3), and rhombic (G1>G2≥G3).	(Appelhans et al., 2010)

TABLE 5.2 (Continued)

Sl no	Method	Examples	Brief description	Ref
		Fluorescence	Fluorescence spectroscopy measures both excitation and emission spectra of a sample. For example, the conformational changes of bovine serum albumin were evaluated by the measurement of intrinsic fluorescence intensity of protein tryptophan residues before and after addition of dendrimers. It was evident that full-generation dendrimers have the stronger effect on biomolecules than the half-generation ones.	(Klajnert and Bryszewska, 2002)
		Chirality, Optical rotation, Circular Dichroism (CD)	The combined techniques are widely used to characterize chiral dendrimers. For example, the decrease in specific rotation is accompanied with an escalation in the number of generation for each group of polyether dendritic bromide coupled with (R)-(+)-1,1′-bi-2,2′-naphthol. CD spectra showed two strong Cotton effects (change in optical circular dichroism curve), which are characteristic of the ^1B_b transition of the naphthalene chromophore. Further CD revealed the absolute configuration of the chiral dendrimers retains 'R' when conjugated with the chiral (R)-naphthol. The dendrimers also exhibited significant dihedral angle due to the loss of an intramolecular hydrogen bond and the increasing steric hindrance.	(Chen et al., 1998)
		X-ray Diffraction (XRD)	XRD is a versatile tool for analyzing the atomic and molecular structure of crystals. The XRD studies are mainly carried out on liquid crystalline dendrimers. For example, XRD experiments have been performed using nickel-filtered copper Kα radiation on various generations of liquid crystalline dendrimers based on the carbosilane dendritic matrix. The study revealed that diffraction pattern the mesogenic units (cyanobiphenyl and methoxyphenyl benzoate) form smectic layers that overcome the restrictions imposed by the globular dendritic core of the molecule.	(Richardson et al., 1999)

TABLE 5.2 (Continued)

Sl no	Method	Examples	Brief description	Ref
		Mass Spectrometry (MS)	Low molecular weight dendrimers are analyzed by chemical ionization or fast atom bombardment techniques whereas for high to extreme molecular weight dendrimers, matrix assisted laser desorption ionization time of flight (MALDI TOF); ion trap tandem MS and electrospray ionization (ESI) are used. A polyphenylene G9 dendrimer having theoretical molecular weight of 1.9 MDa was characterized by MALDI–TOF MS. The experimental weight was found to be approximately 1.5 MDa with a polydispersity of 1.003 and an average of 1.005 for the synthesized homologous series (G1-G9). However, MS data signifying the purity of dendrimer are required to be interpreted with utmost caution as very often falsely negative results are also obtained.	(Baytekin et al., 2006; Räder et al., 2014)
2.	Scattering techniques	Small Angle X-ray Scattering (SAXS)	The average radius of gyration (Rg) of dendrimers in solution is analysed by SAXS. For example, solution small-angle x-ray scattering (SAXS) provides direct structural information of the enzyme-linked DNA dendrimer nanosensor structure. The scattering data for DNA dendrimers with butyrylcholinesterase attached in phosphate buffered saline pH 7.4 suggests Rg of 55Å. Further, Kratky plot of scattering data indicates the heterogeneous internal substructure of the DNA dendrimers.	(Walsh et al., 2015)
		Small-Angle Neutron Scattering (SANS)	The technique is very similar to SAXS and provides greater details of internal dendrimer structure. A SANS analysis to characterize the solution behavior of charged carboxylic acid terminated "cascade" dendrimers revealed that an increase in dendrimer concentration leads to a single broad peak in the scattering profile arising from interdendrimer interaction (IDI). The IDI peak also arises from dissociation of terminal carboxylate groups.	(Huang et al., 2005)
		Dynamic Light Scattering (DLS)	It is utilized for studying dendrimers structure and employed for calculating the diffusion coefficient (D), and the hydrodynamic radius (R_H) is then calculated from D. The R_H derived is comparable with SAXS derived Rg. The reported mean diameter of PAMAM dendrimers as determined by DLS was found to be around 200 nm. An alternate technique to determine the R_H is by laser light scattering (LLS).	(Wang et al., 2012)

TABLE 5.3 Microscopic, Size Exclusion Chromatographic, and Electroanalytical Method of Characterization Including Physical Properties

Sl no	Method	Examples	Brief description	Ref
1	Microscopic	Scanning or Atomic Force Microscopy (SFM/AFM)	In addition to high-resolution imaging, AFM has been utilized for estimating important numerical molecular parameters, such as volume and molecular weight of dendrimers. Tapping mode AFM has successfully imaged individual dendrimer molecules of the higher generations poly(amidoamine) (G5-G10) spread on a mica surface. It also offers a powerful tool for investigating morphology, molecular packing, and molecular dynamics of dendrimers.	(Li et al., 2000; Schlüte and Rabe, 2000)
2	Size exclusion chromatography (SEC)	Gel permeation chromatography	The technique allows for the separation of molecules according to their size and when suitable standards are used, provides relative molar mass distribution of polymers. However, the method should be used with caution with linear polymer standard. Nevertheless, the method is being used more often for determining polydispersity, molar weight distribution (Mw/Mn) and ensuring purity of dendrimers. A G5 star-branched dendrimer with Mw/Mn values <1.03 obtained via SEC showed good congruence with calculated molecular weights and those determined.	(Hirao and Yoo, 2011; Tomalia et al., 2012)
3	Electro-analytical	Coulometry and Cyclic Voltammetry (CV)	Exhaustive coulometry measures the number of electroactive groups in a dendrimer. The technique is mainly used to characterize ferrocenyl termini or in general metallocenes linked to the surface of various dendrimers. Newer coulometric technique successfully probes the mechanisms of interactions (electron transfer) between redox active guest molecules (metallocene) encapsulated in carbon nanotubes. Analysis of such host–guest interactions enables controlling the oxidation state of guest molecules. Alternatively, CV is utilized to investigate the extent of electrostatic binding of metal complexes and dendrimers with respect to size and pH. It also detects the extent of electroactive groups burying (e.g., ferrocyanide) inside the dendrimers.	(Kulczynska et al., 2006; McSweeney et al., 2016; Ornelas et al., 2009)
		Gel Electrophoresis (GE)	Polyacrylamide (PAGE) is routinely used to establish the purity and homogeneity of hydrophilic dendrimers. The technique has been exploited for separation of water soluble dendrimers e.g. PAMAM with high resolution and sensitivity at acidic pH. A combination of PAGE and capillary zone electrophoresis provide an efficient way to determine the electrophoretic mobilities and efficiently separate PAMAM and a core-shell tecto dendrimer with succinamic acid terminal groups based on charge/mass ratio.	(Sharma et al., 2003; Shi et al., 2005)

TABLE 5.3 (Continued)

Sl no	Method	Examples	Brief description	Ref
4	Rheology and physical properties	Intrinsic viscosity	Determination of intrinsic velocity (η) is an important property to characterize the rheological behavior of dendrimer solutions. For example, dilute solution polyether dendrimers exhibit a maximum viscosity as a function of generations and depend on the density of branches and molecular characteristics.	(Lu et al., 2012; Mourey et al., 1992)
		Differential Scanning Calorimetry (DSC)	The technique detects the glass transition temperature (Tg) as a function of branching and terminal group functionality and molecular weight of dendrimers. For example, the longer the graft chain on the surface of polyamidoamine-g-poly(N,N-dimethylaminoethyl methacrylate), the lower is the Tg.	(Gill et al., 2010)
		Dielectric Spectroscopy (DS)	DS provides information about the molecular dynamics of dendrimers. For example PPI dendrimers show three relaxation processes α, d (d for dendrimer), and β. The DS α-relaxation values are consistent to those obtained *via* DSC measurement. The d-relaxation is independent of the number of generations.	(Mohamed et al., 2015)

also efficient in overcoming the pharmacokinetic challenges encountered during delivery of drug molecules as a result of low solubility and limited tissue permeability (Twibanire and Grindley, 2014). The nanoarchitecture of dendrimers found a broad range of application for controlled drug deliveries of drugs to molecular probes, nanoscaffolds, nanodrugs, nanoscale containers, and as investigation tools for biodistribution studies. The section gives an overview of various applications of dendrimers that have revolutionized drug delivery, biomedical, and diagnostic advances.

5.7.1.1   Dendrimers in Cancer Targeting

The nonspecific nature of chemotherapeutic agents to selectively target tumor cells leads to various unwanted toxicity. Most anticancer drugs have a variety of toxic side effects, and this substantially limit their utility (Bharali, 2011; Mousa, 2011; Yingchoncharoen et al., 2016). The approaches for site-specific anticancer drug targeting are considered as a potential challenge in cancer chemotherapy (Tekade et al., 2009a; Tekade and Chougule, 2013; McLornan et al., 2014). Nanotechnology has the prospect to overcome the obstacles in cancer treatment because of the unique size and characteristic bioactivity of nanomaterials. Dendrimers can be conjugated with drugs to discharge the attached compounds for internalization within specific neoplastic cells (Liang et al., 2010; Tekade et al., 2015a). Efficient exploitation of dendrimers for various anticancer drug deliveries and as theranostics in cancer therapy provides multifunctional platforms for cancer detection, treatment, and monitoring for maximum therapeutic effectiveness. Dendrimers conjugated with anticancer drugs have the ability to evade efflux transporter and to enhance bioavailability of the loaded guest molecule (Fernandez et al., 2011; Singh et al., 2016). In light of the above discussion, Parsian et al. successfully synthesized half generations (between G4.5 and G7.5) PAMAM dendrimers-coated magnetic nanoparticle (DCMN) for target delivery of gemcitabine. The G5.5 having the surface modified -COOH groups conjugated with gemcitabine achieved highest drug loading. Gemcitabine nanoparticles were 6- and 3- fold more toxic on SKBR-3 and MCF-7 breast cancer cell lines, respectively, compared to free gemcitabine. They attributed the higher activity of DCMN due to a more efficient intracellular uptake via endocytosis at the target site (Parsian et al., 2016). A recent study reported the encapsulation of 3,4-difluorobenzylidene diferuloylmethane

(CDF), a nontoxic analog of curcumin with PAMAM dendrimer conjugated with folic acid (FA) using equilibrium dialysis method. The focus of the study was to enhance the aqueous solubility and to target FA-PAMAM-CDF particularly to overexpressed folate receptor cervical cancer cells (HeLa) and ovarian cancer cells (SKOV3). The targeted nanoformulation (FA-PAMAM-CDF) exhibited significant anticancer activity compared to the non-targeted PAMAM-CDF due to preferential internalization into HeLa and SKOV3 cells. The IC_{50} were found to be 1.47 μM, 1.70 μM, and 3.60 μM for CDF, FA-PAMAM-CDF, and PAMAM-CDF, respectively, in HeLa cells. Further, the IC_{50} values were 0.49 μM, 0.85 μM, and 3.0 μM for CDF, FA-PAMAM-CDF, and PAMAM-CDF, respectively, in SKOV3 cells (Luong et al., 2016a). In an interesting study, Zhou et al. described the synthesis of more than 1,500 modular degradable dendrimers with wide structural diversity (cores, peripheries, and generations) using sequential, orthogonal click reactions. 5A2-SC8, a dendrimer with amine core and thiol periphery formulated as dendrimer-microRNA (*let*-7) nanoparticle as tumor suppressors, showed high cytotoxic potency to tumors and minimal hepatotoxic liability. It effectively inhibits the growth of aggressive liver tumors in an *MYC*-driven transgenic liver cancer model (Zhou et al., 2016). Therefore, with reasonable modification of dendrimers or decorating with certain properties could make the dendrimer-based nanoformulation target-specific for better therapeutic outcomes and minimizing the toxicity.

5.7.1.2   Dendrimers in Antimicrobial Delivery

The emergence of antimicrobial resistance and concomitant irrational uses endanger the efficacies of most of the available antibiotics. Mining of new antimicrobials by target-based screening of large libraries of synthetic compounds also failed (Lewis, 2013). Further, limited drug discovery and development by the pharmaceutical industries on behalf of reduced economic incentives and challenging regulatory requirements jeopardize the scenario (Ventola, 2015). The crisis of antimicrobial resistance occurs due to increased efflux, decreased influx, target modification, target amplification, repair of a damaged target, enzymatic inactivation, sequestration of antibiotic, target bypass (acquisition of alternative metabolic pathways), protection of target, and intracellular localization among other mechanisms (Yilmaz and Özcengiz, 2016). Considering the setbacks, adopting newer formulation strategies that focus particularly on increasing their delivery efficiencies could improve the antimicrobial efficacy. Dendrimer nanofor-mulations encapsulating a great variety of antimicrobials together with den-drimer drug conjugation enhanced the therapeutic effectiveness and could pave the way for treating recalcitrant infectious diseases (Sharma et al., 2012). A pan-DR-binding epitope peptide derivatized dendrimer complexed with liposomal amphotericin B (PDD-LAmB) was utilized as an adju-vanted nanocarrier to efficiently deliver LAmB to the antigen presenting cells (APC) of Leishmania major *in vivo* mouse model for cutaneous infec-tion. A cross-linking G5 PAMAM dendrimer with the peptide generated the PDD. The nanocarrier enhanced the drug efficacy by 83% and attained 10-fold drug APC targeting. It significantly reduced parasite burden and toxicity and additionally induced antiparasite immunity (Daftarian et al., 2013). Co-conjugated G5 PAMAM dendrimer (*d* 5.4 nm) nanoplatform specifically target lipopolysaccharide molecules present on the surface of gram-negative bacteria. The dendrimer having PMB-mimicking dendritic branch (ethanolamine) conjugated with polymyxin B (PMB) showed potent bactericidal activity *in vitro* against *Escherichia coli* (Wong et al., 2015). Sonawane et al. recently reported a new lipid dendrimer hybrid nanopar-ticle (LDHN) delivery system. Nanosized G4 PAMAM-succinamic (SA) acid dendrimer loaded with vancomycin (VCM) followed by lipidic-based shell encapsulation comprising of Compritol® 888 ATO leads to the forma-tion of LDHNs. The nanoparticles inside a nanoparticle system effectively released VCM over the period of 72 h and followed zero-order kinetics. The core-shell type LDHN achieved drug release in a sustained manner and improved the antibacterial activity of VCM against *Staphylococcus aureus* (sensitive and resistant). At time periods of 18 h and 36 h, VCM loaded LDHNs exhibited higher activity against methicillin-resistant *S. aureus* with a minimum inhibitory concentration of 7.81 µg/ml and 31.25 µg/ml, respectively. Molecular modeling studies also revealed an increase in drug entrapment capacity of the LDHN-based nanoformulation (Sonawane et al., 2016). Therefore, dendrimer-based delivery systems provide practical applications to combat antimicrobial resistance.

5.7.1.3 Dendrimers in Transdermal Drug Delivery (TDD)

Apart from other chemical enhancers, few dendrimers also increase the skin permeability of various drugs. They also facilitate drug solubiliza-tion in the formulation and drug partitioning into the skin for an efficient and safer method of drug delivery. TDD of therapeutics by surface engineered dendrimers too demonstrated immense potentialities as the degree of skin permeability enhancement depends on the dendrimer surface charge (Prausnitz and Langer, 2008; Maheshwari et al., 2012; Yang et al., 2015; Hsu et al., 2016). The feasibility attributed for TDD potential of dendrimer nanocarriers on account of smaller size, extreme molecular uniformity, and functional surfaces (Sun et al., 2005). Various TDD systems were formulated using PAMAM dendrimers to assist the delivery of anti-inflammatory drugs (Chauhan et al., 2003; Yiyun et al., 2007). The relationship between the structure of a PAMAM dendrimer and skin permeability showed that cationic dendrimers have higher skin penetration than neutral and anionic dendrimers, and the penetration (passive and iontophoretic) was inversely proportional to their molecular weight (Venuganti et al., 2011). Yang et al. proposed three hypotheses for skin penetrability of PAMAM-based dendrimers based on the effect of surface charge, size, and hydrophobicity. They highlighted that the smaller the dendrimers, the better is the penetration. Further, surface-modified G2 PAMAM dendrimers either augment or alter skin permeability. The permeation efficiency of the dendrimer–drug conjugates also depends on the partition coefficient (hydrophobicity) (Yang et al., 2012). Recently, Huang et al. reported a TDD incorporating G3 PAMAM dendrimer-based diclofenac gel coupled with sonophoresis. The system showed a 16.5-fold increase in penetration on hairless male Wistar rat skin. The *in vitro* cumulative drug permeated through the skin was found to be 257.3 $\mu g/cm^2$ and 935.21 $\mu g/cm^2$ for DF-dendrimer gel without and with sonophoresis-treated skin, respectively (Huang et al., 2015). Alternatively, for psoriasis treatment, dithranol (DIT) encapsulated within G5 PPI dendrimer synthesized via the divergent method showed a skin penetration of 95.33% in 24 h. The plain DIT only achieved 34.20% penetration. The hyperbranched dendritic nanocarriers displayed slight erythema as determined by the Draize patch test on the skin of albino rats. The permeation rate constant of PPI-DIT was 11.61 $\mu g/cm^2/h$, while the dendrimer-free DIT solution exhibited 2.72 $\mu g/cm^2/h$ with considerable skin irritation (Agrawal et al., 2013).

5.7.1.4   Dendrimers in Oral Drug Delivery

Oral route is the most preferred for polymer-drug conjugates, and this route administers a variety of drugs. However, the utilization of polymer-drug conjugates is marred by limited oral bioavailability because of their large size and molecular weight (Noriega-Luna et al., 2014). The high degree of modifiable surface functionality and encapsulations within the internal cavities augments gastrointestinal absorption. PAMAM dendrimers at low concentrations facilitated the absorption of drugs via small intestine and used as GI penetration enhancers, drug solubilizers, and carriers for oral delivery. The G4 and G3.5 PAMAMs effectively solubilized camptothecin in simulated gastric fluid, and oral absorption increased by 2.3-fold (Lin et al., 2011; Sadekar et al., 2013). They also have the capability to increase the gastrointestinal residence time due to their bioadhesive properties (Dening et al., 2016). However, the oral delivery utilizing PAMAM dendrimers is somewhat compromised by their toxicity and biocompatibility. A strategic manipulation of the surface functionality might pacify the toxicity and PAMAMs could emerge as potential carriers clinically (Yellepeddi and Ghandehari, 2016). Ma et al. carried out an interesting study to enhance the oral absorption of probucol (PB) using nanoliposomes-based (polyethylene glycol) PEGylated G5 PAMAM dendrimer. The particular formulation showed considerable inhibition of plasma total cholesterol and triglyceride elevation in mice. Moreover, the nanoliposomes achieved a greater level of the plasma drug concentration than conventional PB or marketed tablets (PB). The water solubility of PB was improved by the positively charged $-NH_2$ groups of PEGylated-G5 PAMAM than the G4.5 and G5 PAMAM that bears negatively charged -COOH or neutral $-OH$ group, respectively (Ma et al., 2015). Work along similar lines further revealed that liposomal solubilization and encapsulation proved to be far more efficient in increasing oral bioavailability than sole G5 PAMAM–NH_2 dendrimer-drug molecular complex. Simvastatin (SMV)-liposomes increased the C_{max} of SMV by a factor of 3.8, and oral bioavailability by a factor of 3.7 times when compared with SMV (Qi et al., 2015). Alternatively, Mansuri et al. successfully synthesized chitosan-anchored G5 PPI dendrimer for the preparation of albendazole (ABZ)-loaded mucoadhesive dendrimer. The muco dendrimer increased the $t_{1/2}$ of ABZ approximately by 2-fold as compared to the free drug. The muco-PPI elicits higher C_{max} (2.40 µg/mL) when compared with orally administered free ABZ (0.19 µg/mL) and conventional tablet (0.20 µg/mL) (Mansuri et al., 2016). Dendrimer-based formulations for oral delivery have the potential to increase the solubility, dissolution, permeability, bioavailability, efficacy, and biological half-life of drugs (Prajapati et al., 2009; Jain and Tekade, 2013).

5.7.1.5 Dendrimers in Ocular Drug Delivery

Only less than 5% of the applied dose of conventional, that is, solutions, suspensions, and ointment dosage forms, penetrates the cornea and reaches the intraocular tissues. A substantial portion of the instilled dose gets wasted due to the presence of ocular barriers, rapid elimination from the precorneal area, and nasolacrimal drainage. There is also a possibility of systemic absorption that may result in unwanted side effects or toxicity (Ali et al., 2016). Linear bioadhesive polymers partially resolve the disadvantages of the abovementioned dosage forms. Hyaluronic acid or cellulose derivatives increase the drug residence time on the ocular surface and reduces the systemic drug side effects and the dosing frequency. Nevertheless, the associated high viscosity of the polymers induces discomfort and blurred vision. Similarly, the dendritic structures also showed potential in increasing the residence time of drugs via secure attachments and reducing systemic absorption of locally applied ocular drugs. Additionally, aqueous solutions of dendrimer at low concentration attain similar viscosity as water that might nullify the associated discomfort. Eye drops incorporated with dendrimers also facilitate the formation of a spreadable aqueous layer with sufficiently favored by its tensioactive nature (Yavuz et al., 2013; Bravo-Osuna et al., 2016; Tekade and Tekade, 2016; Villanueva et al., 2016). Mishra and Jain synthesized G5 PPI with EDA core and load acetazolamide (AZ). The AZ-encapsulated dendritic nanoarchitectures achieved 56% of maximum drug entrapment efficiency. The cumulative percentage of drug release of the dendrimer formulation showed 83% and 80.4% in pH 7.4 phosphate buffer and simulated tear fluid, respectively, thereby exhibiting a sustained release. ACZ-G5 PPI offers better intraocular pressure lowering efficacy for glaucoma with minimal ocular irritancy (Mishra and Jain, 2014). Conjugates of dexamethasone (DEX) with G3.5 and G4.5 PAMAM dendrimers for retinal delivery also provide improved ocular permeability after subconjunctival and intravitreal applications. *In vitro* studies incorporating ARPE 19 (human corneal epithelium) cell line using {3-(4,5-dimethylthiazol-2-yl)-2,5-diphenyltetrazolium bromide} (MTT) assay revealed no significant toxicity. The hydrolysis and clearance time of the formulation were also markedly enhanced (Yavuz et al., 2016). A recent study demonstrated the efficiency of intraocular gene delivery for posterior segment diseases. The noninvasive retinal gene delivery of the gene model red fluorescent protein plasmid (pRFP) incorporates penetratin (a cell-penetrating peptide) and G3 PAMAM that enable a nanoparticulate (~150 nm) compact gene condensation. The intact system penetrated rapidly from the ocular surface into the fundus with a residual time of more than 8 h, which resulted in efficient expression of RFP (Liu et al., 2016).

5.7.1.6   Dendrimers in Pulmonary Delivery

The uses of nanocarriers have been restricted in pulmonary delivery due to its reported toxicity (Kurmi et al., 2010; Kayat et al., 2011; Chougule et al., 2014). Bai et al. reported the first utilization of PAMAM dendrimers for the delivery of enoxaparin (ENP) via the pulmonary route. The electronegative drug enoxaparin formed a complex with the cationic dendrimers through electrostatic interactions. The absorption efficacy of the complex was estimated indirectly by measuring the plasma anti-factor Xa after administration of the formulations into anesthetized rat lungs. Pulmonary delivery increased 40% absorption of the drug ENP. In contrast, negatively charged PAMAM-bearing surface carboxyl groups (G2.5) failed to improve the bioavailability of ENP. The relative efficacy of the formulations comprising G2 (1%) or G3 (0.5%) PAMAM dendrimer–drug complex was similar to enoxaparin administered subcutaneously in preventing deep vein thrombosis (Bai et al., 2007).

PAMAM dendrimers also served as nanocarriers for pulmonary delivery of poorly water soluble antiasthmatic steroid beclomethasone dipropionate (BDP) through nebulization. Considering all the generations studied, the G4(12) PAMAM having hydrophobic 1,12-diamino dodecane core achieved the highest solubility of BDP. However, nebulizer designs affect the aerosol characteristics rather than dendrimers generations. The BDP-G4(12) PAMAM complex showed sustained release profile as less than 35% of the drug got released after 8 h. Pari LC Sprint® (air-jet) and Aeroneb® Pro (active mesh) nebulizers proved advantageous epitomized by high output and superior fine particle fraction and considered suitable for pulmonary delivery of the nanocarrier (Nasr et al., 2014). A new study compared the pulmonary accumulation and retention time of various lipid {liposomes and PEG DSPE (1,2,-distearoyl-sn-glycero-3-phosphoethanol-amine-N-aminopolyethelenglycol) micelle} and non-lipid-based nanocarriers such as PPI dendrimer- small interfering ribonucleic acid (siRNA) complexes nanoparticles, microporous silica nanoparticles, and quantum dots after delivery by inhalation. Inhalation delivery by lipid-based nano-carriers was found to be more proficient than nonlipid-based carriers for pulmonary delivery. However, nonlipid-based nanocarriers achieved longer lung concentration retention time after inhalation delivery when compared to intravenous injection. Further, doxorubicin (DOX)-loaded liposomes also significantly enhanced the anticancer effect of the drug with limited adverse extrapulmonary effects (Garbuzenko et al., 2014). Therefore, to effectively target lung endothelial cells, Khan et al. modified PAMAM and PPI dendrimers by replacing the free amines with C_{15} and C_{14-16} hydrophobic lipid tails, respectively, by a combinatorial approach. The modified dendrimers were then nanoencapsulated with siRNA and found to favorably target Tie2 (angiopoietin tyrosine kinase receptor) expressed by lung endothelial cells (Khan et al., 2015). In recent times, carboxyl-terminated G4 PAMAM-SA dendrimers were conjugated with DOX by acid-labile hydrazone spacers in the form of nanocarriers to achieve a sustained intracellular-triggered release. Subsequent pulmonary delivery of the nanocarriers improved the therapeutics efficacy of DOX assessed *in vivo* lung metastasis mouse model. A greater lung intracellular concentration of DOX and minimum plasma concentration alleviate the cardiotoxic liability of the drug (Zhong et al., 2016).

5.7.1.7   Dendrimers in Vaccination/Immunomodulation

In the field of vaccinations, dendrimers serve as carriers/presenters of small antigens. They also have immunostimulatory (adjuvant) properties and give rise to multimeric antigenic conjugate vaccines when mixed with specific antigens (Heegaard et al., 2010). The widely used approach is the implementation of MAP dendrimer in peptide-based vaccines. The dendrimer substitutes the conventional carrier protein with amino-modified surface functionality to impart antigenicity. The core matrix consists of trifunctional amino acid lysine, and the polylysine core has the capability to amplify the peptide antigens (Sadler and Tam, 2002; Joshi et al., 2013). The MAP dendrimers circumvent the adverse effects associated with conventional vaccines, carrier proteins, and cytotoxic adjuvants. They elicit a robust immune response than small antigenic peptide and are stable to enzymatic degradation. MAP-based vaccines are either developed by addition of T-cell epitopes, cell-penetrating peptides, and lipophilic moieties or by synthesis of self-assembling peptide dendrimers (Fujita and Taguchi, 2011; Apostolopoulos, 2016). Skwarczynski et al. reported an alkyne-functionalized four-arm star dendrimer with polyacrylate core synthesized by successive atom-transfer radical polymerization and CuAAC "click" reaction. The said dendrimer was then conjugated with J14 peptide epitope construct (a chimera of the streptococcal M protein) to obtain a self-adjuvant vaccine nanoparticle of a size 20 nm. Following subcutaneous immunization, female B10.Br (H-2K) mice produced significant p145-specific antibody titers in mice sera that elicit the strong immune response to the group A *Streptococcus pyogenes* M protein (Skwarczynski et al., 2010). Further, DNA vaccines obtained by complexing peptide-conjugated dendrimers such as Pan-DR epitope and a hemagglutinin-derived peptide conjugated PAMAM (PPD and HAPD) dendrimers with negatively charged DNA. The DNA-peptide-dendrimer nanoparticles when targeted to APC *in vivo* stimulate generation of high-affinity T cells and induce a potent humoral response against tumors. Among the two DNA vaccines, PPD demonstrated higher activity in promoting antitumor immunity (Daftarian et al., 2011). Nonpeptide dendrimers like maltose-modified dense shell G4 PPI glycodendrimer (PPI-mDS G4) was found to activate the NF-κB (nuclear factor-kappa light chain enhancer of activated B cells) signaling pathway. Taking into account the potentiality to induce innate immunity with additional immunomodulation, Pawlik et al. suggest that the application of PPI-based glycodendrimer as a possible biocompatible anticancer vaccine adjuvant (Jatczak-Pawlik et al., 2016). Moreover, a modified dendrimer nanoparticle (MDNP) obtained via reaction between G1 PAMAM dendrimer and 2-tridecyloxirane was nanoencapsulated with replicon mRNA vaccine that encodes protein antigens of influenza virus and Ebola virus. Additionally, a multiplexed MDNP was also nanoencapsulated with multiple replicons mRNA encoding proteins and surface antigens of *Toxoplasma gondii*. The nanoformulations were stable against nuclease degradation and, after a single immunization, provide protection against the deadly pathogens under consideration. The vaccines elicit high levels of antibody titer and antigen-specific CD8+ *T*-cell responses against the encoded protein antigens. The synthetic nanoparticle vaccine platform presents a promising candidate to combat against lethal pathogens during a sudden epidemic without added adjuvant (Chahal et al., 2016). As compared to conventional vaccine formulations, dendrimer-based nanoscale vaccines provide advantages in target delivery of antigens and also function as self-adjuvant for immune potentiation.

5.7.2   BIOMEDICAL APPLICATIONS

Current investigations on dendrimers have widely focused on their utilities in different biomedical applications. The possibilities of varying degree of control over the dendrimer design makes them as promising materials for a range of applications. By accommodating various targeting groups or functionalities, dendrimers provide a flexible architecture that find applications in drug and gene delivery, high-resolution magnetic resonance imaging (MRI), contrast agents, nano-scaffolds/containers, tissue engineering, and molecular probes, etc. Some of the most important biomedical applications of functional dendrimers are summarized as follows:

5.7.2.1   Dendrimers in Gene Delivery

The selection of appropriate systems to deliver therapeutic genes to the disease sites without affecting non-target cells is considered as the cardinal feature for a successful gene therapy (Maheshwari et al., 2015a; Tekade, 2015; Tekade et al., 2015b, 2016). Among various other interventions, gene therapy mainly focuses on gene transfection and correction of genetic mutational disorders, regeneration of tissues by activated stem cells, and facilitation of immune responses to fight cancer (Waehler et al., 2007; Youngren et al., 2013a; Gandhi et al., 2014; Naldini, 2015). Gene therapy either involves delivery of plasmid DNA that increases the activity of the target to produce therapeutic proteins or delivery of genetic material, namely, antisense oligonucleotides and siRNA/DNAzyme, which leads to a reduced activity of the target. The safe and efficient target delivery of therapeutic genes rely on design of improved viral vector or synthetic carrier molecules. Dendrimers also found applications in delivering genes because of their nanoparticulate nature and showed considerable effectiveness for intracellular gene delivery in addition to other synthetic carriers (Dinçer et al., 2005; Dufès et al., 2005). Bifunctional hydroxyl-terminated PAMAM dendrimers (BiD) functionalized with approximately 20 amine groups out of the 64 hydroxyl end groups was efficiently complexed with pBAL plasmid DNA using Fmoc protection/deprotection chemistry. Further, triamcinolone acetonide (TA), a nuclear localization enhancer, was conjugated with the BiD via glutaric acid as a spacer to produce BiD-TA-based gene vectors. BiD-TA-based gene vectors showed improved stability and efficiency in carrying the cargo to the human retinal pigment epithelial cell line (ARPE-19). The complex also elicits enhanced cellular uptake and transfection as determined by the luciferase activity. In addition to dendrimer-TA for DNA complexation, shielding the surface of gene vector by PEGylated dendrimer improved the stability of the system (Mastorakos et al., 2015).

An extensive review reported the importance of PEGylation of PAMAM dendrimers for increasing the efficacy and alleviating the toxicity for gene delivery (Luong et al., 2016b). Very recently, Wang and Cheng reported a library of fluorinated dendrimers with different generations and degree of fluorination obtained by treatment with the heptafluorobutyric anhydride. Subsequently, DNA and siRNA transfection efficacy of the dendrimers (complexed with DNA EGFP [Enhanced Green Fluorescent Protein] plasmid and siRNA targeting firefly luciferase) were assessed on HeLa cells and HeLa-luc cells, respectively. Amongst the EDA-cored G4-G7 fluorinated PAMAM dendrimers, the G5-F7$_{74}$ and G5-F7$_{82}$ revealed the highest efficacy for EGFP transfection and silencing luciferase gene, respectively with improved cellular uptake. Furthermore, the G5 PAMAM dendrimer-based gene delivery systems exhibited limited toxicity on the transfected cells and were more efficacious than the commercial transfection reagent Lipofectamine® 2000 (Wang and Cheng, 2016). Similarly, β-cyclodextrin (β-CD) surface-grafted G5 PAMAM dendrimers entrapped with gold nanoparticles (Au DENPs-β-CD) also showed the ability to compact plasmid DNA at 5:1 nitrogen to phosphate ratio and are less cytotoxic than Au DENPs without β-CD conjugation. Au DENPs-β-CD nanoparticles (Au core size: 2.9 nm) exhibited promising nonviral vectors for efficient gene delivery and could transfect both EGFP and Luc genes to the targeted cells (293 T cells, a human embryonic kidney cell line) (Jieru et al., 2016). In addition to surface functionalization and conjugation with localization enhancer and PEGylation, Hu et al., highlighted that coherent manipulation of dendrimer core also holds promise toward increasing the transfection efficacy and biocompatibility of dendrimers for gene delivery (Hu et al., 2016).

5.7.2.2   Dendrimers in Biomimetic Artificial Proteins

Based on the functionalization of the terminal groups, dendrimers may be covalently linked to each other to act as biomimetic artificial proteins. Chen et al. studied the interaction of dendrimers that mimic the natural dentin matrix protein phosphophoryn on hydroxyapatite crystals (HAC). Among the three G7 PAMAM dendrimers with -COOH, -NH$_2$, and -NHC(O)CH$_3$ capped surfaces, the positively charged dendrimer (-NH2-capped) displayed strong binding capacity with HAC and could control crystal nucleation and growth during biomineralization (Chen et al., 2003). A few peptide dendrimers were also found to function as enzyme catalyst when screened for their esterase activity. In particular, RMG3 {((AcTyrThr)$_8$(DapTrpGly)$_4$-(DapArgSerGly)$_2$DapHisSerNH$_2$) efficiently facilitate an ester hydrolysis reaction of 1-acyloxypyrene-3,6,8-trisulfonates with a catalytic efficiency (K$_{cat}$/K$_M$) of 860 M^{-1} min^{-1}. RMG3 as an enzyme mimic could adopt a globular conformation similar to a molten globular protein and comprises single histidine as the catalytic core and peripheral hydrophobic aromatic amino acids as dendritic branches (S-2,3-diaminopropanoic acid (Dap) as the branching point) (Javor et al., 2007). Similarly, PAMAM dendrimers with a pyridoxamine core and terminal chiral amino groups also served as enzyme mimic that converts the α-keto acids, phenylpyruvic acid and pyruvic acid to L-isomer of phenylalanine and alanine enantioselectively, thereby exhibiting transaminase activity. Molecular modeling studies revealed that the terminal N,N-dimethyl-L-phenylalanine catalytic group backfolded to the core for the enantioselective catalysis. Therefore, dendrimers can attain flexibility similar to protein folding in an aqueous environment (Wei et al., 2009). Furthermore, certain peptide dendrimers also behave as pseudochaperonins that promote refolding of thermal and urea-denatured lipases. Dendrimers provide multivalent surfaces that assist molecular interactions between the dendrimers and the unfolded proteins to refold via noncovalent interactions (Dubey et al., 2013). Notwithstanding, a majority of the research on dendrimer as an artificial protein currently focused on the applicability of phosphorylated PAMAMs as biomimetic remineralization and restoration of dentine enamel (Chen et al., 2015; Wang et al., 2015a; Zhang et al., 2015).

5.7.2.3 Dendrimers as Nanoscale Containers and Nanoscaffolds

Nanoscale containers are molecular containers with inner cavities having nanoscale size in one dimension. They have the capacity to hold nanogram quantities of materials, and one of the potential uses is drug delivery such as gene delivery; conjugate drug delivery; boron neutron capture therapy; molecular recognition; dendritic boxes; etc. (Aulenta et al., 2003; Zhao et al., 2013). According to Vutukuri et al., a nanoscale container constructed of an amphiphilic dendrimer is capable of forming either hydrophobic or hydrophilic cavity and depends on the solvent environment. Moreover, the unimolecular or aggregate nature of the containers is influenced by the size of the interior core in addition to overall dendrimer size (Vutukuri et al., 2004). The interior repeating units of G4 and higher-generation PAMAM dendrimers contribute the characteristic nanocontainer properties exemplified by the fact (Menjoge et al., 2010). A G5 PPI dendrimer-based nanoscale container in which the terminal amino groups conjugated with *t*-Boc (*tert*-butyloxycarbonyl) glycine (TPPI) or mannose (MPPI) showed a marked reduction of hemolytic activity compared to native PPI. Both TPPI and MPPI nanoscale containers were loaded with antiretroviral drug efavirenz to target the human phagocytic system *in vitro*. The cellular uptake of MPPI-loaded drug was 12 times higher than that of free drug and 5.5 times higher than TPPI (Dutta et al., 2007). The dendritic nanoscale container exhibit unimolecular encapsulations with host–guest interactions and apart from therapeutic drugs, they also encapsulate vitamins, metal salts, and diagnostic imaging moieties (Tomalia et al., 2007; Abbasi et al., 2014). In a recent study, the host–guest interaction demonstrated promising delivery of doxorubicin (DOX) to the targeted cancer cells with αvβ3 integrin overexpression. The study adopted a multifunctional G5 PAMAM dendrimer, wherein the terminal amines conjugated with cyclic arginine-glycine-aspartic acid peptide, fluorescein isothiocyanate, and acetylation of unconjugated amine. The interior cavity of the surface-modified dendrimers showed the ability to reduce the degree of freedom of the encapsulated DOX and suggest an increase in coordination with the drug. The cyclic peptide enhances the cellular uptake of the carrier at the target site and hydrophobic interior nanoscale container achieved sustained release of the drug (He et al., 2015). As observed, dendrimers provide surfaces for the attachment and cell-specific groups' presentation for target delivery. Further, solubility modifiers, camouflaging moieties for decreased immunological interactions, imaging tags, cell-specific ligands are also conjugated with the dendrimer nanoscaffold to accomplish specific goals of the vector. For instance, FA and fluorescein attached to dendrimer's surface binds specifically to folate receptor of cancer cells and has been employed for the *in vitro* and *in vivo* delivery of DOX, methotrexate, protein toxins, imaging agents, etc. (van Dongen et al., 2014; Wang et al., 2016). Potential dendrimer nanoscaffold comprising PAMAM, triazine, bis MPA {(2,2-bis(hydroxymethyl) propionic acid}, PPI, PEG core with glycerol and succinic acid, 5ALA (5-aminolevulinic acid), TEA (triethanolamine) core with PAMAM has been developed as theranostic agents for combating prostate cancer (Lo et al., 2013).

5.7.2.4   Microvascular Extravasation

Microvascular extravasation is the movement of molecules from systemic circulation into the adjacent interstitial tissues via the endothelial lining of capillary walls. A promising drug delivery system must extravasate efficiently to reach the target site of action. Sayed et al. reported that dendrimer permeation across completely differentiated Caco-2 cells is inversely proportional to the dendrimer generation as lower-generation dendrimers displayed higher permeation and vice versa (El-Sayed et al., 2001; El-Sayed et al., 2002). Kitchens et al. investigated the influence of molecular weight and size of $-NH_2$-terminated G0–G4 PAMAM dendrimers on extravasation across the microvascular endothelium. The extravasation time order for the PAMAM dendrimers under study was G0<G1<G2<G3<G4 with a range of 143.9 to 422.7 s. The study revealed that the higher the size, the greater is the exclusion of the dendrimers from the endothelial pores of radius 4 to 5 nm. Linear PEG of molecular weight 6000 Da eluted faster than all the cationic and spherical PAMAM dendrimers due to greater hydrodynamic volume and charge neutrality. Further, interactions (electrostatic) between the cationic PAMAMs and the glycocalyx (bearing negatively charged sulfated glycosaminoglycan) lined microvascular endothelium assist in faster extravasation than PEG (Kitchens et al., 2005). Alternatively, during targeted drug delivery to tumor cells, the rate constant of extravasation ($k_{extravasation}$) is difficult to manipulate as it influenced by the tumor size, the convective flow to the tumor, and the vascular permeability. Nevertheless, the elimination and extravasation constants predict the fraction of the dendrimer that reaches the tumor site over time (Lee et al., 2005). Therefore, nanoparticulate molecular cargo-loaded dendrimer must extravasate and distribute efficiently to reach the target cells for a viable therapeutic outcome, but specific literatures on dendrimer extravasation studies have been limited.

5.7.2.5 Dendrimers as Tissue Regenerators

With progress in the pathophysiological knowledge of the tissue damage and regeneration together with advancement in stem cell research, particularly mesenchymal stem cells (MSCs), tissue regeneration is no longer considered a farfetched dream. Tissue reconstructions rely on supportive cells that recruit and differentiate stem cells (native) to the site of injured tissue. The supportive cells typically have a defined extracellular matrix (ECM) that includes myofibroblasts, macrophages, and other immune cells that potentially stimulate tissue repair (DiMarino et al., 2013; Forbes and Rosenthal, 2014). One of the first and pivotal steps toward tissue repair and regeneration is the appropriate selection of scaffold material. It encapsulates cells to regenerate the native ECM followed by overall self-replacement. The versatile nature of dendrimers makes them ideal candidates for incorporation into tissue engineering scaffolds exemplified by the amine-terminated dendrimers that demonstrated as perfect crosslinking agents in collagen scaffolds for corneal tissue engineering (Joshi and Grinstaff, 2008; Noriega-Luna, et al., 2014). Further, dendrimer nanoparticles (NPs) loaded with drugs, growth factors, and genes are utilized to facilitate bone tissue regeneration at the sites of pathology and fracture (Walmsley et al., 2015). For example, ex vivo pre-incubation of dexamethasone (Dex)-loaded carboxymethylchitosan/ poly(amidoamine) dendrimer NPs combined with hydroxyapatite/starch-polycaprolactone scaffolds with rBMSC (rat bone-marrow-derived stromal cells) promote osteogenic differentiation and mineralization of the ECM. The system also enables ectopic bone formation on the back of rats. The delivery of Dex inside the stromal cells by the dendrimer exerts a direct influence on the cellular fate of the stromal cells and makes it potentially suitable for tissue engineering (Oliveira et al., 2009; Monteiro et al., 2015). The MSCs are increasingly becoming a valuable tool as potential targets for genetic manipulation in tissue regeneration and engineering. Taking into account the ability of MSCs toward self-renewal and differentiation into numerous types of cell, Bae et al. reported the transfection efficiency of PAMAM-conjugated arginine (R) (PAMAM H-R), lysine (K) (PAMAM H-K), and ornithine (O) (PAMAM H-O) linked through histidine (H) to human adipose-derived MSCs (AD-MSCs). The conjugated PAMAM derivatives displayed significantly improved transfection efficiency compared to native PAMAM. They also do not interfere with the AD-MSCs differentiation into osteogenic, adipogenic, and chondrogenic lineages. Additionally, PAMAM-H-K and

PAMAM-H-O also exhibited enhanced osteogenic differentiation with low cytotoxicity (Bae et al., 2016). Therefore, these conjugated PAMAMs might prove useful in bone regeneration or repair.

5.7.2.6   Research Tool for Biodistribution Studies

In the case of nanoparticle-based drug-delivery system, the biodistribution studies (BS) assist in overcoming undesirable tissue distribution or excessively low serum half-lives by modulating the particle size, coatings, or other features in early stage formulation and development (Lee et al., 2010). Radiolabelled dendrimers are particularly considered important research tools, and they are used widely for BS. Dendrimers conjugated to radioisotopes such as ^{3}H, ^{14}C, ^{88}Y, ^{111}In, and ^{125}I have been used widely for animal-based research. The information garnered can then be employed to adjust distribution to or away from specific organs for target delivery (Sampathkumar et al., 2007).

Sadekar et al. carried out comparative BS between a linear N(2-hydroxylpropyl)methacrylamide (HPMA) copolymers series and G5-G7 PAMAM dendrimers comprising surface hydroxyl groups (PAMAM-OH) in orthotopic tumor (ovarian) bearing mice. The ^{125}I-labeled polymers were injected in mice, and detection in all the critical organ systems was performed by measuring the γ emission of the ^{125}I. The studies revealed large extent hepatic accumulation of PAMAM dendrimers than the linear HPMA copolymers. Moreover, among all the polymers, the G7 PAMAM-OH exhibited prolonged plasma circulation, improved accumulation in tumors, and highest tumor/blood ratio. The G7 PAMAM-OH with a 4-nm hydrodynamic radius displayed augmented plasma exposure, approximately 25 times greater than the G5 PAMAM-OH and HPMA copolymer (26 kDa) (Sadekar et al., 2011). Therefore, the phenomenon indicates a strong relationship between the plasma exposure and hydrodynamic sizes of the polymer. Thus, the BS suggest that specific polymers could be selected based on their size (hydrodynamic) and architecture for drug targeting and delivery by virtue of preferential organ accumulation (Opina et al., 2015).

Ghai et al. reported a G5 PAMAM dendrimer conjugated with radioactive Gd-BnDOTA (^{153}Gadolinium-benzyl-1,4,7,10-tetraazacyclododecane-tetraacetic acid) (G5-Gd-BnDOTA) via thiourea linkage. Administration of G5-Gd-BnDOTA demonstrated a quick uptake in the deep lymphatic system of mice and enhanced MRI of the internal iliac node of the monkey after injection into the vaginal mucosa. However, a 90-day BS showed that the Gd conjugate undergoes slow elimination from the liver and other organs. Although the conjugate proved efficient lymphatic imaging, the slow clearance from the body and off-target localization might hinder its clinical utility (Ghai et al., 2015). In contrast, another DOTA-based dendrimer obtained by conjugating DOTA-NHS (*N*-hydroxysuccinimide ester) to G4 PAMAM dendrimer and radiolabeled with ^{68}Gallium {[^{68}Ga] DOTA–PAMAM}. The BS of the conjugate showed sufficient accumulation within the tumor tissues and only 2.0% of the injected dose accumulated in the kidneys. Distribution in the lungs, spleen, and liver was comparatively lower than that in the kidneys and demonstrated very negligible uptake in the heart and brain, thus showing limited cytotoxicity. Radiolabeling efficiency together with rapid blood clearance via urinary excretion proved that the [^{68}Ga] DOTA-PAMAM is a promising candidate in the molecular visualization of tumor angiogenesis (Lesniak et al., 2013). Apart from radiolabeled dendrimers, a thorough BS for understanding organ, tissue, cellular uptake, and retention by fluorescence-quantification-based assay also displayed similar efficiency with excellent sensitivity (Michael, 2001). Overall, rational dendrimer conjugation and modifying its structural features have the capabilities to improve the tissue distribution profiles and reduce their toxicity significantly.

5.7.3   DIAGNOSTIC AND MISCELLANEOUS APPLICATIONS

5.7.3.1   Dendrimers in X-Ray Computed Tomography (CT) Contrast Agents

X-ray CT, an imaging technique, has the capability to produce images of transaxial planes through the human body. The development of X-ray CT in the form of multislice helical X-ray CT affords improved contrast image of bodily tissues compare to conventional radiograph (Lusic and Grinstaff, 2013). However, difficulties related to imaging and interface identification between two adjacent tissues or soft tissues imaging necessitate the use of contrast imaging agents for better tissue visualization by X-ray CT (Liu et al., 2012). The exploitation of nanoparticulate (NP) delivery systems allows efficient delivery of the contrast agents and overcome the demerits associated with commonly used iodinated agents. They provide target-specific imaging and angiography, biocompatibility and low toxicity, favorable contrast efficacy, prolonged circulation time, and physiological stability (Kojima et al., 2010). Many research reports indicate that dendrimers, in particular, gold NP-loaded PEGylated dendrimers hold great promise as X-ray CT contrast agents (Peng et al., 2012; Lee et al., 2013; Cai et al., 2015). The PEGylated dendrimers extend the circulation time and tumor accumulation of contrast agents. Furthermore, PAMAM dendrimer-entrapped gold NPs along with iodinated contrast agent as a single entity exhibit a 2-fold increase in the contrasting properties compared to that of clinically available hydrophilic iodinated molecules like Omnipaque® (Li et al., 2014). Innovations in a dendrimer research by modification of dendrimer surface with different imaging agents (fluorescent dyes, traditional small molecular contrast agents, or metal ion/chelator complexes, etc.) intensify the number of unique and promising contrast agents for X-ray CT imaging (Qiao and Shi, 2015). More recently, You et al. revealed contrast agents obtained by covalent attachment of multiple copies of tetraiodophthaloyl (TIP) groups to the amine surface of partially PEGylated PAMAM dendrimers. The four iodinated dendrimer NPs with the diameter in the range of 13 to 22 nm demonstrated enhanced imaging capacities, excellent *in vivo* safety profile, water solubility, and extended blood half-lives (You et al., 2016).

5.7.3.2 Dendrimers as MRI Imaging Agents

MRI, a noninvasive technique for imaging, is presently routinely used for disease diagnosis. It provides detailed anatomical information of organs, soft tissues, bone including other bodily structures. Nevertheless, associated low sensitivity demands the use of contrast agents for high sensitivity imaging with better resolution. One of the widely used contrast agents is Gd(III) complexes, administered in high dosage due to low efficiency and rapid renal excretion. Herein, the applicability of polymer scaffold exists for tagging multiple agents to intensify the contrast and subsequent dose reduction. However, contrary to dendrimers, linear and other hyperbranched polymers failed due to toxicity and incomplete renal excretion. Dendrimers provide an alternative synthetic scaffold as biocompatible carriers of MRI contrast agents. For example, Gadomer® 17, a polylysine dendrimer combined with 24 Gd DOTA complexes, is under clinical trials. They are also equally efficient in reducing the dose of Gd(III) complexes due to high relaxivity on account of slower particle tumbling rates in solution (Langereis et al., 2007; Longmire et al., 2014). Biodegradable gadolinium-conjugated PEGylated G3 PAMAM polydisulfide dendrimer (thiolated) nanoclusters displayed bright image contrast as observed in the MRI of the abdominal aorta. Although the CA undergoes slow degradation in the serum via a thiol-disulfide exchange reaction, nonetheless it showed prolonged blood circulation time with a half-life of less than 1.6 h in mice. The PEGylation enhanced the water solubility and facilitated the renal clearance (Huang et al., 2012). In an additional study, Miyake et al. reported few novel chiral and racemic dendrimer-triamine coordinated Gd complexes (DTG) with polyaminoalcohol end groups along with chiral DTG complexes with polyol end groups MRI CA. Further conjugation of the CA with PEG resulted in low toxicity probably due to free amino groups masking. Among the DTG CA, the chiral DTG with polyaminoalcohol end groups exhibited five times higher longitudinal relaxivity than that of clinically used Gd-DTPA (diethylenetriamine-pentaacetic acid) (Miyake et al., 2015). Consequently, recent efforts directed toward the development of dendrimer-based CA have obtained promising results. Various studies revealed that they prolong the half-life in the blood, pacify the toxicity of conventional Gd complexes, augment the longitudinal relaxivity, and improve the sensitivity/specificity of MRI.

5.7.3.3   Dendrimers as Molecular Probes

Biosensors such as DNA microarrays (DM) finds application in cancer diagnostic and predictions, gene expression studies, detection of nucleotide mutations, genotyping of individuals, etc. (Ntzani and Ioannidis, 2003; Rosi and Mirkin, 2005; Govindarajan et al., 2012). Amino-modified oligonucleotides linked with the amine terminals of the immobilized dendrons using di(N-succinimidyl)carbonate as linker provides sufficient space for hybridization probe DNA with incoming DNAs. The generated DM facilitates the detection of single nucleotide variations in the p53 gene in genomic DNAs from cancer cell lines (Caminade et al., 2006). Covering of bio-barcode DNA probe with immobilized G4-PAMAM dendrimers on gold NPs loaded with the chemiluminescent reagent N-(4-aminobutyl)-N-ethylisoluminol enables precise and direct detection of extremely low concentration DNA (detection limit: 6 fM (femtomolar)L^{-1} of target DNA) (Liu et al., 2011). A avidin biotin-G2 PAMAM dendrimer Gd(III) supramolecular complex–based molecular probe showed high affinity for cell surface lectins found in ovarian cancer cells. Intraperitoneal administration of the probe in mice specifically target and delivered chelated Gd(III) and rhodamine green to the ovarian tumors using the SHIN3 ovarian cancer model. Dual modality MRI and fluorescence imaging may utilize this particular tumor-targeted molecular probe (Xu et al., 2007). Further, conjugation of multiple organic cyanine dyes to G5 and G6 PAMAM dendrimers afforded the synthesis of dye-conjugated fluorescent dendritic nanoprobes (FDNs). The single-molecule fluorescence measurements of FDNs demonstrated enhanced photostability and high-resolution biological imaging technique compared to single organic dye molecules probes (Kim et al., 2013). The DNA dendrimer nanocarriers prepared from Y-DNAs building blocks are used to deliver functional nucleic acids (FNAs) in situ intracellular for molecular probing applications such as monitoring of biological molecules (e.g., histidine and ATP) in living cells. DNA dendrimer scaffold embedded with FNAs such as DNAzyme and aptamer self-deliver FNAs into living cells coupled with target recognition capabilities (Meng et al., 2014). The DNA dendrimer scaffolds possess larger surface area to volume ratio that provides the advantage for strategic placement of sensing components and fluorescent probes within the dendrimers. For example, nanoscale DNA dendrimer scaffolds incorporated with butyrylcholinesterase and fluorescein detect acetylcholine and can be utilized to quantify the spatio-temporal fluctuations of neurotransmitter release (Walsh et al., 2015). The inherent properties of dendrimers that combine large surface area and versatile surface functionalities allow them to integrate various biosensing molecular probes for diagnostic and other biomedical applications.

5.7.3.4   Dendrimers in Modified Electrodes

Various types of molecular building blocks are employed for modification of electrode surface to achieve specific goals, namely selective absorption and detection of analytes. Particular sensors to detect biologically significant analytes such as DNA, dopamine, histamine, and glucose, etc., employed modified electrodes (MEs). Bonné et al. reported that nanocomposite membrane comprising boronic acid G1 PAMAM (ethylenediamine core, eight amino surface groups) dendrimer-modified nanofibrillar cellulose creates stable membranes for spectrophotometric or electrochemical detection. The modified electrode reversibly reduced 2-electron 2-proton of immobilized probe molecule alizarin red S that occur in approximately 60-nm zone close to the electrode surface. The presence of dendrimer renders the boronic acid-binding sites immobile for strong interaction with alizarin red S dye. The ME may find utility for *in vivo* sensing of D-glucose and similar physiological analytes (Bonne et al., 2010; Sun and James, 2015). Similarly, 6-(ferrocenyl) hexanethiol (FcSH)-modified electrode was fabricated by integrating gold nanoparticles (AuNPs) on the surface of PAMAM-modified gold electrode for glucose analysis. Additionally, immobilization of a self-assembled mono-layer of glucose oxidase onto the gold surface via FcSH and cysteamine was carried out. The electrochemical response of the ME toward glucose substrate revealed a detection limit of 0.6 mM (Karadag et al., 2013). In recent times, Li et al. reported a glass carbon (GC) electrode modified with hybrid multiwalled carbon nanotube (MWCNT)-PAMAM dendrimer that incorporates methylene blue (MB) as the redox indicator. Oligonucleotide capture probes were immobilized onto the ME (MB/probe/MWCNT-PAMAM/GC) for electrochemical detection of micro RNA24 extracted from HeLa cell. With a low detection limit of 0.5 fM, the ME showed improved signal to MB reduction when compared to bare GC electrode (Li et al., 2015). In general, the exploitation of PAMAM dendrimer in MEs for probe immobilization displayed electrochemical detection of analytes with high sensitivity. Apart from PAMAM ME, PPI dendrimer-based gold NP modified exfoliated graphite electrode was studied for electrochemical detection of *o*-nitrophenol (Ndlovu et al., 2014). Additionally, PAMAM dendrimer-based graphite MEs have also been reported for the detection of an anticancer drug (Erdem et al., 2011), and biocompatible nanosized PAMAM dendrimer films were used for the fabrication of protein-modified electrodes (Shen, 2004; Hu, et al., 2004).

5.7.3.5   Dendrimers in Cosmetics

Many cosmetic companies like L'Oréal, Dow Chemical Company, Unilever, and Revlon have several patents on dendrimer-based cosmetics for various applications in hair, skin, and nail care products (Singh and Sharma, 2016). A few patents currently applied and/or granted for possible utilization of dendrimers in cosmetic preparation are summarized in tabular form (Table 5.4).

TABLE 5.4 Dendrimers Patented for Cosmetic Uses

Publication number	Applicant	Publication Year	Brief Description	Ref.
US9260607 B2	Dow Corning Toray Co., Ltd.	2016	Carbosiloxane dendrimer as surface-treatment agent for powders comprising of skin care, hair care, antiperspirant, deodorant product or a UV-ray protective agent.	Iimura and Furukawa, 2016
WO2016160796 A1	Revlon Consumer Products Corporation	2016	A keratin restorative composition comprising of keratin-peptide/dendrimer for ameliorating nail delamination, brittleness, cracking, chipping, splitting, dryness, and texturing.	Valia et al., 2016
WO2015092632 A3	L'Oréal	2015	Polymer bearing a carbosiloxane dendrimer unit for skin preparations as water-in-oil emulsion.	Khachikian and Ricard, 2015
WO2014030771 A3	Dow Corning Toray Co., Ltd.	2014	Copolymer having carbosiloxane dendrimer that has good compatibility with hardly soluble ultraviolet absorbers.	Iimura and Hori, 2014
WO2009112682 A1	The National Center for Scientific Research, France	2009	Encapsulation of Vitamin C into water soluble dendrimers or the preparation of a cosmetic anti-aging or wrinkle.	Astruc et al., 2009
WO2005092275 A1	L'Oréal	2005	A dendritic polymer with peripheral fatty chains comprising of surfactant and cosmetic agent for washing and conditioning hairs.	Vic and Samain, 2005

TABLE 5.4 (Continued)

Publication number	Applicant	Publication Year	Brief Description	Ref.
US6582685 B1	Unilever Home & Personal Care Usa, Division Of Conopco, Inc.	2003	A hydroxyl-functionalized dendritic macromolecule constructed from polyester units for hair styling composition such as spray, gel or mousse.	Adams et al., 2003
US6395867 B1	L'Oréal	2002	Dendrimer with thiol function used as the antioxidant preservative in haircare, skin care and other makeup products.	Maignan, 2002

5.8 SURFACE ENGINEERED DENDRIMERS

Dendrimers are gaining popularity owing to their number of associated advantages, unique properties, and proved versatility. Dendritic carriers have gained significant interest in diverse areas of gene and drug delivery. Despite several pharmaceutical applications, the associated toxicity issues, drug leakage, immunogenicity, and uptake by reticuloendothelial system restrict the clinical applications of these macromolecules. The toxicity is primarily associated with the cationic amine groups present on the surface (Satija et al., 2007). To overcome these shortcomings, some safe and modified dendrimers have been developed by surface engineering of the parent dendrimers with the rationale of decreasing toxicity, preventing the uptake by the reticuloendothelial system, increasing targeting efficiency and cellular uptake of therapeutic agents as well as enhancing gene transfection efficiency (Tekade et al., 2008; Yellepeddi et al., 2009; Dhakad et al., 2013). Surface engineering for drug targeting involves the modification of dendrimer surface with specific targeting ligands like FA, biotin, transferrin, cell penetrating peptides, antibodies etc. In an investigation, FA-conjugated poly(propyleneimine) (PPI) dendrimer was designed for the delivery of melphalan to target the cancer cells (Kesharwani et al., 2015c). In another study, PAMAM dendrimer-transferrin conjugate was developed to improve the gene delivery to the cancer cells. The formulation was reported to exhibit enhanced efficacy toward transfection of HepG2, HeLa, and CT26 cell lines (Urbiola et al., 2015)]. Pu et al. in 2013 synthesized a polyhedral oligomeric silsesquioxane (POSS)-centered poly(L-glutamic acid) dendrimer engineered with biotin and doxorubicin on the surface. Surface functionalization with biotin significantly enhanced the *in vitro* inhibition activity of drug-dendrimer conjugates (Pu et al., 2013). In another investigation, Sialyl Lewis X antibodies-conjugated PAMAM dendrimer were synthesized to specifically target the colon cancer HT29 cells. Antibody-conjugated dendrimer drug complex shows enhanced and concentration-dependent capture of HT29 cells (Xie et al., 2015). Narsireddy et al. synthesized a functionalized dendrimer conjugate for targeted photodynamic therapy (PDT). PAMAM dendrimer (G4) was conjugated with a photosensitizer and a nitrilotriacetic acid (NTA) group. The conjugation of NTA group was done to specifically bind a peptide specific to human epidermal growth factor 2. The peptide-modified dendrimer conjugate was reported to be more effective than free photosensitizer in PDT-mediated cell death assays in HER2-positive cells, SK-OV-3 (Narsireddy et al., 2015).

PEGylation of dendrimers is one of the strategies of surface engineering to subdue the limitations related to dendrimers, especially toxicity and stability. PEGylation is the linking or conjugation of polyethylene glycol with dendrimer system. PEGylated dendrimers have been reported to show less toxicity than cationic dendrimers in various hemolytic and cell line studies (Gajbhiye et al., 2007). In research, doxorubicin-loaded PEGylated polyester dendrimers were synthesized. *In vitro* studies suggest the enhanced stability of the PEGylated dendrimer conjugate with long circulation half-life in mice. It was also reported that the formulation delivers comparable doxorubicin as that of clinically used PEGylated liposomal formulation DOXIL (Guillaudeu et al., 2008). In another study, doxorubicin-conjugated G5 polylysine PEGylated dendrimers were designed. The conjugate shows lower systemic toxicity and antitumor efficacy equivalent to the liposomal or solution-based doxorubicin formulations (Kaminskas et al., 2012). A number of studies indicated that PEGylation is the efficient and simplest technique for the modification of dendrimer surface to enhance the stability with better biocompatibility and increase the encapsulation of drug as well as targeting ability (Thakur et al., 2013, 2015).

In addition, various other surface engineering strategies can also be employed to improve efficiency of dendrimers, such as functionalization with oligoamine (Uchida et al., 2011), chitosan (Tekade et al., 2014, 2015c;

Maheshwari et al., 2015b), imidazolium (Luo et al., 2011), guanidium (Tziveleka et al., 2007), phosphonium (Ornelas-Megiatto et al., 2012), dexamethasone (Jeon et al., 2015), triamcinolone acetonide (Ma et al., 2009), FA (Ohyama et al., 2016), nucleobase (Wang et al., 2015b), saponins (glycyrrhizin) (Chopdey et al., 2015), luteinizing hormone-releasing hormone (LHRH) peptide (Ghanghoria et al., 2016a, 2016b), neomycin (Yadav et al., 2016), and paromomycin (Ghilardi et al., 2013).

5.9 BIOCOMPATIBILITY AND TOXICITY OF DENDRIMERS

Extensive applications of dendritic scaffolds have been reported in the pharmaceutical field, a glimpse of which is shown in this chapter. However, the associated toxicity of dendrimers restricts their use, which is mainly attributed to the surface cationic charge of the dendrimers exemplified by the presence of $-NH_2$ groups at the periphery that often interact with biological membranes, which are negatively charged (Jain et al., 2010). Biocompatibility is also one of the important attributes related to the compatibility of the biological systems with dendritic carriers, as the interaction of amino groups of dendrimers with the cell membrane considerably influence the permeability and stability of cell membranes, which subsequently leads to structural disruptions (Tekade et al., 2009b; Sebestik et al., 2012; Youngren et al., 2013b). Cationic dendrimers have been reported to show concentration-dependent cytotoxicity and hemolytic activity (Fischer et al., 2003; Chen et al., 2004). In comparison with low-generation dendrimers, the higher-generation G4-G8 PAMAM dendrimers have been reported to show toxicity on account of high cationic charge density (Shah et al., 2011). In an investigation, low-generation dendrimers and Arg–Gly–Asp conjugated dendrimers were screened for toxicity in a zebrafish embryo model. Owing to the presence of amino groups, G4 dendrimers were found to be toxic to zebrafish embryos. On the other hand, carboxylic acid terminal G3.5 dendrimers were reported to be nontoxic in the same model (Heiden et al., 2007). Likewise, several other studies suggest the associated toxicities of dendrimers, which depend upon the surface terminal groups as well as dendrimer generation (Chauhan et al., 2010; Albertazzi et al., 2013). Further, the consideration of toxic nature of cationic dendrimers is extremely important for successful use in gene delivery and transfection (Dufès et al., 2005) where the surface groups with the positive charge are essential for ionic complexation with DNA. These groups also work as a proton buffer that subsequently causes endosome disruption (Tang et al., 1996; Sonawane et al., 2003).

In comparison with other delivery systems (drug and gene), dendritic carriers have demonstrated lower toxicities, efficient drug delivery, and targeting as well as successful gene transfection, but further research is required to optimize the toxicity of these novel carriers to be utilized in clinical conditions (McNerny et al., 2010). It has been thoroughly investigated that cationic dendrimers can be modified by the surface functionalization to avoid toxicity issues and to prevent their accumulation in the liver (Malik et al., 2000; Jevprasesphant et al., 2003). Dendrimers conjugated with amino acids like phenylalanine and glycine have been reported to show a considerable reduction in toxicities (Agashe et al., 2006). Ihre et al. in 2002 designed and synthesized polyester dendrimers (Ihre et al., 2002), which were found to be nontoxic to cells and exhibit specific accumulation in the tumor cells of the mice-bearing subcutaneous B16F10 tumor (Gillies et al., 2005). In another investigation, polyether imine dendrimers were synthesized and evaluated for *in vitro* cytotoxicity in different cell lines. Findings suggest that the synthesized dendrimers do not show significant cytotoxicity in the cell lines under study (Krishna et al., 2005). Polyether copolyester dendrimers were synthesized (Dhanikula and Hildgen, 2007a) and evaluated for cytotoxicity on RAW 264.7 cell lines, which revealed the biocompatibility of the dendrimers as no cell death was observed up to the concentration of 250 g/ml (Dhanikula and Hildgen, 2007b). Similarly, several other modified dendrimers like melamine dendrimers, peptide dendrimers, triazine dendrimers, etc., were synthesized and evaluated for toxicity. These dendrimers were reported to be comparatively less toxic (Chen et al., 2004; Agrawal et al., 2007; Chouai and Simanek, 2008). Surface engineered dendrimers including PEGylated dendrimers have been discussed in this chapter under previous headings.

5.10 CONCLUSION AND FUTURE PERSPECTIVES

Due to the incredible properties of dendrimers, they have found a wide range of applications in drug delivery, therapeutics, and diagnostics aids. Controlled synthesis and polymerization techniques impart unique architecture and precise control over the dendrimer structure, which provides an exceptional domain for the attachment of drugs or genes. Surface functionality, void spaces, and structural uniformity of the dendrimers allow targeting moieties and therapeutic agents with diverse structures to be encapsulated in the internal voids, conjugated on the surface, or attached through non-covalent interactions, hydrogen bonding, electrostatic interactions, or chemical bonding. This chapter focuses on the enormous utilitarian perspective that a dendritic carrier presents in consideration of vast expansion in the field of nanotechnology. Requisite properties for nanocarriers such as nontoxicity, biocompatibility, biodegradation, and nonimmunogenicity can be accomplished with dendritic carriers also by adopting controlled synthesis conditions and by utilizing the selective moieties and spacers during the synthesis depending upon the characteristics desired. This chapter also attempts to throw the light on the various procedures discovered to synthesize different types of dendrimers with variable nomenclature and properties.

With the discovery of these magnificent macromolecules, scientists have explored numerous of their applications with an especial attention focused on the delivery of therapeutic agents including drugs as well as genes. In this chapter, an emphasis has been given on the therapeutic, biomedical and diagnostic applications of dendrimers. The ability of dendritic scaffolds to cross highly specific biological barriers to deliver the therapeutic agents into the targeted cells or intracellular compartments has gained ample of interest of pharmaceutical scientists to explore the novel properties of this versatile carrier. Surface engineering or surface functionality modification of dendrimers presents plethora of opportunities for the researchers to fabricate these macromolecules with specificity for particular targeting ligands. The literature covered in this chapter focuses on the various aspects of surface engineering of dendrimers that reduce the toxicity associated with native dendrimers and to enhance the target specificity and accumulation in the cells. The high drug entrapment efficiency and flexibility of dendritic carriers for surface functionalization by anchoring with the number of ligands has exponentially increased the number of investigations and researches on these excellent carriers. Although several *in vitro* studies were performed to evaluate the dendrimers for their efficacy and safety. Further, *in vivo* studies along with more persuasive and comprehensive data are needed to establish these novel constructs for clinical usage.

KEYWORDS

- applications of dendrimers
- biocompatibility
- dendrimer generations
- evaluation of dendrimers
- hyperbranched structures
- synthesis of dendrimers

REFERENCES

Abbasi, E., Aval, S. F., Akbarzadeh, A., Milani. M., Nasrabadi, H. T., Joo, S. W., Hanifehpour, Y., Koshki, K. N., & Asl, R. P., (2014). Dendrimers: synthesis, applications, and properties. *Nanoscale Res. Lett.*, *9*, 247.

Abd-El-Aziz, A. S., Agatemor, C., Etkin, N., Overy, D. P., Lanteigne, M., McQuillan, K., & Kerr, R. G., (2015). Antimicrobial organometallic dendrimers with tunable activity against multidrug-resistant bacteria. *Biomacromolecules.*, *16*, 3694–3703.

Abid, C. K. Z., Jackeray, R., Jain, S., Chattopadhyay, S., Asif, S., & Singh, H., (2016). Antimicrobial efficacy of synthesized quaternary ammonium polyamidoamine dendrimers and dendritic polymer network. *J. Nanosci. Nanotechnol.*, *16*, 998–1007.

Adams, G., Ruth Ashton, M., & Khoshdel, E., (2003). *Hydroxyl-Functionalized Dendritic Macromolecules in Topical Cosmetic and Personal Care Compositions.* U. S. Patent, 6, *582*, 685 B1, June 24.

Agashe, H. B., Dutta, T. D., Garg, M., & Jain, N. K., (2006). Investigations on the toxicological profile of functionalized fifth-generation poly (propylene imine) dendrimer. *J. Pharm. Pharmacol.*, *58*, 1491–1498.

Agrawal, P., Gupta, U., & Jain, N. K., (2007). Glycoconjugated peptide dendrimers-based nanoparticulate system for the delivery of chloroquine phosphate. *Biomaterials.*, *28*, 3349–3359.

Agrawal, U., Mehra, N. K., Gupta, U., & Jain, N. K., (2013). Hyperbranched dendritic nanocarriers for topical delivery of dithranol. *J. Drug Target.*, *21*, 497–506.

Albertazzi, L., Gherardini, L., Brondi, M., Sulis Sato, S., Bifone, A., Pizzorusso, T., Ratto, G. M., & Bardi, G., (2013). *In vivo* distribution and toxicity of PAMAM dendrimers in the central nervous system depend on their surface chemistry. *Mol. Pharm.*, *10*, 249–260.

Ali, J., Fazil, M., Qumbar, M., Khan, N., & Ali, A., (2016). Colloidal drug delivery system: amplify the ocular delivery. *Drug Deliv.*, *23*, 710–726.

Alper, J., (1991). Rising chemical "stars" could play many roles. *Science*, 251, 1562–1664.

Apostolopoulos, V., (2016). Vaccine delivery methods into the future. *Vaccines (Basel)*, *4*, doi: 10. 3390/vaccines4020009.

Appelhans, D., Oertel, U., Mazzeo, R., Komber, H., Hoffmann, J., Weidner, S., Brutschy, B., Voit, B., & Ottaviani, M. F., (2010). Dense-shell glycodendrimers: UV/Vis and electron paramagnetic resonance study of metal ion complexation. *Proc. R. Soc. A.*, *466*, 1489–1513.

Astruc, D., Ruiz, A. J., & Boisselier, E., (2009). Encaps*ulation of Vitamin C into Water Soluble Dendrimers*. W. O. Patent, *112*, 682 A1, September 17.

Aulenta, F., Hayes, W., & Rannard, S., (2003). Dendrimers: a new class of nanoscopic containers and delivery devices. *Eur. Polym. J.*, *39*, 1741–1771.

Bae, Y., Lee, S., Green, E. S., Park, J. H., Ko, K. S., Han, J., & Choi, J. S., (2016). Characterization of basic amino acids-conjugated PAMAM dendrimers as gene carriers for human adipose-derived mesenchymal stem cells. *Int. J. Pharm.*, *501*, 75–86.

Bai, S., Thomas, C., & Ahsan, F., (2007). Dendrimers as a carrier for pulmonary delivery of enoxaparin, a low-molecular weight heparin. *J. Pharm. Sci.*, *96*, 2090–2106.

Balaji, B. S., & Lewis, M. R., (2009). Double exponential growth of aliphatic polyamide dendrimers via AB2 hypermonomer strategy. *Chem. Commun.*, 4593–4595.

Baytekin, B., Werner, N., Luppertz, F., Engeser, M., Brüggemann, J., Bitter, S., Henkel, R., Felder, T., & Schalley, C. A., (2006). How useful is mass spectrometry for the characterization of dendrimers? "Fake defects" in the ESI and MALDI mass spectra of dendritic compounds. *Int. J. Mass Spectrom.*, *249–250*, 138–148.

Bhattacharya, P., Geitner, N. K., Sarupria, S., & Ke, P. C., (2013). Exploiting the physico-chemical properties of dendritic polymers for environmental and biological applications. *Phys. Chem. Chem. Phys.*, *15*, 4477–4490.

Biricova, V., & Laznickova, A., (2009). Dendrimers: Analytical characterization and applications. *Bioorg. Chem.*, *37*, 185–192.

Boas, U., & Heegaard, P. M. H., (2004). Dendrimers in drug research. *Chem. Soc. Rev.*, *33*, 43–63.

Boas, U., Christensen, J. B., & Heegaard, P. M. H., (2006). Dendrimers: design, synthesis and chemical properties. *J. Mater. Chem.*, *16*, 3785–3798.

Bonne, M. J., Galbraith, E., James, T. D., Wasbrough, M. J., Edler, K. J., Jenkins, A. T. A., Helton, M., McKee, A., Thielemans, W., Psillakis, E., & Marken, F., (2010). Boronic acid dendrimer receptor modified nanofibrillar cellulose membranes. *J. Mater. Chem.*, *20*, 588–594.

Bravo-Osuna, I., Vicario-de-la-Torre, M., Andrés-Guerrero, V., Sánchez-Nieves, J., Guzmán-Navarro, M., de la Mata, F. J., Gómez, R., de Las Heras, B., Argüeso, P., Ponchel, G., Herrero-Vanrell, R., & Molina-Martínez, I. T., (2016). Novel water-soluble mucoadhesive carbosilane dendrimers for ocular administration. *Mol. Pharm.*, *13*, 2966–2976.

Buhleier, E., Wehner, W., F., & Vogtle, F., (1978). "Cascade"- and "nonskid-chain-like" syntheses of molecular cavity topologies. *Synthesis*, 155–158.

Cai, H., Li, K., Li, J., Wen, S., Chen, Q., Shen, M., Zheng, L., Zhang, G., & Shi, X., (2015). Dendrimer-assisted formation of Fe3O4/Au nanocomposite particles for targeted dual mode CT/MR imaging of tumors. *Small*, *11*, 4584–4593.

Caminade, A. M., (2011). Methods of characterization of dendrimers. In *Dendrimers: Towards Catalytic, Material and Biomedical Uses*, Caminade, A. M., Turrin, C. O., Laurent, R., Ouali, A., Nicot, B. D. Eds., John Wiley & Sons, Ltd, Chichester, UK, pp. 35–66.

Caminade, A. M., Laurent, R., & Majoral, J. P., (2005). Characterization of dendrimers. *Adv. Drug Deliv. Rev.*, *57*, 2130–2146.

Caminade, A. M., Padié, C., Laurent, R., Maraval, A., & Majoral, J. P., (2006). Uses of dendrimers for DNA microarrays. *Sensors.*, *6*, 901–914.

Cao, W., & Zhu, L., (2011). Synthesis and unimolecular micelles of amphiphilic dendrimer-like Star polymer with various functional surface groups. *Macromolecules.*, *44*, 1500–1512.

Carlmark, A., Hawker, C., Hult, A., & Malkoch, M., (2009). New methodologies in the construction of dendritic materials. *Chem. Soc. Rev.*, *38*, 352–362.

Chahal, J. S., Khan, O. F., Cooper, C. L., McPartlan, J. S., Tsosie, J. K., Tilley, L. D., Sidik, S. M., Lourido, S., Langer, R., Bavari, S., Ploegh, H. L., & Anderson, D. G., (2016). Dendrimer-RNA nanoparticles generate protective immunity against lethal Ebola, H1N1 influenza, and *Toxoplasma gondii* challenges with a single dose. *Proc. Natl. Acad. Sci. USA.*, *113*, E4133–E4142.

Chauhan, A. S., Jain, N. K., & Diwan, P. V., (2010). Pre-clinical and behavioural toxicity profile of PAMAM dendrimers in mice. *Proc. R. Soc. A.*, *466*, 1535–1550.

Chauhan, A. S., Sridevi, S., Chalasani, K. B., Jain, A. K., Jain, S. K., Jain, N. K., & Diwan, P. V., (2003). Dendrimer-mediated transdermal delivery: enhanced bioavailability of indomethacin. *J. Control. Release*, *90*, 335–343.

Chen, H. T., Neerman, M. F., Parrish, A. R., & Simanek, E. E., (2004). Cytotoxicity, hemolysis, and acute *in vivo* toxicity of dendrimers based on melamine, candidate vehicles for drug delivery. *J. Am. Chem. Soc.*, *126*, 10044–10048.

Chen, H., Holl, M. B., Orr, B. G., Majoros, I., & Clarkson, B. H., (2003). Interaction of dendrimers (artificial proteins) with biological hydroxyapatite crystals. *J. Dent. Res.*, *82*, 443–448.

Chen, L., Yuan, H., Tang, B., Liang, K., & Li, J., (2015). Biomimetic remineralization of human enamel in the presence of polyamidoamine dendrimers *in vitro*. *Caries Res.*, *49*, 282–290.

Chen, Y. M., Chen, C. F., & Xi, F., (1998). Chiral dendrimers with axial chirality. *Chirality.*, *10*, 661–666.

Choi, Y., Thomas, T., Kotlyar, A., Islam, M. T., & Baker, J. R., (2005). Synthesis and functional evaluation of DNA-assembled polyamidoamine dendrimer clusters for cancer cell-specific targeting. *Chem. Biol.*, *12*, 35–43.

Chopdey, P. K., Tekade, R. K., Mehra, N. K., Mody, N., & Jain, N. K., (2015). Glycyrrhizin conjugated dendrimer and multi-walled carbon nanotubes for liver specific delivery of doxorubicin. *J. Nanosci. Nanotechnol.*, *15*, 1088–1100.

Chouai, A., & Simanek, E. E., (2008). Kilogram-scale synthesis of a second-generation dendrimer based on 1, 3, 5-triazine using green and industrially compatible methods with a single chromatographic step. *J. Org. Chem.*, *73*, 2357–2366.

Chougule, M. B., Tekade, R. K., Hoffmann, P. R., Bhatia, D., Sutariya, V. B., & Pathak, Y., (2014). Nanomaterial-based gene and drug delivery: pulmonary toxicity considerations. In: *Biointeractions of Nanomaterials*, Sutariya, V. B., Pathak, Y., (eds.), CRC Press, Taylor & Francis Group, Florida, pp. 225–248.

Ciolkowski, M., Petersen, J. F., Ficker, M., Janaszewska, A., Christensen, J. B., Klajnert, B., & Bryszewska, M., (2012). Surface modification of PAMAM dendrimer improves its biocompatibility. *Nanomedicine*, *8*, 815–817.

Daftarian, P. M., Stone, G. W., Kovalski L. Kumar, M., Vosoughi, A., Urbieta, M., Blackwelder, P., Dikici, E., Serafini, P., Duffort, S., Boodoo, R., Rodríguez-Cortés, A., Lemmon, V., Deo, S., Alberola, J., Perez, V. L., Daunert, S., & Ager, A. L., (2013). A targeted and adjuvanted nanocarrier lowers the effective dose of liposomal amphotericin B and enhances adaptive immunity in murine cutaneous Leishmaniasis. *J. Infect. Dis.*, *208*, 1914–1922.

Daftarian, P., Kaifer, A. E., Li, W., Blomberg, B. B., Frasca, D., Roth, F., Chowdhury, R., Berg, E. A., Fishman, J. B., Al Sayegh, H. A., Blackwelder, P., Inverardi, L., Perez, V. L., Lemmon, V., & Serafini, P., (2011). Peptide-conjugated PAMAM dendrimer as a universal DNA vaccine platform to target antigen-presenting cells. *Cancer Res., 71*, 7452–7462.

Dening, T. J., Rao, S., Thomas, N., & Prestidge, C. A., (2016). Oral nanomedicine approaches for the treatment of psychiatric illnesses. *J. Control. Release, 223*, 137–156.

Dhakad, R. S., Tekade, R. K., & Jain, N. K., (2013). Cancer targeting potential of folate targeted nanocarrier under comparative influence of tretinoin and dexamethasone. *Curr. Drug Deliv., 10*, 477–491.

Dhanikula, R. S., & Hildgen, P., (2007a). Conformation and distribution of groups on the surface of amphiphilic polyether-co-polyester dendrimers: effect of molecular architecture. *J. Colloid. Interface Sci., 311*, 52–58.

Dhanikula, R. S., & Hildgen, P., (2007b). Influence of molecular architecture of polyetherco-polyester dendrimers on the encapsulation and release of methotrexate. *Biomaterials, 28*, 3140–3152.

DiMarino, A. M., Caplan, A. I., & Bonfield, T. L., (2013). Mesenchymal stem cells in tissue repair. *Front Immunol., 4*, 201.

Dinçer, S., Türk, M., & Piskin, E., (2005). Intelligent polymers as nonviral vectors. *Gene Ther., 12*, S139–S145.

Donnio, B., Buathong, S. W., Bury, I., & Guillon, D., (2007). Liquid crystalline dendrimers. *Chem. Soc. Rev., 36*, 1495–1513.

Dubey, P., Gautam, S., Kumar, P. P. P., Sadanandan, S., Haridas, V., & Gupta, M. N., (2013). Dendrons and dendrimers as pseudochaperonins for refolding of proteins. *RSC Adv., 3*, 8016–8020.

Dufès, C., Uchegbu, I. F., & Schätzlein, A. G., (2005). Dendrimers in gene delivery. *Adv. Drug Deliv. Rev., 57*, 2177–2202.

Dutta, T., Agashe, H. B., Garg, M., Balakrishnan, P., Kabra, M., & Jain, N. K., (2007). Poly (propyleneimine) dendrimer based nanocontainers for targeting of efavirenz to human monocytes/macrophages *in vitro. J. Drug Target, 15*, 89–98.

Dvornic, P. R., (2006). PAMAMOS: the first commercial silicon-containing dendrimers and their applications. *J. Polym. Sci. A. Polym. Chem, 44*, 2755–2773.

El-Sayed, M., Ginski, M., Rhodes, C., & Ghandehari, H., (2002). Transepithelial transport of poly (amidoamine) dendrimers across Caco-2 cell monolayers. *J. Control. Release, 81*, 355–365.

El-Sayed, M., Kiani, M. F., Naimark M. D., Hikal, A. H., & Ghandehari, H., (2001). Extravasation of poly(amidoamine) (PAMAM) dendrimers across microvascular network endothelium, *Pharm. Res., 18*, 23–28.

Erdem, A., Karadeniz, H., & Caliskan, A., (2011). Dendrimer modified graphite sensors for detection of anticancer drug Daunorubicin by voltammetry and electrochemical impedance spectroscopy. *Analyst., 136*, 1041–1045.

Fernandez, F. A., Manchanda, R., & McGoron, A. J., (2011). Theranostic applications of nanomaterials in cancer: drug delivery, image-guided therapy, and multifunctional platforms. *Appl. Biochem. Biotechnol., 165*, 1628–1651.

Fischer, D., Li, Y., Ahlemeyer, B., Krieglstein, J., & Kissel, T., (2003). *In vitro* cytotoxicity testing of polycations: influence of polymer structure on cell viability and hemolysis. *Biomaterials, 24*, 1121–1131.

Fischer, M., & Vögtle, F., (1999). Dendrimers: from design to application-a progress report. *Angew. Chem. Int. Ed., 38*, 884–905.

Flory, P. J., (1941). Molecular size distribution in three dimensional polymers. III. tetrafunctional branching units. *J. Am. Chem. Soc.*, *63*, 3096–3100.

Forbes, S. J., & Rosenthal, N., (2014). Preparing the ground for tissue regeneration: from mechanism to therapy. *Nat. Med.*, *20*, 857–869.

Fréchet, J. M. J., (1994). Functional polymers and dendrimers: reactivity, molecular architecture, and interfacial energy. *Science*, *263*, 1710–1715.

Frechet, J. M. J., (2002). Dendrimers and supramolecular chemistry. *Proc. Natl. Acad. Sci. USA.*, *99*, 4782–4787.

Friedhofen, J. H., & Vogtle, F., (2006). Detailed nomenclature for dendritic molecules. *New J. Chem.*, *30*, 32–43.

Fujita, Y., & Taguchi, H., (2011). Current status of multiple antigen-presenting peptide vaccine systems: application of organic and inorganic nanoparticles. *Chem. Cent. J.*, *5*, 48.

Furera, V. L., Vandyukovb, A. E., Padiec, C., Majoralc, J. P., Caminadec, A. M., & Kovalenkob, V. I., (2015). Raman spectroscopy studies of phosphorus dendrimers with phenoxy and deuterophenoxy terminal groups. *Vib. Spectrosc.*, *80*, 17–23.

Gajbhiye, V., Kumar, P. V., Tekade, R. K., & Jain, N. K., (2007). Pharmaceutical and biomedical potential of PEGylated dendrimers. *Curr. Pharm. Des.*, *13*, 415–429.

Gajbhiye, V., Palanirajan, V. K., Tekade, R. K., & Jain, N. K., (2009b). Dendrimers as therapeutic agents: a systematic review. *J. Pharm. Pharmacol.*, *61*, 989–1003.

Gajbhiye, V., Vijayaraj Kumar, P., Tekade, R. K., & Jain, N. K., (2009a). PEGylated PPI dendritic architectures for sustained delivery of H2 receptor antagonist. *Eur. J. Med. Chem.*, *44*, 1155–1166.

Gandhi, N. S., Tekade, R. K., & Chougule, M. B., (2014). Nanocarrier mediated delivery of siRNA/miRNA in combination with chemotherapeutic agents for cancer therapy: current progress and advances. *J. Control. Release*, *194*, 238–256.

Garbuzenko, O. B., Mainelis, G., Taratula, O., & Minko, T., (2014). Inhalation treatment of lung cancer: the influence of composition, size and shape of nanocarriers on their lung accumulation and retention. *Cancer Biol. Med.*, *11*, 44–55.

Ghai, A., Singh, B., Hazari, P. P., Schultz, M. K., Parmar, A., Kumar, P., Sharma, S., Dhawan, D., & Mishra, A. K., (2015). Radiolabeling optimization and characterization of 68Ga labeled DOTA-polyamido-amine dendrimer conjugate-animal biodistribution and PET imaging results. *Appl. Radiat. Isot.*, *105*, 40–46.

Ghanghoria, R., Kesharwani, P., Tekade, R. K., & Jain, N. K., (2016a). Targeting luteinizing hormone-releasing hormone: A potential therapeutics to treat gynecological and other cancers. *J. Control. Release*, doi: 10. 1016/j. jconrel. *11*. 002.

Ghanghoria, R., Tekade, R. K., Mishra, A. K., Chuttani, K., & Jain, N. K., (2016b). Luteinizing hormone-releasing hormone peptide tethered nanoparticulate system for enhanced antitumoral efficacy of paclitaxel. *Nanomedicine*, *11*, 797–816.

Ghilardi, A., Pezzoli, D., Bellucci, M. C., Malloggi, C., Negri, A., Sganappa, A., Tedeschi, G., Candiani, G., & Volonterio, A., (2013). Synthesis of multifunctional PAMAM-aminoglycoside conjugates with enhanced transfection efficiency. *Bioconjug. Chem.*, *24*, 1928–1936.

Gill, P., Moghadam, T. T., & Ranjbar, B., (2010). Differential scanning calorimetry Techniques: applications in biology and nanoscience. *J. Biomol. Tech.*, *21*, 167–193.

Gillies, E. R., Dy, E., Frechet, J. M. J., & Szoka, F. C., (2005). Biological evaluation of polyester dendrimer: poly(ethylene oxide) "Bow-Tie" hybrids with tunable molecular weight and architecture. *Mol. Pharm.*, *2*, 129–138.

Govindarajan, R., Duraiyan, J., Kaliyappan, K., & Palanisamy, M., (2012). Microarray and its applications. *J. Pharm. Bioallied. Sci.*, *4*, 310–312.

Grayson, S. M., & Fréchet, J. M. J., (2001). Convergent dendrons and dendrimers: From synthesis to applications. *Chem. Rev.*, *101*, 3819–3868.

Guerra, S., Nguyen, T. L. A., Furrer, J., Nierengarten, J. F., Barbera, J., & Deschenaux, R., (2016). Liquid-crystalline dendrimers designed by click chemistry. *Macromolecules*, *49*, 3222–3231.

Guillaudeu, S. J., Fox, M. E., Haidar, Y. M., Dy, E. E., Szoka, F. C., & Frechet, J. M., (2008). PEGylated dendrimers with core functionality for biological applications. *Bioconjug. Chem.*, *19*, 461–469.

Hawker, C., & Frechet, J. M. J., (1969). A new convergent approach to monodisperse dendritic macromolecules. *J. Chem. Soc. Chem. Commun.*, 1010–1013.

He, X., Alves, C. S., Oliveira, N., Rodrigues, J., Zhu, J., Bányai, I., Tomás, H., & Shi, X., (2015). RGD peptide-modified multifunctional dendrimer platform for drug encapsulation and targeted inhibition of cancer cells. *Colloids Surf. B. Biointerfaces*, *125*, 82–89.

Heegaard, P. M. H., Boas, U., & Sorensen, N. S., (2010). Dendrimers for vaccine and immunostimulatory uses. A review. *Bioconjugate Chem.*, *21*, 405–418.

Heiden, T. C., Dengler, E., Kao, W. J., Heideman, W., & Peterson, R. E., (2007). Developmental toxicity of low generation PAMAM dendrimers in zebrafish. *Toxicol. Appl. Pharmacol.*, *225*, 70–79.

Heise, R. L., Blakeney, B. A., & Pouliot, R. A., (2015). Polymers in tissue engineering, In: *Advanced Polymers in Medicine*, Puoci, F., (ed.), Springer International Publishing, Switzerland, pp. 195–196.

Hirao, A., & Yoo, H. S., (2011). Dendrimer-like star-branched polymers: Novel structurally well-defined hypcrbranched polymers. *Polymer J.*, *43*, 2–17.

Hsu, H. J., Bugno, J., Lee, S. R., & Hong, S., (2016). Dendrimer-based nanocarriers: A versatile platform for drug delivery. *Wiley Interdiscip. Rev. Nanomed. Nanobiotechnol.*, doi: 10. 1002/wnan. 1409.

Hu, J., Hu, K., & Cheng, Y., (2016). Tailoring the dendrimer core for efficient gene delivery. *Acta. Biomater.*, *35*, 1–11.

Huang, B., Dong, W. J., Yang, G. Y., Wang, W., Ji, C. H., & Zhou, F. N., (2015). Dendrimer-coupled sonophoresis-mediated transdermal drug-delivery system for diclofenac. *Drug Des. Devel. Ther.*, *9*, 3867–3876.

Huang, C. H., Nwe, K., Zaki, A. A., Brechbiel, M. W., & Tsourkas, A., (2012). Biodegradable polydisulfide dendrimer nanoclusters as MRI contrast agents. *ACS Nano.*, *6*, 9416–9424.

Huang, Q. R., Dubin, P. L., Lal, J., Moorefield, C. N., & Newkome, G. R., (2005). Small-angle neutron scattering studies of charged carboxyl-terminated dendrimers in solutions. *Langmuir.*, *21*, 2737–2742.

Ihre, H. R., Padilla De Jesus, O. L., Szoka, F. C., & Frechet, J. M. J., (2002). Polyester dendritic systems for drug delivery applications: design, synthesis, and characterization. *Bioconj. Chem.*, *13*, 443–452.

Iimura, T., & Furukawa, H., (2016). Surface-Treatment agent forpowder for use in cosmetic and cosmetic containing powder treated with the same. *U. S. Patent 9, 260*, 607 B2.

Iimura, T., & Hori, S., (2014). Copolymer having carbosiloxane dendrimer structure and composition and cosmetic containing the same. W. O. Patent, 030, 771 A3.

Jain, K., Kesharwani, P., Gupta, U., & Jain, N. K., (2010). Dendrimer toxicity: let's meet the challenge. *Int. J. Pharm.*, *394*, 122–142.

Jain, N. K., & Tekade, R. K., (2013). Dendrimers for enhanced drug solubilization. In: *Drug Delivery Strategies for Poorly Water-Soluble Drugs*, Douroumis, D., Fahr, A., (eds.), John Wiley & Sons Ltd, Oxford, UK, pp. 373–409.

Jatczak-Pawlik, I., Gorzkiewicz, M., Studzian, M., Appelhans, D., Voit, B., Pulaski, L., & Maculewicz, B., K., (2016). Sugar-modified poly(propylene imine) dendrimers stimulate the NF-κB pathway in a myeloid cell line. *Pharm. Res.*, doi: 10. 1007/s11095–016–2049–3.

Javor, S., Delort, E., Darbre, T., & Reymond, J. L., (2007). A peptide dendrimer enzyme model with a single catalytic site at the core. *J. Am. Chem. Soc.*, *129*, 13238–13246.

Jeon, P., Choi, M., Oh, J., & Lee, M., (2015). Dexamethasone-conjugated polyamidoamine dendrimer for delivery of the Heme Oxygenase-1 gene into the ischemic brain. *Macromol. Biosci.*, *15*, 1021–1028.

Jeong, Y., Kim, S. T., Jiang, Y., Duncan, B., Kim, C. S., Saha, K., Yeh, Y. C., Yan, B., Tang, R., Hou, S., Kim, C., Park, M. H., & Rotello, V. M., (2016). Nanoparticle-dendrimer hybrid nanocapsules for therapeutic delivery. *Nanomedicine*, *11*, 1571–1578.

Jevprasesphant, R., Penny, J., Jalal, R., Attwood, D., McKeown, NB., & D'Emanuele, A., (2003). The influence of surface modification on the cytotoxicity of PAMAM dendrimers. *Int. J. Pharm.*, *252*, 263–266.

Jieru, Q., Lingdan, K., Xueyan, C., Aijun, L., Hongru, T., & Xiangyang, S., (2016). Dendrimer-entrapped gold nanoparticles modified with β-cyclodextrin for enhanced gene delivery applications. *RSC Adv.*, *6*, 25633–25640.

Joshi, N., & Grinstaff, M., (2008). Applications of dendrimers in tissue engineering. *Curr. Top. Med. Chem.*, *8*, 1225–1236.

Joshi, V. G., Dighe, V. D., Thakuria, D., Malik, Y. S., & Kumar, S., (2013). Multiple antigenic peptide (MAP): a synthetic peptide dendrimer for diagnostic, antiviral and vaccine strategies for emerging and re-emerging viral diseases. *Indian J. Virol.*, *24*, 312–320.

Kalomiraki, M., Thermos, K., & Chaniotakis, N. A., (2016). Dendrimers as tunable vectors of drug delivery systems and biomedical and ocular applications. *Int. J. Nanomedicine.*, *11*, 1–12.

Kaminskas, L. M., McLeod, V. M., Kelly, B. D., Cullinane, C., Sberna, G., Williamson, M., Boyd, B. J., Owen, D. J., & Porter, C. J., (2012). Doxorubicin-conjugated PEGylated dendrimers show similar tumoricidal activity but lower systemic toxicity when compared to PEGylated liposome and solution formulations in mouse and rat tumor models. *Mol. Pharm.*, *9*, 422–432.

Karadag, M., Geyik, C., Demirkol, D. O., Ertas, F. N., & Timur, S., (2013). Modified gold surfaces by 6-(ferrocenyl)hexanethiol/dendrimer/gold nanoparticles as a platform for the mediated biosensing applications. *Mater. Sci. Eng. C Mater. Biol. Appl.*, *33*, 634–640.

Kaur, A., Jain, K., Mehra, N. K., & Jain, N. K., (2017). Development and characterization of surface engineered PPI dendrimers for targeted drug delivery. *Artif. Cells Nanomed. Biotechnol.*, *45*(3), 414–425.

Kavyani, S., Amjad-Iranagh, S., Dadvar, M., & Modarress, H., (2016). Hybrid dendrimers of PPI(core)-PAMAM(shell): a molecular dynamics simulation study. *J. Phys. Chem. B.*, *120*, 9564–9575.

Kawaguchi, T., Walker, K. L., Wilkins, C. L., & Moore, J. S., (1995). Double exponential dendrimer growth. *J. Am. Chem. Soc.*, *117*, 2159–2165.

Kayat, J., Gajbhiye, V., Tekade, R. K., & Jain, N. K., (2011). Pulmonary toxicity of carbon nanotubes: a systematic report. *Nanomedicine*, *7*, 40–49.

Kesharwani, P., Banerjee, S., Gupta, U., Amin, M. C. I. M., Padhye, S., Sarkar, F. H., & Iyer, A. K., (2015b). PAMAM dendrimers as promising nanocarriers for RNAi therapeutics. *Materials Today*, *18*, 565–572.

Kesharwani, P., Tekade, R. K., & Jain, N. K., (2015a). Dendrimer generational nomenclature: the need to harmonize. *Drug Discov. Today*, *20*, 497–499.

Kesharwani, P., Tekade, R. K., & Jain, N. K., (2015c). Generation dependent safety and efficacy of folic acid conjugated dendrimer based anticancer drug formulations. *Pharm. Res.*, *32*, 1438–1450.

Khachikian, H., & Ricard, A., (2015). *Cosmetic Composition Comprising a Polymer Bearing a Carbosiloxane Dendrimer Unit and Expanded Polymer Particles.* W. O. Patent 2015, *092*, 632 A3, September 24.

Khan, O. F., Zaia, E. W., Jhunjhunwala, S., Xue, W., Cai, W., Yun, D. S., Barnes, C. M., Dahlman, J. E., Dong, Y., Pelet, J. M., Webber, M. J., Tsosie, J. K., Jacks, T. E., Langer, R., & Anderson, D. G., (2015). Dendrimer-inspired nanomaterials for the *in vivo* delivery of siRNA to lung vasculature. *Nano Lett.*, *15*, 3008–3016.

Kim, Y., Kim, S. H., Tanyeri, M., Katzenellenbogen, J. A., & Schroeder, C. M., (2013). Dendrimer probes for enhanced photostability and localization in fluorescence imaging. *Biophys. J.*, *104*, 1566–1575.

Kitchens, K. M., El-Sayed, M. E. H., & Ghandehari, H., (2005). Transepithelial and endothelial transport of poly (amidoamine) dendrimer. *Adv. Drug Deliv. Rev.*, *57*, 2163–2176.

Klajnert, B., & Bryszewska, M., (2002). Fluorescence studies on PAMAM dendrimers interactions with bovine serum albumin. *Bioelectrochemistry.*, *55*, 33–35.

Kojima, C., Kono, K., Maruyama, K., & Takagishi, T., (2000). Synthesis of polyamidoamine dendrimers having poly(ethylene glycol) grafts and their ability to encapsulate anticancer drugs. *Bioconjug. Chem.*, *11*, 910–917.

Kojima, C., Umeda, Y., Ogawa, M., Harada, A., Magata, Y., & Kono, K., (2010). X-ray computed tomography contrast agents prepared by seeded growth of gold nanoparticles in PEGylated dendrimer. *Nanotechnology*, *21*, 245104.

Kolb, H. C., Finn, M. G., & Sharpless, K. B., (2001). Click chemistry: diverse chemical function from a few good reactions. *Angew. Chem. Int. Ed. Engl.*, *40*, 2004–2021.

Krishna, T. R., Jain, S., Tatu, U. S., & Jayaraman, N., (2005). Synthesis and biological evaluation of 3-amino-propan-1-ol based poly(ether imine) dendrimers. *Tetrahedron.*, *61*, 4281–4288.

Kulczynska, A., Frost, T., & Margerum, L. D., (2006). Effect of PAMAM dendrimer size and pH on the electrostatic binding of metal complexes using cyclic voltammetry. *Macromolecules*, *39*, 7372–7377.

Kurmi, B. D., Kayat, J., Gajbhiye, V., Tekade, R. K., & Jain, N. K., (2010). Micro- and nanocarrier-mediated lung targeting. *Expert Opin. Drug Deliv.*, *7*, 781–794.

Langereis, S., Dirksen, A., Hackeng, T. M., Van Genderena, M. H. P., & Meijer, E. W., (2007). Dendrimers and magnetic resonance imaging. *New J. Chem.*, *31*, 1152–1160.

Lazniewska, J., Milowska, K., & Gabryelak, T., (2012). Dendrimers-revolutionary drugs for infectious diseases. *Wiley Interdiscip. Rev. Nanomed. Nanobiotechnol.*, *4*, 469–491.

Lee, C. C., MacKay, J. A., Fréchet, J. M. J., & Szoka, F. C., (2005). Designing dendrimers for biological applications. *Nat. Biotechnol.*, *23*, 1517–1526.

Lee, M. J., Veiseh, O., Bhattarai, N., Sun, C., Hansen, S. J., Ditzler, S., Knoblaugh, S., Lee, D., Ellenbogen, R., Zhang, M., & Olson, J. M., (2010). Rapid pharmacokinetic and biodistribution studies using cholorotoxin-conjugated iron oxide nanoparticles: a novel non-radioactive method. *PLoS One.*, *5*, e9536.

Lee, N., Choi, S. H., & Hyeon, T., (2013). Nano-sized CT contrast agents. *Adv. Mater.*, *25*, 2641–2660.

Lesniak, W. G., Mishra, M. K., Jyoti, A., Balakrishnan, B., Zhang, F., Nance, E., Romero, R., Kannan, S., & Kannan, R. M., (2013). Biodistribution of fluorescently labeled PAMAM dendrimers in neonatal rabbits: effect of neuroinflammation. *Mol. Pharm.*, *10*, 4560–4571.

Lewis, K., (2013). Platforms for antibiotic discovery. *Nat. Rev. Drug Discov.*, *12*, 371–387.

Li, F., Peng, J., Zheng, Q., Guo, X., Tang, H., & Yao, S., (2015). Carbon nanotube-polyamidoamine dendrimer hybrid-modified electrodes for highly sensitive electrochemical detection of microRNA24. *Anal. Chem.*, *87*, 4806–4813.

Li, J., Piehler, L. T., Qin, D., Baker, J. R., & Tomalia, D. A., (2000). Visualization and characterization of poly(amidoamine) dendrimers by atomic force microscopy. *Langmuir.*, *16*, 5613–5616.

Li, X., Anton, N., Zuber, G., & Vandamme, T., (2014). Contrast agents for preclinical targeted X-ray imaging. *Adv. Drug Deliv. Rev.*, *76*, 116–133.

Liang, X. J., Chen, C., Zhao, Y., & Wang, P. C., (2010). Circumventing tumor resistance to chemotherapy by nanotechnology. *Methods Mol. Biol.*, *596*, 467–488.

Liko, F., Hindré, F., & Fernandez-Megia, E., (2016). Dendrimers as innovative radiopharmaceuticals in cancer radionanotherapy. *Biomacromolecules*, *17*, 3103–3114.

Lim, J., Kostiainen, M., Maly, J., Da Costa, V. C. P., Annunziata, O., Pavan, G. M., & Simanek, E. E., (2013). Synthesis of large dendrimers with the dimensions of small viruses. *J. Am. Chem. Soc.*, *135*, 4660–4663.

Lin, Y., Fujimori, T., Kawaguchi, N., Tsujimoto, Y., Nishimi, M., Dong, Z., Katsumi, H., Sakane, T., & Yamamoto, A., (2011). Polyamidoamine dendrimers as novel potential absorption enhancers for improving the small intestinal absorption of poorly absorbable drugs in rats. *J. Control. Release*, *149*, 21–28.

Liu, C., Jiang, K., Tai, L., Liu, Y., Wei, G., Lu, W., & Pan, W., (2016). Facile non-invasive retinal gene delivery enabled by penetratin. *ACS Appl. Mater. Interfaces*, *8*, 19256–19267.

Liu, P., Hun, X., & Qing, H., (2011). Dendrimer-based biosensor for chemiluminescent detection of DNA hybridization. *Microchim. Acta.*, *175*, 201–207.

Liu, Y., Ai, K., & Lu, L., (2012). Nanoparticulate X-ray computed tomography contrast agents: from design validation to *in vivo* applications. *Acc. Chem. Res.*, *45*, 1817–1827.

Lizama, O. V., Vilos, C., & Lara, E. D., (2016). Techniques of structural characterization of dendrimers. *Curr. Org. Chem.*, *20*, 2591–2605.

Lo, S. T., Kumar, A., Hsieh, J. T., & Sun, X., (2013). Dendrimer nanoscaffolds for potential theranostics of prostate cancer with a focus on radiochemistry. *Mol. Pharm.*, *10*, 793–812.

Longmire, M. R., Ogawa, M., Choyke, P. L., & Kobayashi, H., (2014). Dendrimers as high relaxivity MR contrast agents. *Nanomed. Nanobiotechnol.*, *6*, 155–162.

Lu, Y. Y., Shi, T. F., An, L. J., & Wang, Z. G., (2012). Intrinsic viscosity of polymers: from linear chains to dendrimers. *EPL.*, *97*, 64003.

Lu, Y., Slomberg, D. L., Shah, A., & Schoenfisch, M. H., (2013). Nitric oxide-releasing amphiphilic poly(amidoamine) (PAMAM) dendrimers as antibacterial agents. *Biomacromolecules*, *14*, 3589–3598.

Luo, K., Li, C., Wang, G., Nie, Y., He, B., Wu, Y., & Gu, Z., (2011). Peptide dendrimers as efficient and biocompatible gene delivery vectors: synthesis and *in vitro* characterization. *J. Control. Release*, *155*, 77–87.

Luong, D., Kesharwani, P., Deshmukh, R., Amin, M. C. I. M., Gupta, U., Greish, K., & Iyer, A. K., (2016b). PEGylated PAMAM dendrimers: Enhancing efficacy and mitigating toxicity for effective anticancer drug and gene delivery. *Acta Biomater*, *43*, 14–29.

Luong, D., Kesharwani, P., Killinger, B. A., Moszczynska, A., Sarkar, F. H., Padhye, S., Rishi, A. K., & Iyer, A. K., (2016a). Solubility enhancement and targeted delivery of a potent anticancer flavonoid analogue to cancer cells using ligand decorated dendrimer nano-architectures. *J. Colloid Interface Sci.*, *484*, 33–43.

Lusic, H., & Grinstaff, M. W., (2013). X-Ray computed tomography contrast agents. *Chem. Rev.*, *113*, 1641–1666.

Ma, K., Hu, M. X., Qi, Y., Zou, J. H., Qiu, L. Y., Jin, Y., Ying, X. Y., & Sun, H. Y., (2009). PAMAM-triamcinolone acetonide conjugate as a nucleus-targeting gene carrier for enhanced transfer activity. *Biomaterials*, *30*, 6109–6118.

Ma, Q., Han, Y., Chen, C., Cao, Y., Wang, S., Shen, W., Zhang, H., Li, Y., Van Dongen, M. A., He, B., Yu, M., Xu, L., Holl, M. M. B., Liu, G., Zhang, Q., & Qi, R., (2015). Oral absorption enhancement of probucol by PEGylated G5 PAMAM dendrimer modified nanoliposomes. *Mol. Pharm.*, *12*, 665–674.

Madaan, K., Kumar, S., Poonia, N., Lather, V., & Pandita, D., (2014). Dendrimers in drug delivery and targeting: Drug-dendrimer interactions and toxicity issues. *J. Pharm. Bioallied. Sci.*, *6*, 139–150.

Maheshwari, R., Tekade, M., Sharma, P. A., & Tekade, R. K., (2015a). Nanocarriers assisted siRNA gene therapy for the management of cardiovascular disorders. *Curr. Pharm. Des.*, *21*, 4427–4440.

Maheshwari, R., Thakur, S., Singhal, S., Patel, R. P., Tekade, M., & Tekade, R. K., (2015b). Chitosan encrusted nonionic surfactant based vesicular formulation for topical administration of ofloxacin. *Sci. Adv. Mater.*, *7*, 1163–1176.

Maignan, J., (2002). *Polymers with Thiol Terminal Function*. U. S. Patent, 6, 395, 867 B1, May 28.

Maiti, P. K., Cagın, T., Wang, G., & Goddard, W. A., (2004). Structure of PAMAM Dendrimers: Generations 1 through 11. *Macromolecules*, *37*, 6236–6254.

Malik, N., Wiwattanapatapee, R., Klopsch, R., Lorenz, K., Frey, H., Weener, J. W., Meijer, E. W., Paulus, W., & Duncan, R., (2000). Dendrimers: relationship between structure and biocompatibility *in vitro*, and preliminary studies on the biodistribution of 125I-labelled polyamidoamine dendrimers *in vivo*. *J. Control. Release*, *65*, 133–148.

Mansuri, S., Kesharwani, P., Tekade, R. K., & Jain, N. K., (2016). Lyophilized mucoadhesive-dendrimer enclosed matrix tablet for extended oral delivery of albendazole. *Eur. J. Pharm. Biopharm.*, *102*, 202–213.

Maraval, V., Pyzowski, J., Caminade, A. M., & Majoral, J. P., (2003). "Lego" chemistry for the straight forward synthesis of dendrimers. *J. Org. Chem.*, *68*, 6043–6046.

Mastorakos, P., Kambhampati, S. P., Mishra, M. K., Wu, T., Song, E., Hanes, J., & Kannan, R. M., (2015). Hydroxyl PAMAM dendrimer-based gene vectors for transgene delivery to human retinal pigment epithelial cells. *Nanoscale*, *7*, 3845–3856.

McLornan, D. P., List, A., & Mufti, G. J., (2014). Applying synthetic lethality for the selective targeting of cancer. *N. Engl. J. Med.*, *371*, 1725–1735.

McNerny, D. Q., Leroueil, P. R., & Baker, J. R., (2010). Understanding specific and nonspecific toxicities: a requirement for the development of dendrimer-based pharmaceuticals. *Wiley Interdiscip. Rev. Nanomed. Nanobiotechnol.*, *2*, 249–259.

McSweeney, R. L., Chamberlain, T. W., Baldoni, M., Lebedeva, M. A., Davies, E. S., Besley, E., & Khlobystov, A. N., (2016). Direct measurement of electron transfer in nanoscale host-guest systems: metallocenes in carbon nanotubes. *Chem. Eur. J.*, *22*, 13540–13549.

Mekelburger, H., Jaworek, W., & Vogtle, F., (1992). Dendrimers, arborols, and cascade molecules: breakthrough into generations of new materials. *Angew. Chem. Int. Ed.*, *31*, 1571–1576.

Meng, H. M., Zhang, X., Lv, Y., Zhao, Z., Wang, N. N., Fu, T., Fan, H., Liang, H., Qiu, L., Zhu, G., & Tan, W., (2014). DNA dendrimer: an efficient nanocarrier of functional nucleic acids for intracellular molecular sensing. *ACS Nano.*, *8*, 6171–6181.

Menjoge, A. R., Kannan, R. M., & Tomalia, D. A., (2010). Dendrimer-based drug and imaging conjugates: design considerations for nanomedical applications. *Drug Discov. Today*, *15*, 171–185.

Michael, G., (2001). X-ray computed tomography. *Phys. Educ.*, *36*, 442–451.

Michigan Nanotechnology Institute for Medicine and Biological Sciences. http://www. nano. med. umich. edu/Platforms/Tecto-Dendrimers. html (accessed Dec 21, 2016).

Milhem, O. M., Myles, C., McKeown, N. B., Attwood, D., & D'Emanuele, A., (2000). Polyamidoamine starburst dendrimers as solubility enhancers. *Int. J. Pharm.*, *197*, 239–241.

Mishra, V., & Jain, N. K., (2014). Acetazolamide encapsulated dendritic nano-architectures for effective glaucoma management in rabbits. *Int. J. Pharm.*, *461*, 380–390.

Miyake, Y., Kimura, Y., Orito, N., Imai, H., Matsuda, T., Toshimitsu, A., & Kondo, T., (2015). Synthesis and functional evaluation of chiral dendrimer-triamine-coordinated Gd complexes with polyaminoalcohol end groups as highly sensitive MRI contrast agents. *Tetrahedron.*, *71*, 4438–4444.

Mody, N., Tekade, R. K., Mehra, N. K., Chopdey, P., & Jain, N. K., (2014). Dendrimer, liposomes, car nanotubes and PLGA nanoparticles: one platform assessment of drug delivery potential. *AAPS Pharm. Sci. Tech.*, *15*, 388–399.

Mohamed, F., Hofmann, M., Pötzschner, B., Fatkullin, N., & Rössler, E. A., (2015). Dynamics of PPI dendrimers: a study by dielectric and ^{2}H NMR spectroscopy and by field-cycling ^{1}H NMR relaxometry. *Macromolecules*, *48*, 3294–3302.

Monteiro, N., Martins, A., Reis, R. L., & Neves, N. M., (2015). Nanoparticle-based bioactive agent release systems for bone and cartilage tissue engineering. *Regener. Ther.*, *1*, 109–118.

Mourey, T. H., Turner, S. R., Rubinstein, M., Frechet, J. M. J., Hawker, C. J., & Wooley, K. L., (1992). Unique behavior of dendritic macromolecules: intrinsic viscosity of polyether dendrimers. *Macromolecules*, *25*, 2401–2406.

Mousa, S. A., & Bharali, D. J., (2011). Nanotechnology-based detection and targeted therapy in cancer: nano-bio paradigms and applications. *Cancers*, *3*, 2888–2903.

Naldini, L., (2015). Gene therapy returns to centre stage. *Nature*, *526*, 351–360.

Nanjwade, B. K., Bechra, H. M., Derkar, G. K., Manvi, F. V., & Nanjwade, V. K., (2009). Dendrimers: emerging polymers for drug-delivery systems. *Eur. J. Pharm. Sci.*, *38*, 185–196.

Narsireddy, A., Vijayashree, K., Adimoolam, M. G., Manorama, S. V., & Rao, N. M., (2015). Photosensitizer and peptide-conjugated PAMAM dendrimer for targeted *in vivo* photodynamic therapy. *Int. J. Nanomedicine.*, *10*, 6865–6878.

Nasr, M., Najlah, M., D'Emanuele A., & Elhissi, A., (2014). PAMAM dendrimers as aerosol drug nanocarriers for pulmonary delivery via nebulization. *Int. J. Pharm.*, *461*, 242–250.

Ndlovu, T., Arotiba, O. A., & Mamba, B. B., (2014). Poly(propyleneimine) dendrimer-gold nanoparticle modified exfoliated graphite electrode for the electrochemical detection of o-nitrophenol. *Int. J. Electrochem. Sci.*, *9*, 8330–8339.

Newkome, G. R., Baker, G. R., Young, J. K., & Traynhama, J. G., (1993). Systematic nomenclature for cascade polymers. *J. Polym. Sci. A. Polym. Chem.*, *31*, 641–651.

Noriega-Luna, B., Godinez, L. A., Rodríguez, F. J., Rodríguez, A., Zaldivar-Lelo de Larrea, G., Sosa-Ferreyra, C. F., Mercado-Curiel, R. F., Manriquez, J., & Bustos, E., (2014). Applications of dendrimers in drug delivery agents, diagnosis, therapy and detection. *J. Nanomater.*, doi:10. 1155/2014/507273.

Ntzani, E. E., & Ioannidis, J. P., (2003). Predictive ability of DNA microarrays for cancer outcomes and correlates: an empirical assessment. *Lancet.*, *362*, 1439–1444.

Ohyama, A., Higashi, T., Motoyama, K., & Arima, H., (2016). *In vitro* and *in vivo* tumor-targeting siRNA delivery using folate-PEG-appended dendrimer (G4)/α-cyclodextrin conjugates. *Bioconjug. Chem.*, *27*, 521–532.

Oliveira, J. M., Sousa, R. A., Kotobuki, N., Tadokoro, M., Hirose, M., Mano, J. F., Reis, R. L., & Ohgushi, H., (2009). The osteogenic differentiation of rat bone marrow stromal cells cultured with dexamethasone-loaded carboxymethylchitosan/poly(amidoamine) dendrimer nanoparticles. *Biomaterial*, *30*, 804–813.

Opina, A. C., Wong, K. J., Griffiths, G. L., Turkbey, B. I., Bernardo, M., Nakajima, T., Kobayashi, H., Choyke, P. L., & Vasalatiy, O., (2015). Preparation and long-term biodistribution studies of a PAMAM dendrimer G5–Gd-BnDOTA conjugate for lymphatic imaging. *Nanomedicine*, *10*, 1423–1437.

Ornelas-Megiatto, C., Wich, P. R., & Fréchet, J. M., (2012). Polyphosphonium polymers for siRNA delivery: an efficient and nontoxic alternative to polyammonium carriers. *J. Am. Chem. Soc.*, *134*, 1902–1905.

Ornelas, C., Ruiz, J., Belin, C., & Astruc, D., (2009). Giant dendritic molecular electrochrome batteries with ferrocenyl and pentamethylferrocenyl termini. *J. Am. Chem. Soc.*, *131*, 590–601.

Pande, S., & Crooks, R. M., (2011). Analysis of poly(amidoamine) dendrimer structure by UV–Vis spectroscopy. *Langmuir.*, *27*, 9609–9613.

Panicker, R. K. G., & Krishnapillai, S., (2014). Synthesis of on resin poly(propylene imine) dendrimer and its use as organocatalyst. *Tetrahedron Lett.*, *55*, 2352–2354.

Parsian, M., Mutlub, P., Yalcin, S., Tezcaner, A., & Gunduz, U., (2016). Half generations magnetic PAMAM dendrimers as an effective system for targeted gemcitabine delivery. *Int. J. Pharm.*, *515*, 104–113.

Patel, H. K., Gajbhiye, V., Kesharwani, P., & Jain, N. K., (2016). Ligand anchored poly (propyleneimine) dendrimers for brain targeting: comparative *in vitro* and *in vivo* assessment. *J. Colloid. Interface. Sci.*, *482*, 142–150.

Peng, C., Zheng, L., Chen, Q., Shen, M., Guo, R., Wang, H., Cao, X., Zhang, G., & Shi, X., (2012). PEGylated dendrimer-entrapped gold nanoparticles for *in vivo* blood pool and tumor imaging by computed tomography. *Biomaterials*, *33*, 1107–1119.

Popescu, M. C., Filip, D., Vasile, C., Cruz, C., Rueff, J. M., Marcos, M., Serrano, J. L., & Singurel, G., (2006). Characterization by fourier transform infrared spectroscopy (FT-IR) and 2D IR correlation spectroscopy of PAMAM dendrimer. *J. Phys. Chem. B.*, *110*, 14198–14211.

Prajapati, R. N., Tekade, R. K., Gupta, U., Gajbhiye, V., & Jain, N. K., (2009). Dendrimer-mediated solubilization, formulation development and *in vitro-in vivo* assessment of piroxicam. *Mol. Pharm.*, *6*, 940–950.

Prausnitz, M. R., & Langer, R., (2008). Transdermal drug delivery. *Nat. Biotechnol.*, *26*, 1261–1268.

Pu, Y., Chang, S., Yuan, H., Wang, G., He, B., & Gu, Z., (2013). The anti-tumor efficiency of poly(L-glutamic acid) dendrimers with polyhedral oligomeric silsesquioxane cores. *Biomaterials*, *34*, 3658–3666.

Qi, R., Zhang, H., Xu, L., Shen, W., Chen, C., Wang, C., Cao, Y., Wang, Y., van Dongen, B., He, M. A., Wang, S., Liu, G., Holl, M. M. B., & Zhang, Q., (2015). G5 PAMAM dendrimer versus liposome: a comparison study on the *in vitro* transepithelial transport and *in vivo* oral absorption of simvastatin. *Nanomedicine.*, *11*, 1141–1151.

Qiao, Z., & Shi, X., (2015). Dendrimer-based molecular imaging contrast agents, *Prog. Polym. Sci.*, *44*, 1–27.

Räder, H. J., Nguyen, T. T. T., & Müllen, K., (2014). MALDI–TOF Mass spectrometry of polyphenylene dendrimers up to the megadalton range. Elucidating structural integrity of macromolecules at unrivaled high molecular weights. *Macromolecules*, *47*, 1240–1248.

Richardson, R. M., Whitehouse, I. J., Ponomarenko, S. A., Boiko, N. I., & Shibaev, V. P., (1999). XRay diffraction from liquid crystalline carbosilane dendrimers. *Mol. Cryst. Liq. Cryst.*, *330*, 167–174.

Ritzén, A., & Frejd, T., (1999). Synthesis of a chiral dendrimer based on polyfunctional amino acids. *Chem. Commun.*, 207–208.

Rosi, N. L., & Mirkin, C. A., (2005). Nanostructures in biodiagnostics. *Chem. Rev.*, *105*, 1547–1562.

Sadekar, S., Ray, A., Janàt-Amsbury, M., Peterson, C. M., & Ghandehari, H., (2011). Comparative biodistribution of PAMAM dendrimers and HPMA copolymers in ovarian tumour-bearing mice. *Biomacromolecules*, *12*, 88–96.

Sadekar, S., Thiagarajan, G., Bartlett, K., Hubbard, D., Ray, A., McGill, L. D., & Ghandehari, H., (2013). Poly(amido amine) dendrimers as absorption enhancers for oral delivery of camptothecin. *Int. J. Pharm.*, *456*, 175–185.

Sadler, K., & Tam, J. P., (2002). Peptide dendrimers: applications and synthesis. *J. Biotechnol.*, *90*, 29–195.

Samanta, K., Jana, P., Bäcker, S., Knauer, S., & Schmuck, C., (2016). Guanidiniocarbonyl pyrrole (GCP) conjugated PAMAM-G2, a highly efficient vector for gene delivery: the importance of DNA condensation. *Chem. Commun.*, *52*, 12446–12449.

Sampathkumar, S. G., & Yarema, K. J., (2007). Dendrimers in cancer treatment and diagnosis. In *Nanotechnologies for the Life Sciences*, Kumar, C. S. R., (ed.), Wiley-VCH Verlag GmbH & Co. KGaA, Weinheim, pp. 28.

Saovapakhiran, A., D'Emanuele A., Attwood, D., & Penny, J., (2009). Surface modification of PAMAM dendrimers modulates the mechanism of cellular internalization. *Bioconjug. Chem.*, *20*, 693–701.

Satija, J., Gupta, U., & Jain, N. K., (2007). Pharmaceutical and biomedical potential of surface engineered dendrimers. *Crit. Rev. Ther. Drug Carrier Syst.*, *24*, 257–306.

Schlüter, A. D., & Rabe, J. P., (2000). Dendronized polymers: synthesis, characterization, assembly at interfaces, and manipulation. *Angew. Chem. Int. Ed. Engl.*, *39*, 864–883.

Sebestik, J., Reinis, M., & Jezek, J., (2012). Biocompatibility and toxicity of dendrimers. In *Biomedical Applications of Peptide-, Glyco- and Glycopeptide Dendrimers, and Analogous Dendrimeric Structures*, Springer: Vienna, pp. 111–114.

Shah, N., Steptoe, R. J., & Parekh, H. S., (2011). Low-generation asymmetric dendrimers exhibit minimal toxicity and effectively complex DNA. *J. Pept. Sci.*, *17*, 470–478.

Sharma, A., Arya, D. K., Dua, M., Chhatwal, G. S., & Johri, A. K., (2012). Nano-technology for targeted drug delivery to combat antibiotic resistance. *Expert. Opin. Drug Deliv.*, *9*, 1325–1332.

Sharma, A., Mohanty, D. K., Desai, A., & Ali, R., (2003). A simple polyacrylamide gel electrophoresis procedure for separation of polyamidoamine dendrimers. *Electrophoresis.*, *24*, 2733–2739.

Sharma, P. A., Maheshwari, R., Tekade, M., & Tekade, R. K., (2015). Nanomaterial Based Approaches for the Diagnosis and Therapy of Cardiovascular Diseases. *Curr. Pharm. Des.*, *21*, 4465–4478.

Shen, L., & Hu, N., (2004). Heme protein films with polyamidoamine dendrimer: direct electrochemistry and electrocatalysis. *Biochim. Biophys. Acta.*, *1608*, 23–33.

Shi, X., Patri, A. K., Lesniak, W., Islam, M. T., Zhang, C., Baker, J. R., & Balogh, L. P., (2005). Analysis of poly(amidoamine)-succinamic acid dendrimers by slab-gel electrophoresis and capillary zone electrophoresis. *Electrophoresis.*, *26*, 2960–2967.

Shiao, T. C., & Roy, R., (2012). Glycodendrimers as functional antigens and antitumor vaccines. *New J. Chem.*, *36*, 324–339.

Singh, J., Jain, K., Mehra, N. K., & Jain, N. K., (2016). Dendrimers in anticancer drug delivery: mechanism of interaction of drug and dendrimers. *Artif. Cells Nanomed. Biotechnol.*, *44*, 1626–1634.

Singh, T. K., & Sharma, N., (2016). Nano Biomaterials in cosmetics: Current status and future prospects. In: *Nanobiomaterials in Galenic Formulations and Cosmetics: Applications of Nanobiomaterials*, Grumezescu, A. M., (ed.), William Andrew, Elsevier, Oxford, UK, p 159.

Skwarczynski, M., Zaman, M., Urbani, C. N., Lin, I. C., Jia, Z., Batzloff, M. R., Good, M. F., Monteiro, M. J., & Toth, I. T., (2010). Polyacrylate dendrimer nanoparticles: a self-adjuvanting vaccine delivery system. *Angew. Chem. Int. Ed. Engl.*, *49*, 5742–5745.

Sonawane, N. D., Szoka, F. C. Jr., & Verkman, A. S., (2003). Chloride accumulation and swelling in endosomes enhances DNA transfer by polyamine-DNA polyplexes. *J. Biol. Chem.*, *278*, 44826–44831.

Sonawane, S. J., Kalhapure, R. S., Rambharose, S., Mocktar, C., Vepuri, S. B., Soliman, M., & Govender, T., (2016). Ultra-small lipid-dendrimer hybrid nanoparticles as a promising strategy for antibiotic delivery: *in vitro* and *in silico* studies. *Int. J. Pharm.*, *504*, 1–10.

Sun, M., Fan, A., Wang, Z., & Zhao, Y., (2012). Dendrimer-mediated drug delivery to the skin. *Soft. Matter*, *8*, 4301–4305.

Sun, X., & James, T. D., (2015). Glucose sensing in supramolecular chemistry. *Chem. Rev.*, *115*, 8001–8037.

Svenson, S., & Tomalia, D. A., (2005). Dendrimers in biomedical applications-reflections on the field. *Adv. Drug Deliv. Rev.*, *57*, 2106–2129.

Szulc, A., Pulaski, L., Appelhans, D., Voit, B., & Maculewicz, B. K., (2016). Sugar-modified poly(propylene imine) dendrimers as drug delivery agents for cytarabine to overcome drug resistance. *Int. J. Pharm.*, *513*, 572–583.

Tam, J. P., & Spetzler, J. C., (2001). Synthesis and application of peptide dendrimers as protein mimetics. *Curr. Protoc. Protein Sci.*, *17*, 1851–1853.

Tang, M. X., Redemann, C. T., & Szoka, F. C. Jr., (1996). *In vitro* gene delivery by degraded polyamidoamine dendrimers. *Bioconjug. Chem.*, *7*, 703–714.

Tang, Y. H., Huang, A. Y., Chen, P. Y., Chen, H. T., & Kao, C. L., (2011). Metallodendrimers and dendrimer nanocomposites. *Curr. Pharm. Des.*, *17*, 2308–2330.

Taratula, O., Garbuzenko, O. B., Kirkpatrick, P., Pandya, I., Savla, R., Pozharov, V. P., He, H., & Minko, T., (2009). Surface-engineered targeted PPI dendrimer for efficient intracellular and intratumoral siRNA delivery. *J. Control. Release, 140*, 284–293.

Tekade, R. K., (2015). Editorial: contemporary siRNA therapeutics and the current state-of-art. *Curr. Pharm. Des., 21*, 4527–4528.

Tekade, R. K., & Chougule, M. B., (2013). Formulation development and evaluation of hybrid nanocarrier for cancer therapy: Taguchi orthogonal array based design. *BioMed. Res. Int., 2013*, 712678.

Tekade, R. K., & Tekade, M., (2016). Ocular bioadhesives and their applications in ophthalmic drug delivery. In *Nano-Biomaterials for Ophthalmic Drug Delivery*, Springer International Publishing, Switzerland, pp. 211–230.

Tekade, R. K., Dutta, T., Gajbhiye, V., & Jain, N. K., (2009b). Exploring dendrimer towards dual drug delivery: pH responsive simultaneous drug-release kinetics. *J. Microencapsul., 26*, 287–296.

Tekade, R. K., Dutta, T., Tyagi, A., Bharti, A. C., Das, B. C., & Jain, N. K., (2008). Surface-engineered dendrimers for dual drug delivery: a receptor up-regulation and enhanced cancer targeting strategy. *J. Drug Target, 16*, 758–772.

Tekade, R. K., Kumar, P. V., & Jain, N. K., (2009a). Dendrimers in oncology: an expanding horizon. *Chem. Rev., 109*, 49–87.

Tekade, R. K., Maheshwari, R. G., Sharma, P. A., Tekade, M., & Chauhan, A. S., (2015b). siRNA therapy, challenges and underlying perspectives of dendrimer as delivery vector. *Curr. Pharm. Des., 21*, 4614–4636.

Tekade, R. K., Tekade, M., Kesharwani, P., & D'Emanuele A., (2016). RNAi-combined nano-chemotherapeutics to tackle resistant tumors. *Drug Discov. Today, 21*, 1761–1774.

Tekade, R. K., Tekade, M., Kumar, M., & Chauhan, A. S., (2015a). Dendrimer-stabilized smart-nanoparticle (DSSN) platform for targeted delivery of hydrophobic antitumor therapeutics. *Pharm. Res., 32*, 910–928.

Tekade, R. K., Youngren-Ortiz, S. R., Yang, H., Haware, R., & Chougule, M. B., (2014). Designing hybrid onconase nanocarriers for mesothelioma therapy: a Taguchi orthogonal array and multivariate component driven analysis. *Mol. Pharm., 11*, 3671–3683.

Tekade, R. K., Youngren-Ortiz, S. R., Yang, H., Haware, R., & Chougule, M. B., (2015c). Albumin-chitosan hybrid onconase nanocarriers for mesothelioma therapy. *Cancer Res., 75*, 3680–3680.

Thakur, S., Kesharwani, P., Tekade, R. K., & Jain, N. K., (2015). Impact of pegylation on biopharmaceutical properties of dendrimers. *Polymer., 59*, 67–92.

Thakur, S., Tekade, R. K., Kesharwani, P., & Jain, N. K., (2013). The effect of polyethylene glycol spacer chain length on the tumor-targeting potential of folate-modified PPI dendrimers. *J. Nanopart. Res., 15*, 1625.

Tintaru, A., Ungaro, R., Liu, X., Chen, C., Giordano, L., Peng, L., & Charles, L., (2015). Structural characterization of new defective molecules in poly(amidoamine) dendrimers by combining mass spectrometry and nuclear magnetic resonance. *Anal. Chim. Acta., 853*, 451–459.

Tomalia, D. A., & Fréchet, J. M. J., (2002). Discovery of dendrimers and dendritic polymers: a brief historical perspective. *J. Polym. Sci. A. Polym. Chem., 40*, 2719–2728.

Tomalia, D. A., Baker, H., Dewald, J., Hall, M., Kallos, G., Martin, S., Roeck, J., Ryder, J., & Smith, P., (1985). A new class of polymers: starburst-dendritic macromolecules. *Polym. J., 17*, 117–132.

Tomalia, D. A., Christensen, J. B., & Boas, U., (2012). Homogeneity and molecular weight. In *Dendrimers, Dendrons, and Dendritic Polymers: Discovery, Applications, and the Future*, Cambridge University Press, UK, pp. 170–171.

Tomalia, D. A., Reyna, L. A., & Svenson, S., (2007). Dendrimers as multi-purpose nanodevices for oncology drug delivery and diagnostic imaging. *Biochem. Soc. Trans., 35,* 61–67.

Twibanire, J. K., & Grindley, T. B., (2014). Polyester dendrimers: smart carriers for drug delivery. *Polymers., 6,* 179–213.

Tziveleka, L. A., Psarra, A. M., Tsiourvas, D., & Paleos, C. M., (2007). Synthesis and characterization of guanidinylatedpoly(propylene imine) dendrimers as gene transfection agents. *J. Control. Release., 117,* 137–146.

Uchida, H., Miyata, K., Oba, M., Ishii, T., Suma, T., Itaka, K., Nishiyama, N., & Kataoka, K., (2011). Odd-even effect of repeating aminoethylene units in the side chain of N-substituted polyaspartamides on gene transfection profiles. *J. Am. Chem. Soc., 133,* 15524–15532.

Uppuluri, S., Swanson, D. R., Piehler, L. T., Li, J., Hagnauer, G. L., & Tomalia, D. A., (2000). Core–shell tecto (dendrimers): I. synthesis and characterization of saturated shell models. *Adv. Mater., 12,* 796–800.

Urbiola, K., Blanco-Fernandez, L., Navarro, G., Rodl, W., Wagner, E., Ogris, M., Tros de, & Ilarduya, C., (2015). Evaluation of improved PAMAM-G5 conjugates for gene delivery targeted to the transferrin receptor. *Eur. J. Pharm. Biopharm., 94,* 116–122.

Valia, D., Kaiser, M., Conger, C., Scheibert, S. E., & Keratin, T. W. O., (2016). Patent, 2016, *160,* 796 A1, October 6.

Van Dongen, M. A., Silpe, J. E., Dougherty, C. A., Kanduluru, A. K., Choi, S. K., Orr, B. G., Low, P. S., & Holl, M. M. B., (2014). Avidity mechanism of dendrimer–folic acid conjugates. *Mol. Pharm., 11,* 1696–1706.

Ventola, C. L., (2015). The antibiotic resistance Crisis: part 1: causes and threats. *P. T., 40,* 277–283.

Venuganti, V. V., Sahdev, P., Hildreth, M., Guan, X., & Perumal, O., (2011). Structure-skin permeability relationship of dendrimers. *Pharm. Res., 28,* 2246–2260.

Vic, G., & Samain, H., (2005). *Cosmetic Composition Comprising a Dendritic Polymer with Peripheral Fatty Chains, a Surfactant and a Cosmetic Agent, and Uses Thereof.* W. O. Patent, 092, 275 A1, October 6.

Villanueva, J. R., Navarro, M. G., & Villanueva, L. R., (2016). Dendrimers as a promising tool in ocular therapeutics: latest advances and perspectives. *Int. J. Pharm., 511,* 359–366.

Vutukuri, D. R., Basu, S., & Thayumanavan, S., (2004). Dendrimers with both polar and apolar nanocontainer characteristics. *J. Am. Chem. Soc., 126,* 15636–15637.

Waehler, R., Russell, S. J., & Curiel, D. T., (2007). Engineering targeted viral vectors for gene therapy. *Nat. Rev. Genet., 8,* 573–587.

Walmsley, G. G., McArdle, A., Tevlin, R., Momeni, A., Atashroo, D., Hu, M. S., Feroze, A. H., Wong, V. W., Lorenz, P. H., Longaker, M. T., & Wan, D. C., (2015). Nanotechnology in bone tissue engineering. *Nanomedicine., 11,* 1253–1263.

Walsh, R., Morales, J. M., Skipwith, C. G., Ruckh, T. T., & Clark, H. A., (2015). Enzyme-linked DNA dendrimer nanosensors for acetylcholine. *Sci. Rep., 5,* 14832.

Wang, B., Lu, Z. R., & Tan, M., (2016). Development of dendrimer-based nanomaterials for diagnostic and therapeutic applications. In *Nanomaterials in Pharmacology (Methods in Pharmacology and Toxicology)*, Lu, Z. R., Sakuma S., (eds.), Springer Science and Business Media, New York, pp. 47–63.

Wang, H., Wei, H., Huang, Q., Liu, H., Hu, J., Cheng, Y., & Xiao, J., (2015b). Nucleobase-modified dendrimers as nonviral vectors for efficient and low cytotoxic gene delivery. *Colloids Surf. B Biointerfaces.*, *136*, 1148–1155.

Wang, M., & Cheng, Y., (2016). Structure-activity relationships of fluorinated dendrimers in DNA and siRNA delivery. *Acta Biomater.*, *46*, 204–210.

Wang, T., Yang, S., Wang, L., & Feng, H., (2015a). Use of multifunctional phosphorylated PAMAM dendrimers for dentin biomimetic remineralization and dentinal tubule occlusion. *RSC Adv.*, *5*, 11136–11144.

Wang, X., Guerrand, L., Wu, B., Li, X., Boldon L., Chen, W. R., & Liu, L., (2012). Characterizations of polyamidoamine dendrimers with scattering techniques. *Polymers.*, *4*, 600–616.

Wei, G. Z., Kui, L., Chuan, S. W., Yao, W., & Bin, H., (2010). New-generation biomedical materials: Peptide dendrimers and their application in biomedicine. *Sci. China Chem.*, *53*, 458–478.

Wei, S., Wang, J., Venhuizen, S., Skouta, R., & Breslow, R., (2009). Dendrimers in solution can have their remote catalytic groups folded back into the core: enantioselective transaminations by dendritic enzyme mimics-II. *Bioorg. Med. Chem. Lett.*, *19*, 5543–5546.

Wei, T., Chen, C., Liu, J., Liu, C., Posocco, P., Liu, X., Cheng, Q., Huo, S., Liang, Z., Fermeglia, M., Pricl, S., Liang, X. J., Rocchi, P., & Peng, L., (2015). Anticancer drug nanomicelles formed by self-assembling amphiphilic dendrimer to combat cancer drug resistance. *Proc. Natl. Acad. Sci. USA*, *112*, 2978–2983.

Wong, P. T., Tang, S., Tang, K., Coulter, A., Mukherjee, J., Gam, K., Baker, J. R., & Choi, S. K., (2015). Lipopolysaccharide binding heteromultivalent dendrimer nanoplatform for Gram negative cell targeting. *J. Mater. Chem. B.*, *3*, 1149–1156.

Wrońska, N., Felczak, A., Zawadzka, K., Poszepczyńska, M., Różalska S., Bryszewska, M., Appelhans, D., & Lisowska, K., (2015). Poly(Propylene Imine) dendrimers and amoxicillin as dual-action antibacterial agents. *Molecules*, *20*, 19330–19342.

Xie, J., Wang, J., Chen, H., Shen, W., Sinko, P. J., Dong, H., Zhao, R., Lu, Y., Zhu, Y., & Jia, L., (2015). Multivalent conjugation of antibody to dendrimers for the enhanced capture and regulation on colon cancer cells. *Sci. Rep.*, *5*, 9445.

Xiong, Z., Wang, Y., Zhu, J., He, Y., Qu, J., Effenberg, C., Xia. J., Appelhans, D., & Shi, X., (2016). Gd-Chelated poly(propylene imine) dendrimers with densely organized maltose shells for enhanced MR imaging applications. *Biomater. Sci.*, *4*, 1622–1629.

Xu, H., Regino, C. A. S., Koyama, Y., Hama, Y., Gunn, A. J., Bernardo, M., Kobayashi, H., Choyke, P. L., & Brechbiel, M. W., (2007). Preparation and preliminary evaluation of a biotin-targeted, lectin-targeted dendrimer-based probe for dual-modality magnetic resonance and fluorescence imaging. *Bioconjugate Chem.*, *18*, 1474–1482.

Yadav, S., Deka, S. R., Jha, D., Gautam, H. K., & Sharma, A. K., (2016). Amphiphilic azobenzene-neomycin conjugate self-assembles into nanostructures and transports plasmid DNA efficiently into the mammalian cells. *Colloids Surf. B Biointerfaces.*, *148*, 481–486.

Yang, J., Hu, J., He, B., & Cheng, Y., (2015). Transdermal delivery of therapeutic agents using dendrimers (US20140018435A1): a patent evaluation. *Expert Opin. Ther. Pat.*, *25*, 1209–1214.

Yang, Y., Sunoqrot, S., Stowell, C., Ji, J., Lee, C. W., Kim, J. W., Khan, S. A., & Hong, S., (2012). Effect of size, surface charge, and hydrophobicity of poly(amidoamine) dendrimers on their skin penetration. *Biomacromolecules, 13*, 2154–2162.

Yavuz, B., Pehlivan, S, B., Bolu, S. B., Sanyal, R. N., Vural, I., & Ünlü, N., (2016). Dexamethasone-PAMAM dendrimer conjugates for retinal delivery: preparation, characterization and *in vivo* evaluation. *J. Pharm. Pharmacol., 68*, 1010–1020.

Yavuz, B., Pehlivan, S. B., & Ünlü, N., (2013). Dendrimeric systems and their applications in ocular drug delivery. *Scientific World J., 732340.*

Yellepeddi, V. K., & Ghandehari, H., (2016). Poly (amidoamine) dendrimers in oral delivery. *Tissue Barriers., 4*, e1173773.

Yellepeddi, V. K., Kumar, A., & Palakurthi, S., (2009). Surface modified poly(amido)amine dendrimers as diverse nanomolecules for biomedical applications. *Expert Opin. Drug Deliv., 6*, 835–850.

Yilmaz, C., & Özcengiz, G., (2016). Antibiotics: pharmacokinetics, toxicity, resistance and multidrug efflux pumps. *Biochem. Pharmacol.,* doi: 10. 1016/j. bcp. 2016. 10. 005.

Yingchoncharoen, P., Kalinowski, D. S., & Richardson, D. R., (2016). Lipid-based drug delivery systems in cancer therapy: what is available and what is yet to come. *Pharmacol. Rev., 68*, 701–787.

Yiyun, C., Na, M., Tongwen, X., Rongqiang, F., Xueyuan, W., Xiaomin, W., & Longping, W., (2007). Transdermal delivery of nonsteroidal anti-inflammatory drugs mediated by polyamidoamine (PAMAM) dendrimers. *J. Pharm. Sci., 96*, 595–602.

You, S., Jung, H. Y., Lee, C., Choe, Y. H., Heo, J. Y., Gang, G. T., Byun, S. K., Kim, W. K., Lee, C. H., Kim, D. E., Kim, Y. I., & Kim, Y., (2016). High-performance dendritic contrast agents for X-ray computed tomography imaging using potent tetraiodobenzene derivatives. *J. Control. Release, 226*, 258–267.

Youngren, S. R., Tekade, R. K., Gustilo, B., Hoffmann, P. R., & Chougule, M. B., (2013a). STAT6 siRNA matrix-loaded gelatin nanocarriers: formulation, characterization, and ex vivo proof of concept using adenocarcinoma cells. *Bio. Med. Res. Int., 858946.*

Youngren, S. R., Tekade, R. K., Hoffmann, P. R., & Chougule, M. B., (2013b). Biocompatible nanocarrier mediated delivery of STAT-6 siRNA to cancer cells. *Cancer Res., 73*, 3313–3313.

Zhang, H., Yang, J., Liang, K., Li, J., He, L., Yang, X., Peng, S., Chen, X., Ding, C., & Li, J., (2015). Effective dentin restorative material based on phosphate-terminated dendrimer as artificial protein. *Colloids Surf. B Biointerfaces., 128*, 304–314.

Zhang, W., & Simanek, E. E., (2000). Dendrimers based on melamine. Divergent and orthogonal, convergent syntheses of a G3 dendrimer. *Org. Lett., 2*, 843–845.

Zhao, X., Meng, G., Han, F., Li, X., Chen, B., Xu, Q., Zhu, X., Chu, Z., Kong, M., & Huang, Q., (2013). Nanocontainers made of various materials with tunable shape and size. *Sci. Rep., 3*, 2238.

Zhong, Q., Bielski, E. R., Rodrigues, L. S., Brown, M. R., Reineke, J. J., da Rocha, S. R., (2016). Conjugation to poly(amidoamine) dendrimers and pulmonary delivery reduce cardiac accumulation and enhance antitumor activity of doxorubicin in lung metastasis. *Mol. Pharm., 13*, 2363–2375.

Zhou, K., Nguyen, L. H., Miller, J. B., Yan, Y., Kos, P., Xiong, H., Li, L., Hao, J., Minnig, J. T., Zhu, H., & Siegwart, D. J., (2016). Modular degradable dendrimers enable small RNAs to extend survival in an aggressive liver cancer model. *Proc. Natl. Acad. Sci. USA., 113*, 520–525.

CHAPTER 6

DENDRIMERS FOR CONTROLLED RELEASE DRUG DELIVERY

PHUNG NGAN LE, DAI HAI NGUYEN, CUU KHOA NGUYEN, and NGOC QUYEN TRAN

Department of Materials and Pharmaceutical Chemistry, Vietnam Academy of Science and Technology, HCMC70000, Vietnam, E-mail: tnquyen@iams.vast.vn

CONTENTS

ABSTRACT

Dendrimers are one of the most studied nanocarriers for drug-delivery systems because they can be prepared with predetermined nanosize and drugs are loaded in their inner cavities or functionalized on their surface via (non) covalent interactions. These novel properties of dendrimers enable to carry drugs, proteins, and genes. Therefore, it was understandable that Tomalia, who first introduced and laid a sustainable foundation of dendrimer to the biomaterial field, was predicted to be one of the candidates for Nobel Prize by

Thomson Reuter in 2011. At present, therapeutic dendrimer-based products are being been commercially used in biomedical applications. The dendritic nanostructures are potentially employed in the high encapsulation together with maintaining suitable drug release rate. Besides, the facile modification by subsequent corporation with smart polymeror hybrid dendrimer could considerably reduce the side-effects of nanocarrier and improve their bio-compatibility. This review exclusively focuses on the hyperbranched dendrimers and multifunctional dendritic drug-delivery systems. Some recent approaches in the synthesis of many specific agents-modified dendrimers are also reported.

6.1 INTRODUCTION

The bottleneck of drug pipeline has steadily increased with time (Lipinski, 2000; Lipinski et al., 2001). Particularly, some drug discovery studies indicated that while an estimated two-fifth of current drugs on commercials were poorly soluble based on the biopharmaceutical classification system (BCS), the majority of drugs in the development phase were characterized as BCS Class II or Class IV drugs (Savjani et al., 2012; Kalepuet et al., 2015). To overcome the limitation of some hydrophobic drugs and increment in therapeutic effectiveness, the advent of tailor-made nanocarriers involving wisely engineered as well as precisely designed nanostructures has paved way for drug delivery. Some novel drug-delivery systems (DDS) based on organic, inorganic, and hybrid materials have been developed to maintain the drug concentration in targeted organs/tissues for a longer period of time and reduce the related side effects (Safraetal, 2000; Soppimath et al., 2001; Abraham et al., 2005; Ding et al., 2011; Gardikis et al., 2010; Spence et al., 2015). Among them, dendrimer has emerged as one of the excellent candidates for the development of controlled DDS over other conventional nanoscale carriers (Svenson et al., 2005; Gupta et al., 2006). Markedly, the unique features of dendrimers and their ease of functionalization enabled to design outstanding hybrid inorganic/organic systems that articulated multiple functions into a single scaffold (Shi et al., 2006; Mehdipoor et al., 2011). This chapter details the recent progresses made in this swiftly evolving field of some typical dendrimer-based nanomaterials in the "smart" DDS, and especially spotlights the development of model system from monofunctional to multifunctional dendritic nanocarriers in this regard. Moreover, some recent approaches in the preparation of many bulky dendrimers-based materials are also reported for DDS.

6.2 NANOCARRIER-BASED DENDRIMER IN DDS

Dendrimers are highly hyperbranched macromolecules that emanate from an initiator core and then symmetrically branch with repetitive structures made of monomer to develop their generations (Tomalia, 1985; Pushkar et al., 2006). The molecules with star-like architecture have well-defined nano-size dimensions as well as molecular weights. Drug molecules can be conjugated to the surface group or entrapped inside the core. With time, the class of dendrimers has been differentiated depending upon the desired applications. The majority of popular outstanding dendrimers in drug delivery were categorized as poly(amidoamide) (PAMAM) (Tomalia, 1985; Hawker et al., 1990; Esfand et al., 2001; Yavuz et al., 2015), poly(propyleneimine) (PPI) (Berg et al., 1993; Richter et al., 2001; Taratula, 2009; Kaura et al., 2016), and poly(L-lysine) (PLL) (Al-Hamra et al., 2005; Al-Jamal et al., 2013; Rahimi et al., 2016). Yet, tailor-made dendrimers generally had a well-predetermined structure with a multi-reactive-site central core, layer-by-layer branches possibly upgrading to higher generations, and surface end groups to attach external ligands (Bosman et al., 1999; Frechet et al., 2001; Sowinska et al., 2014). The most appealing structure of starburst dendrimers has provided manifold special opportunity for the chemical interaction between host and guests and are well equipped to control drug release rate. A dendrimer was facile to interact in several ways with organic compounds, such as drugs via electrostatic interactions and covalent bonds. Exclusively, the presence of easily functionalized end groups significantly induced chemical corporations between dendrimer and drug molecules resulting in improvement in solubility of hydrophobic drugs (Jansen et al., 1999; Beezer et al., 2003; Prajapati et al., 2009; Soto-Castro et al., 2012; Zhou et al., 2013). Moreover, there were chemical linkages and their own mechanism of cleavage, which enable the drug release from dendrimer. Some of them were profoundly impacted by the pH (ester, hydrazine,..), while others were under enzymatic control (peptide, amide,...) (Patri et al., 2005; Kurtoglu et al., 2010; Zhang et al., 2014). However, the physical incorporation of drugs into dendrimer was more eminent straight forward and rapid preparation method without adversely impacting drug pharmacological activity and faster release than normal chemical conjugation (Najlah et al., 2006; Cami-nadeet al., 2014). The entrapment of drug molecules into dendrimer resulted in non-bonding interactions. Dendrimers were considered as dendritic structure and unimolecular micelles to capture the external molecules inside their internal voids by host–guest interactions. The increasing generation of dendrimers meant that more drug would be entrapped inside the well-defined interior cavities. Physically incorporated drugs were tracked down from the carrier due to degradation of covalent bond between drugs and dendrimer by appropriate enzymes or undergoing the physical changes as well as stimulus like pH, temperature, etc. (Medina et al., 2009). It was usually swift and depended on several factors such as the drug partition coefficient between hydrophobic and aqueous environments, strength of drug/dendrimer interactions, dendrimer generations, and surface groups (Nanjwade et al., 2009).

Of all, the PAMAM dendrimer has widely been investigated for the drug-delivery applications accounting for the integration to the *in vivo* prerequisite studies (Esfand et al., 2001; Madaan et al., 2014). Take platinum dendrimer complexes as an example. Nguyen and collaborators (Figure 6.1) indicated that together with the enhancement of cisplatin molecules conjugated to nanocomplex, the *in vivo* result also made a considerable contribution to intensive research (Kirkpatrick et al., 2011; Yellepeddi et al., 2011, 2013; Tran et al., 2013; Nguyen et al., 2015).

Furthermore, in comparison with the other rival called "poly (propylene imine) (PPI)" dendrimer in drug delivery, PAMAM has drawn more specific attention notwithstanding the similar open structure and inner pockets encapsulating hydrophobic drugs. The PAMAM dendrimer had a much higher loading ability and showed better-sustained release ability and is less toxic than the PPI dendrimer at high drug feeding ratios (Shao et al., 2011).

In another study relevant to the PAMAM dendrimer, Nguyen et al. prepared a series highly lipophilic pluronics-conjugated PAMAM dendrimer but reported a new conclusion that the hydrophobic drug encapsulation efficacy increases proportionally to the increase in lipophilic pluronics nanocarriers in comparison with other lesser lipophilic groups, which played a key role in delivering poorly water insoluble drugs and application to biomedicine (Nguyen et al., 2016). To extensively develop more applications from homogenous characteristics of PPI and PAMAM dendrimer, Sajjad and coworkers showed the combination of the hybrid dendrimers of

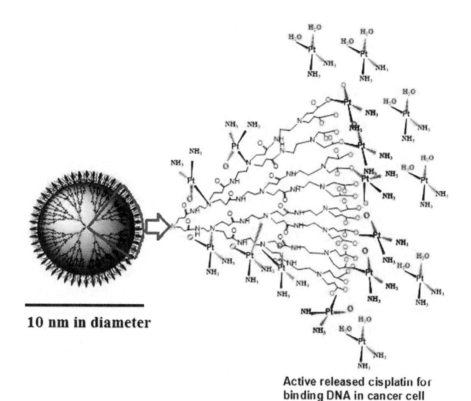

Active released cisplatin for
binding DNA in cancer cell

FIGURE 6.1   Complex formation of aquated cisplatin and carboxylated PAMAM G3.5.

PPI(core)–PAMAM (shell), which significantly multiplied the encapsulation capacity of the hybrid (Kavyani et al., 2016).

6.3   APPROACHES ON MULTIFUNCTIONAL DENDRIMER AS SMART DRUG-DELIVERY SYSTEMS

Along with the expansion of approaches to overcome limitations of nonspecific bio-distribution and achieve the release of payloads at the target sites in a controlled manner, the term "smart" drug-delivery system (DDS) was coined. In other words, nanocarriers were designed with targeting agents and solubilizing polymers for synergistically improving drug effectiveness. The drug-loaded nanoplatform ensured that the drug would not only upsurge the solubility of hydrophobic drugs but also released at the targets where the nanocarriers accumulated by active or passive targeting strategy. These systems could be manipulated by one or more internal and external stimulus.

6.3.1   ENDOGENOUS FACTORS

6.3.1.1   pH-Responsive Dendrimer

pH in the human body varies along the gastrointestinal trac, and also in pathophysiological conditions (such as inflammation and tumors) or in subcellular compartments. Thus, it was considered as an essential factor for the intensive nanocarrier research with pH stimulus. Polymers with changes in conformation in terms of alterations to surface wettability, charge state, or solubility were considered outstanding in DDS application. Notably, dendrimer had an easily swelling nanostructure due to protonization of amine groups in acidic medium, resulting into a higher amount of released drug. The polymer responded to low external pH could maintain the payloadinto the carriers in the bloodstream while causing their dissociation over time, which facilitate the release of encapsulated molecules at target agents in control. Criscionea et al. (2009) demonstrated a spontaneous self-assembly of poly(amidoamine) dendrimers complex nanoscopic and microscopic particulates following partial fluorination of the constituent dendrimer subunits, which has an exclusive capacity in MRI. Significant developments in science inevitably led to combine stimulus of pH applied in smart DDS with other stimuli such as temperature and redox potential to achieve more precise and specific release at the target (Hui et al., 2010; Chen et al., 2011; Cheng et al., 2013).

6.3.1.2   Enzyme-Responsive Dendrimer

Enzyme stimulus was apart from other endogenous factors having ability to respond to a biological molecule to regulate the function of natural materials, whereas the enzyme-sensitive dendrimer aspect has lately been emerging in DDS applications. Lee at al. (2015) synthesized doxorubicin-conjugated dendrimer nanoparticles (DenGDP) using glycyl-phenylalanyl-leucyl-glycine tetrapeptide (Gly-Phe-Leu-Gly) for cathepsin B-responsive drug targeting of cancer cells. It was shown that DenGDP had a more efficiency and bio-compatibility than dendritic complex with peptide conjugation. Similarly, Wang et al. (2016) modified the generation 4 PAMAM dendrimer as a nanocarrier with an enzyme sensitivity. The results indicated that modified enzyme-sensitive dendrimer had a more efficiency in anti-cancer drug entrapment together with the high cytotoxicity in comparison with the free drug The precise control of the initial response time of the systems was still a main obstacle of enzyme-responsive DDSs.

6.3.2 EXOGENOUS FACTORS

6.3.2.1 Temperature-Responsive Dendrimer

Thermal-sensitive dendrimers are interesting due to the variation in temperature applied externally in a noninvasive manner. The polymer-phase transitions in aqueous solutions by temperature, which were accompanied by (partial) dehydration of the polymer chain, has been the most convenient and effective factor since this phenomenon provides high potential for bio-medical applications, such as drug delivery (Schmaljohann et al., 2006). A functionally facile lipoic acid conjugated thermo-platform could undergo slow and sustained release at 37 42°C (Castonguay et al., 2011). Exclusively, the change in polymer conformation during the low temperature solution temperature (LCST) transition characteristics has been investigated more for many years. Le et al. (Figure 6.2) showed that covalent graft of thermo-responsive poly(Nisopropylacrylamide) (PNIPAM) to the periphery of PAMAM dendrimers enhanced the hydrophobic drug payload and triggered to control their drug release profile in tumors (Le et al., 2016).

As a rule, thermal-sensitive nanocarriers were designed to retain their payloads and release the drugs rapidly considering the temperature difference between cancer tissues and normal tissues. Hence, utmost care should be taken to prevent their sensitivity to slight temperature changes (Gerweck et al., 1996; Mura et al., 2013).

6.3.2.2 Magnetic-Responsive Dendrimer

Unlike common agents, magnetic dendrimer was mainly considered as an imaging-capable drug delivery agent in magnetic resonance imaging

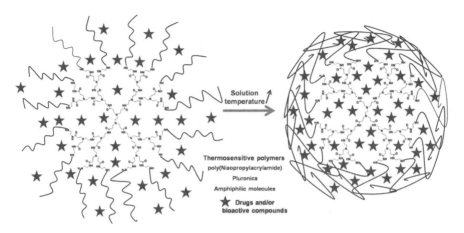

FIGURE 6.2 Mechanism of several drugs encapsulated in an amphiphilic dendrimer.

(MRI). As an example, a paramagnetic-contrast-agent complex with PAMAM dendrimers for MRI was widely used in the diagnosis of both tumor vasculature and lymphatic invasion (Kobayashi et al., 2004; Ye et al., 2013). Broadly, magnetic-sensitive dendrimer approaches could offer the possibility of being systematically administered but directed toward a specific target in the human body by subsequent support of an applied magnetic field.

6.3.2.3 Light-Responsive Dendrimer

Dendritic copolymers formed through self-assembly of photo-responsive block copolymers were utilized as they fulfilled the stringent requirements of smart drug delivery. This specific behavior which manipulated the drug encapsulation by photo-illumination originated from the incorporation of azo-benzene moieties within the dendrimer design. The photo-isomerization of these subunits resulted in a topographical modification, which gave rise to an enclosed dendrimer core, allowing for the physical entrapment of small molecule guests (Nguyen et al., 2011). Recently, Nazemi et al. synthesized a new series of fully photodegradable polyester dendrimer backbone incorporated with photodegradable o-nitrobenzyl esters. Photo-sensitive dendritic nanocarriers could instantaneously release all the payloads at tumor sites once they were irradiated with UV-vis (Nazemi et al., 2013).

6.3.3 OTHER DESIGNS OF SMART DENDRIMERS

Addition to the single responsive dendrimers in the intelligent DDS, the dendritic nanocarriers based on dual stimulus such as pH/temperature, pH/redox, pH/magnetic field, temperature/magnetic field, temperature/enzyme, v/v, etc., were popularly researched (Hui et al., 2005; Wang et al., 2007; Shen et al., 2011; Hu et al., 2014). Markedly, novel environmental stimuli-responsive nanoscale vehicles were designed to control drug delivery and release, leading to superior *in vitro* and/or *in vivo* anti-cancer efficacy. Chandra et al. introduced a multiple sensitive dendrimer, in which loading efficiency was estimated to be roughly 95% with a sustained pH-and-temperature-responsive release along with the addition of enzyme cathepsin B to degrade the dendritic shell and then trigger sustained drug release at target tumors (Chandra et al., 2015).

More importantly, a conventional method based on EPR effect was only effective to tumors with highly permeable solid areas; however, the disparity of malignancies prevented the passive targeting approaches in some cases that had lower permeable adsorption. Hence, to precisely achieve the payloads at target pathophysiological sites, the subsequent conjugation of specific targeting ligands on nanocarrier surfaces could highly facilitate their selective binding to the tumors sites in which the signals of overexpression could be captured. A wide range of external ligands such as folic acid, biotin, amino acids, and peptides was successfully incorporated onto the surface of dendrimer. Zhang et al. employed the third-generation PAMAM dendrimer after modification with polyvalent saccharide into the methotrexate (MTX) delivery system (Zhang et al., 2011). In another case, Mekuria et al. successfully engineered IL-6 antibody and RGD peptide-conjugated PAMAM dendrimer for the active delivery to targeted HeLa cells, which not only upsurged the drug loading but also enhanced drug tracking from carrier corresponded with greater cytotoxicity (Mekuria et al., 2016). Broadly, smart DDSs for controlled drug release (e.g., polymers, liposome, and organic–inorganic hybrid biomaterials) have been a terrain of discussion (De Jong et al., 2008).

6.3.4 DENDRIMER-BASED HYDROGELS FOR CONTROLLING DELIVERY

In the drug-delivery field, the polymeric hydrogel networks have received much attention. Several kinds of physical and chemical hydrogels were reported as versatile platforms for delivering protein, gene, and small bioactive molecules (Tran et al., 2011; Nguyen et al., 2014). The preparation of functional hydrogels play an important role in delivering specific molecules due to their interaction between the hydrogel platform and the guest molecules that could enhance solubility of hydrophobic drugs and delivery efficiency. The utilization of outstanding dendrimer-functional hydrogels has emerged as an advanced approach in the field as the dendrimer macromolecules were externally amine or carboxylic groups that could be functionalized with crosslinking agents to form hydrogel matrix, and these functional groups play a role in controlling drug delivery. Ketoconazole, a well-known antifungal, was conjugated and encapsulated in the PAMAM dendrimer, thereby enhancing its solubility and antifungal activity. The interaction between the dendrimer and ketoconazole formed a conjugation that might affect hydrophobic interaction in the formation of the physical hydrogel (Winnicka et al., 2012). A peptide and vinylsulfone-functionalized acetyl PAMAM dendrimer and thiolated hyaluronic acid were quickly cross-linked to form hydrogel via Michael type reaction. The matrix served as a good well platform for tissue regeneration and drug or cell delivery (Bi et al., 2015). A photo cross-linked polyethylene glycol acrylate-conjugated poly-amidoamine dendrimer was used as a novel platform for ocular drug delivery of brimonidine and timolol maleate (Holden et al., 2012). A PLGA-conjugated dendrimer hydrogel platform also sustainably delivered two brimonidine and timolol maleate antiglaucoma drugs (Yang et al., 2012). The PAMAM dendrimer (Figure 6.3) also performed the role of a cationic precursor in enzymatic preparation of tetronic-dendrimer hydrogel for controlling delivery of heparin anionic drug (Tran et al., 2013). Apart from the abovementioned dendrimer-based hydrogels that performed a significant role in drug delivery, some of functional dendrimer/dendron hydrogels have been developed, which could be potential platforms in the drug-delivery field (Kaga et al., 2016).

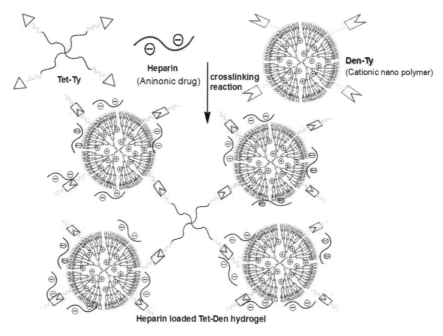

Tet-Ty

Heparin
(Aninonic drug)

crosslinking
reaction

Den-Ty
(Cationic nano polymer)

Heparin loaded Tet-Den hydrogel

FIGURE 6.3 Cationic dendrimer-tetronic hydrogel for delivering heparin.

6.4 FUTURE PERSPECTIVES

The administered delivery systems should retain the drugs in nanocarriers during circulation and release precisely at the target cells. A great number of smart DDSs have been developed and approved in clinical applications such as Doxil (Liposomal DOX) and Lipusu (Liposomal PTX) (Kopeček, 2003; Lavan et al., 2003). To obtain smart properties, most of state-of-art DDSs were designed with sophisticated structures and formulations, but their efficiency was lost while scaling up for industrial production. Hence, they are recommended to be used as simply formulations, and more attentions need to be paid to the development of advanced approaches wherein clinical doses will highly prevent the normal tissue damage and tissue-penetration depth at least. By the expeditious development of material science, pharmaceutical sciences, and biomedical science, controlled release nanomaterials together with nanotechnologies will be used popularly for smart dendrimer-based DDSs and accelerate the translation of smart drugs from the bench to the bedside.

KEYWORDS

- biomaterials
- drug delivery systems
- smart polymers
- starburst dendrimers
- sustained/controlled drug release
- transfection

REFERENCES

Abraham, S. A., Waterhouse, D. N., Mayer, L. D., Cullis P. R., Madden, T. D., & Bally, M. B., (2005). The liposomal formulation of doxorubicin. *Methods Enzymol.*, *391*, 71–97.

Al-Hamra, M., & Ghaddar, T. H., (2005). Facile synthesis of poly-(l-lysine) dendrimers with a pentaaminecobalt(III) complex at the core. *Tetrahedron Lett.*, *46*(34), 5711–5714.

Al-Jamal, K. T., Al-Jamal, W. T., Wang, J. T. W., Rubio, N., Buddle, J., Gathercole, D., Zloh, M., & Kostarelos, K., (2013). Cationic poly-l-lysine dendrimer complexes doxorubicin and delays tumor growth *in vitro* and *in vivo*. *ACS Nano.*, *7*, 1905–1917.

Beezer, A. E., King, A. S., Martin, I. K., Mitchel. J. C., Twyman, L. J., & Wain, C. F., (2003). Dendrimers as potential drug carriers, encapsulation of acidic hydrophobes within water soluble PAMAM derivatives. *Tetrahedron.*, *59*, 3873–3880.

Berg van den, E. M. M., & Meijer, E. W., (1993). Poly(propylene imine) dendrimers: large-scale synthesis by heterogeneously catalyzed hydrogenations. *Angew. Chem. Int. Ed.*, *32*, pp. 1308–1311.

Bi, X., Amie, L. J., Allen, A., Ramaboli, M., Campbell, E., West, D., Maturavongsadit, P., Brummett, K, Wang, Q., (2015). Synthesis of PAMAM dendrimer-based fast cross-linking hydrogel for biofabrication. *J. Biomater. Sci. Polym. Ed.*, *26*(11), 669–682.

Bosman, A. W., Janssen, H. M., & Meijer, E. W., (1999). About dendrimers: Structure, physical properties, and applications. *Chem. Rev.*, *99*(7), 1665–1688.

Caminade, M. A., & Turrin, C. O., (2014). Dendrimers for drug delivery. *J. Mater. Chem. B*, *2*, 4055–4066.

Castonguay, A., Wilson, E., Al-Hajaj, N., Petitjean, L., Paoletti, J., Maysinger, D., & Kakkar, A., (2011). Thermosensitive dendrimer formulation for drug delivery at physiologically relevant temperatures. *Chem. Commun.*, *47*, 12146–12148.

Chandra, S., Noronha, G., Dietrich, S., Lang, H., & Bahadur, D., (2015). Dendrimer-magnetic nanoparticles as multiple stimuli responsive and enzymatic drug delivery vehicle. *[J. Magn. Magn. Mater.*, *380*, 7–12.

Chen, W., Zhong, P., Meng, F., Cheng, R., Deng, C., Feijen, J., & Zhong, Z., (2013). Redox and pH-responsive degradable micelles for dually activated intracellular anticancer drug release. *J. Control Release*, *169*, 171–179.

Cheng, R., Meng, F., Deng, C., Klok, H. A., & Zhong, Z., (2013). Dual and multi-stimuli responsive polymeric nanoparticles for programmed site-specific drug delivery. *Biomaterials., 34*, 3647–3657.

Criscionea, M. J, Leb, B. L., Sterna, E., Brennanc, M., Rahnerd, M., Papademetrisa, X., & Fahmy, M. T., (2009). Self-assembly of pH-responsive fluorinated dendrimer-based particulates for drug delivery and noninvasive imaging. *Biomaterials, 30*(23–24), 3946–3955.

De Jong, W. H., & Borm, P. J., (2008). Drug delivery and nanoparticles: Applications and hazards. *Int. J. Nanomedicine, 3*(2), 133–149.

Ding, D., Zhu, Z., Li. R., Wu, W., Jiang, X., & Liu, B., (2011). Nanospheres-incorporated implantable hydrogel as a trans-tissue drug delivery system. *ACS Nano, 5*(4), 2520–2534.

Esfand, R., & Tomalia, D. A., (2001). Poly(amidoamine) (PAMAM) dendrimers: from biomimicry to drug delivery and biomedical applications. *Drug. Discov. Today., 6*(8), 15, 427–436.

Frechet, J. M. J., & Tomalia, D. A., (2001). *Dendrimers and Other Dendritic Polymers.* Chichester: Wiley, UK.

Gardikis, K., Hatziantoniou, S., Bucos, M., Fessas, D., Signorelli, M., Felekis, T., Zervou, M., Screttas, C. G., Steele, B. R, Ionov, M., Micha-Screttas, M., Klajnert, B., Bryszewska, M., & Demetzos, C., (2010). New drug delivery nanosystem combining liposomal and dendrimeric technology (liposomal locked-in dendrimers) for cancer therapy. *J. Pharm. Sci., 99*(8), 3561–3571.

Gerweck, L. E., & Seetharaman, K., (1996). Cellular pH gradient in tumor versus normal tissue: potential exploitation for the treatment of cancer. *Cancer Res., 56*, 1194–1198.

Gupta, U., Agashe, H. B., Asthana, A., & Jain, N. K., (2006). A review of *in vitro-in vivo* investigations on dendrimers: the novel nanoscopic drug carriers. *Nanomedicine, 2*(2), 66–73.

Hawker, C., Frechet J. M. J., (1990). Preparation of polymers with controlled molecular architecture: A new convergent approach to dendritic macromolecules. *J. Am. Chem. Soc., 112*, 7638–7647.

Holden, A. C., Tyagi, P., Thakur, A., Kadam, R., Jadhav, G., Kompella, B. U., & Yang, H., (2012). Polyamidoamine dendrimer hydrogel for enhanced delivery of antiglaucoma drugs, *Nanomedicine: NBM., 8*, 776–783.

Hu, W., Cheng, L., Cheng, L., Zheng, M., Lei, Q., Hu, Z., Xu, M., Qiu, L., & Chen, D., (2014). Redox and pH-responsive poly (amidoamine) dendrimer-poly (ethylene glycol) conjugates with disulfide linkages for efficient intracellular drug release. *Colloids. Surf. B. Biointerfaces., 123*, 254–263.

Hui, H., Fan, D. X., & Cao, Z. L., (2015). Thermo- and pH-sensitive dendrimer derivatives with a shell of poly(N, N-dimethylaminoethyl methacrylate) and study of their controlled drug release behavior. *Polymer., 46*(22), 9514–9522.

Jansen, B. A. J., Zwan, J. van der., Reedijk, J., Dulk, Den H., & Brouwer, J., (1999). A tetranuclear platinum compound designed to overcome cisplatin resistance. *Eur. J. Inorg. Chem.,* 1429–1433.

Kaga, S., Arslan, M., Sanyal, R., & Sanyal, A., (2016). Dendrimers and dendrons as versatile building blocks for the fabrication of functional hydrogels. *Molecules, 21*, 497.

Kalepu, S., & Nekkanti, V., (2015). Insoluble drug delivery strategies: review of recent advances and business prospects. *Acta. Pharm. Sin. B., 5*(5), 442–453.

Kaur, A., Jain, K., Mehra, N. K., & Jain, N. K., (2016). Development and characterization of surface engineered PPI dendrimers for targeted drug delivery. *Artif. Cells. Nanomed. Biotechnol.*. DOI: 10. 3109/21691401. 2016. 1160912.

Kavyani, S., Amjad, I. S., Dadvar, M., & Modarress, H., (2016). Hybrid dendrimers of PPI(core)–PAMAM(shell): A molecular dynamics simulation study. *J. Phys. Chem. B.*, *120*(36), 9564–9575.

Kesharwani, P., Tekade, R. K., & Jain, N. K., (2014). Formulation development and *in vitro–in vivo* assessment of the fourth-generation PPI dendrimer as a cancer-targeting vector. *Nanomedicine (Lond).*, *9*(15), 2291–308.

Kirkpatrick, G. J., Plumb, J. A., Sutcliffe, O. B., Flint, D. J., & Wheate, N. J., (2011). Evaluation of anionic half generation 3. 5–6. 5 poly(amidoamine) dendrimers as delivery vehicles for the active component of the anticancer drug cisplatin. *J. Inorg. Biochem.*, *105*(9), 1115–1122.

Kobayashi, H., Reijnders, K., English, S., Yordanov, A. T., Milenic, D. E., Sowers, A. L., Citrin, D., Krishna, M. C., Waldmann, T. A., Mitchell, J. B., & Brechbiel, M. W., (2004). Application of a macromolecular contrast agent for detection of alterations of tumor vessel permeability induced by radiation. *Clin. Cancer Res.*, *10*(22), 7712–7720.

Kopecek, J., (2003). Smart and genetically engineered biomaterials and drug delivery systems. *Eur. J. Pharm. Sci.*, *20*, 1–16.

Kurtoglu, Y. E., Mishra, M. K., Kannan, S., & Kannan, R. M., (2010). Drug release characteristics of PAMAM dendrimer-drug conjugates with different linkers. *Int. J. Pharm.*, *384*(1–2), 189–194.

Lavan, D. A., McGuire, T., & Lange, R., (2003). Small-scale systems for *in vivo* drug delivery. *Nat. Biotechnol.*, *21*, 1184–1191.

Le, P. N., Nguyen, N. H, Nguyen, C. K., & Tran, N. Q., (2016). Smart dendrimer-based nanogel for enhancing 5-fluorouracil loading efficiency against MCF7 cancer cell growth. *Bull. Mater. Sci.*, *39*(6) 1493–1500.

Lee, S. J., Jeong, Y. I., Park, H. K., Kang, D. H., Oh, J. S., Lee, S. G., & Lee, H. C., (2015). Enzyme-responsive doxorubicin release from dendrimer nanoparticles for anticancer drug delivery. *Int. J. Nanomedicine.*, *10*, 5489–5503.

Lipinski, C. A., (2000). Drug-like properties and the causes of poor solubility and poor permeability. *J. Pharmacol. Toxicol. Methods*, *44*, 235–249.

Lipinski, C. A., Lombardo, F., Dominy, B. W., & Feeney, P. J., (2001). Experimental and computational approaches to estimate solubility and permeability in drug discovery and development settings. *Adv. Drug Deliv. Rev.*, *46*, 3–26.

Madaan, K., Kumar, S., Poonia, N., Lather, V., & Pandita, D., (2014). Dendrimers in drug delivery and targeting: Drug-dendrimer interactions and toxicity issues. *J. Pharm. Bioallied. Sci.*, *6*(3), 139–150.

Medina, S. H., & El-Sayed, M. E., (2009). Dendrimers as carriers for delivery of chemotherapeutic agents. *Chem. Rev.*, *109*(7), 3141–3157.

Mehdipoor, E., Adeli, M., Bavadi, M., Sasanpourb, P., Bizhan Rashidian, B., (2011). A possible anticancer drug delivery system based on carbon nanotube–dendrimer hybrid nanomaterials. *J. Mater. Chem.*, *21*, 15456–15463.

Mekuria, S. L., Debele, T. A., Chou, H. Y., & Tsai, H. C., (2016). IL-6 Antibody and RGD peptide conjugated poly(amidoamine) dendrimer for targeted drug delivery of hela cells. *J. Phys. Chem. B.*, *120*(1), 123–130.

Minko, T., (2009). Surface-engineered targeted PPI dendrimer for efficient intracellular and intratumoral siRNA delivery. *J. Control. Release*, *140*, 284–293.

Mura, S., Nicolas, J., & Couvreur, P., (2013). Stimuli-responsive nanocarriers for drug delivery. *Nat. Mater*, *12*, 991–1003.

Najlah, M., Freeman, S., Attwood, D. D., & Emanuele, A., (2006). Synthesis, characterization and stability of dendrimer prodrugs. *Int. J. Pharm., 308*, 175–182.

Nanjwade, B. K., Bechra, H. M., Derkar, G. K., Manvi, F. V., & Nanjwade, V. K. (2009). Dendrimers: emerging polymers for drug-delivery systems. *Eur. J. Pharm. Sci. Sa., 38*(3), 185–196.

Nazemi, A., Schon, B. T., & Gillies. R. E., (2013). Synthesis and degradation of backbone photodegradable polyester dendrimers. *Org. Lett.*, *15*(8), 1830–1833.

Nguyen, H., Nguyen, C. K., Nguyen, N. H., & Tran, N. Q., (2015). Improved method for cisplatin-loading dendrimer and behavior of the complex nanoparticles *in vitro* release and cytotoxicity. *J. Nanoscience and Nanotechnol.*, *16*(6), 4106–4110.

Nguyen, M. K., & Alsberg, E., (2014). Bioactive factor delivery strategies from engineered polymer hydrogels for therapeutic medicine. *Chem. Eng. Res.*, *39*(7), 1236–1265.

Nguyen, T. T., Turp, D., Wang, D., Nölscher, B., Laquai, F., & Müllen, K., (2011). A fluorescent, shape-persistent dendritic host with photoswitchable guest encapsulation and intramolecular energy transfer. *J. Am. Chem. Soc.*, *133*(29), 11194–11204.

Patri, A. K., Kukowska-Latallo, J. F., & Baker, J. R. Jr., (2005). Targeted drug delivery with dendrimers: comparison of the release kinetics of covalently conjugated drug and non-covalent drug inclusion complex. *Adv. Drug. Deliv. Rev.*, *57*(15), 2203–2214.

Prajapati, R. N., Tekade, R. K., Gupta, U., Gajbhiye, V., & Jain, N. K., (2009). Dendrimer-mediated solubilization, formulation development and *in vitro-in vivo* assessment of piroxicam. *Mol. Pharm.*, *6*(3), 940–50.

Pushkar, S., Philip, A., Pathak, K., & Pathak, D., (2006). Dendrimers: nanotechnology derived novel polymers in drug delivery. *Indian J. Pharm. Educ. Res., 40*(3), 153–158.

Rahimi, A., Amjad-Iranagh, S., & Modarress, H., (2016). Molecular dynamics simulation of coarse-grained poly (L-lysine) dendrimers. *J. Mol. Model, 22*(3), 59.

Richter-Egger, D. L., Tesfai, A., & Tucker, S. A., (2001). Spectroscopic investigations of poly(Propyleneimine) dendrimers using the solvatochromic probe phenol blue and comparisons to poly(amidoamine) dendrimers. *Anal. Chem.*, *73*, 5743–5751.

Safra, T., Muggia, F., Jeffers, S., Tsao-Wei, D. D., Groshen, S., Lyass, O., Henderson, R., Berry, G., & Gabizon, A., (2000). Pegylated liposomal doxorubicin (Doxil): reduced clinical cardiotoxicity in patients reaching or exceeding cumulative doses of 500 mg/m2. *Ann. Oncol.*, *11*(8), 1029–1033.

Savjani, K. T., Gajjar, A. K., & Savjani, J. K., (2016). Drug solubility: Importance and enhancement techniques. ISRN Pharm. 2012, 2012: 195727. Published online: Jul 5, 2012 DOI: 10.5402/2012/195727.

Schmaljohann, D., (2006). Thermo- and pH-responsive polymers in drug delivery. *Adv. Drug. Delivery. Rev.*, *58*, 1655–1670.

Shao, N., Su, Y., Hu, J., Zhang, J., Zhang, H., & Cheng, Y., (2011). Comparison of generation 3 polyamidoamine dendrimer and generation 4 polypropylenimine dendrimer on drug loading, complex structure, release behavior, and cytotoxicity. *Int. J. Nanomedicine.*, *6*, 3361–3372.

Shen, Y., Ma, X., Zhang, B., Zhou, Z., Sun, Q., Jin, E., Sui, M., Tang, J., Wang, J., & Fan, M., (2011). Degradable dual pH- and temperature-responsive photoluminescent dendrimers. *Chemistry*, *17*(19), 5319–26.

Shi, X. Y., Ganser, T. R., Sun, K., Balogh, L. P., & Baker, J. R. Jr., (2006). Characterization of crystalline dendrimer-stabilized gold nanoparticles. *Nanotechnology*, *17*, 1072–1078.

Soppimath, K. S., Aminabhavi, T. M., Kulkarni, A. R., & Rudzinski, W. E., (2001). Bio-degradable polymeric nanoparticles as drug delivery devices. *J. Control. Release.*, *70*(1–2), 1–20.

Soto-Castro, D., Cruz-Morales, J. A., Ramírez, A. M. T., Guadarrama P., (2012). Solubilization and anticancer-activity enhancement of Methotrexate by novel dendrimeric nanodevices synthesized in one-step reaction. *Bioorg. Chem.*, *41–42*, 13–21.

Sowinska, M., & Urbanczyk-Lipkowska, Z., (2014). Advances in the chemistry of dendrimers. *New J. Chem.*, *38*, 2168–2203.

Spencer, D. S., Puranik, A. S., & Peppas, N. A., (2015). Intelligent nanoparticles for advanced drug delivery in cancer treatment. *Curr. Opin. Chem. Eng.*, *7*, 84–92.

Svenson, S., & Tomalia, D. A., (2005). Dendrimers in biomedical applications–Reflections on the field. *Adv. Drug Deliv. Rev.*, *57*(15), 2106–2129.

Taratula, O., Garbuzenko, O. B., Kirkpatrick, P., Pandya, I., Savla, R., Pozharov, V. P., He, H., & Minko, T., (2009). Surface-engineered targeted PPI dendrimer for efficient intracellular and intratumoral siRNA delivery. *J. Control Release*, *140*(3), 284–293.

Tomalia, D. A., Baker, H., Dewald, J., Hall, M., Kallos, G., Martin, S., & Roeck, J., (1985). A new class of polymers: Starburst-dendritic. *Polymer Journal*, *17*(1), 117–132.

Tong, N. N. A., Nguyen, T. H., Nguyen, D. H., Nguyen, C. K., & Tran, N. Q., (2015). In situ preparation and characterizations of cationic dendrimer-based hydrogels for controlled heparin release, *J. Macromol. Sci. A.*, *52*(10), 830–837.

Tran, N. Q., Choi, J. H., Bae, J. W., Choi, J. W., Joung, Y. K., & Park, K. D., (2011). Self-assembled nanogel of pluronic-conjugated heparin as a versatile drug nanocarrier. *Macromol. Res.*, *19*, 180–188.

Tran, N. Q., Joung, Y. K., Lih, E., & Park, K. D., (2011). *In situ* forming and rutin-releasing chitosan hydrogels as an injectable dressing for dermal wound healing, *Biomacromol.*, *2*, 2872–2880.

Tran, N. Q., Nguyen, C. K., & Nguyen, T. P., (2013). Dendrimer-based nanocarriers demonstrating a high efficiency for loading and releasing anticancer drugs against cancer cells *in vitro* and *in vivo*. *Adv. Nat. Sci: Nanosci. Nanotechnol.*, *4*, 045013.

Wang, S. H., Shi, X., Van Antwerp, M., Cao, Z., Swanson, S. D., Bi, X., et al., (2007). Dendrimer functionalized iron oxide nanoparticles for specific targeting and imaging of cancer cells. *Adv. Funct. Mater.*, *17*(16), 3043–3050.

Wang, Y., Luo, Y., Zhao, Q., Wang, Z., Xu, Z., & Jia, X., (2016). An enzyme-responsive nanogel carrier based on PAMAM dendrimers for drug delivery. *ACS Appl. Mater. Interfaces*, *8*(31) 19899–19906.

Winnicka, K., Wroblewska, M., Wieczorek, P., Sacha, T. P., & Tryniszewska, E., (2012). Hydrogel of ketoconazole and PAMAM dendrimers: Formulation and antifungal activity. *Molecules*, *17*, 4612–4624.

Yang, H., Tyagi, P., Kadam, R. S, Holden, C. A., & Kompella, U. B., (2012). Hybrid dendrimer hydrogel/PLGA nanoparticle platform sustains drug delivery for one week and antiglaucoma effects for four days following one-time topical administration. *ACS Nano*, *6*(9), 7595–606.

Yavuz, B., Pehlivan, S. B., Vural, I., & Unlu, N., (2016). *In vitro/in vivo* evaluation of dexamethasone—pamam dendrimer complexes for retinal drug delivery. *J. Pharm. Sci.*, *104*(11), 3814–23.

Ye, M., Qian, Y., Tang, J., Hu, H., Sui, M., & Shen, Y., (2013). Targeted biodegradable dendritic MRI contrast agent for enhanced tumor imaging. *J. Control Release*, *169*(3), 239–45.

Yellepeddi, V. K., Kumar, A., Maher, D. M., Chauhan, S. C., Vangara, K. K., & Palakurthi, S., (2011). Biotinylated PAMAM dendrimers for intracellular delivery of cisplatin to ovarian cancer: role of SMVT. *Anticancer Res.*, *31*(3), 897–906.

Yellepeddi, V. K., Vangara, K. K., & Palakurthi, S., (2013). Poly(amido)amine (PAMAM) dendrimer–cisplatin complexes for chemotherapy of cisplatin-resistant ovarian cancer cells. *J. Nanopart. Res.*, *15*, 1897. Published online: Aug 09, 2013. DOI: 10.1007/s11051-013-1897-6. http://link. springer. com/ (Accessed on Nov, 2013).

Yuan, H., Luo, K., Lai, Y., Pu, Y., He, B., Wang, G., Wu, Y., & Gu, Z. W., (2010). A novel poly(l-glutamic acid) dendrimer based drug: Delivery system with both pH-sensitive and targeting functions. *Mol. Pharmaceutics.*, *7*(4), 953–962.

Zhang, C., Pan, D., Luo, K., She, W., Guo, C., Yang, Y., & Gu, Z., (2014). Peptide dendrimer-Doxorubicin conjugate-based nanoparticles as an enzyme-responsive drug delivery system for cancer therapy. *Adv. Healthc. Mater.*, *3*(8), 1299–1308.

Zhang, Y., Thomas, T. P., Lee, K. H., Li, M., Zong, H., Desai, A. M., Kotlyar, A., Huang, B., Holl, M. M., & Baker, J. R. Jr., (2011). Polyvalent saccharide-functionalized generation 3 poly(amidoamine) dendrimer-methotrexate conjugate as a potential anticancer agent. *Bioorg. Med. Chem.*, *19*(8), 2557–2264.

Zhou, Z., D'Emanuele, A., & Attwood, D., (2013). Solubility enhancement of paclitaxel using a linear-dendritic block copolymer. *Int. J. Pharm.*, *452*(1–2), 173–179.

CHAPTER 7

DENDRIMERS IN TARGETED DRUG DELIVERY

ANKIT SETH,[1] PIYOOSH A. SHARMA,[1] RAHUL MAHESHWARI,[2]
MUKTIKA TEKADE,[3] SUSHANT K. SHRIVASTAVA,[1]
and RAKESH K. TEKADE[2]

[1]Department of Pharmaceutical Engineering & Technology,
Indian Institute of Technology (Banaras Hindu University),
Varanasi–221005, India

[2]TIT College of Pharmacy, Technocrats Institute of Technology,
Anand Nagar, Bhopal, Madhya Pradesh–462021, India

[3]National Institute of Pharmaceutical Education and Research
(NIPER)– Ahmedabad, Opposite Air Force Station Palaj, Gandhinagar,
Gujarat 382355, India, E-mail: rakeshtekade@gmail.com

CONTENTS

ABSTRACT

Advancement in research leads to the development of novel drug-delivery carriers especially dendrimers, which have emerged as unique vectors in the expanding field of drug targeting and delivery. The variety in their structure offer a wide range of interactions such as surface conjugation, void space entrapment, non-covalent bonding, etc., with drug molecule, ligands, and other moieties, thus making them a subject of the first choice for exploring new methods of drug delivery. Use of dendrimers has granted scientists immense potential to maneuver pharmacokinetic and pharmacodynamic aspects of drugs to achieve desired goals and eliminate disadvantages associated with many drugs. Dendrimers are gaining popularity owing to their modifiable surface that can be engineered with a variety of ligands and other molecules to enhance the targeting potential as well to minimize the toxicity and uptake by the reticuloendothelial system. This chapter focuses on the targeted drug-delivery strategies based on dendrimers and summarizes several investigations on dendrimer-assisted targeting of various sites including the brain, cancer, pulmonary system, cardiovascular system, and segments of the eye.

7.1   INTRODUCTION

The search for new drug-delivery systems and targeting approaches represents one of the important endeavors for pharmaceutical scientists. Drug targeting may be referred as the delivery of the active drug and confining its pharmacological activity specifically to the targeted site with the rationale of optimizing the therapeutic index such that the side effects should be reduced (Safari and Zarnegar, 2014; Tibbitt et al., 2016). In 1906, Paul Ehrlich had envisaged the concept of drug targeting and predicted the usage of carriers intended to deliver the drug to the targeted site (Jain and Jain, 1997). Several important studies have been carried out in the past, which guided the development of versatile carriers for delivering therapeutics to the defined targets. One of the developments includes microencapsulation technique for producing microparticles made up of biocompatible polymers or lipids for entrapping both the hydrophilic as well as hydrophobic drugs (Singh et al., 2010).

Advances in research lead to the development of nanotechnology for delivering the drugs in an innovative way by utilizing particulates of nanosize range. Due to the minute size and high surface area, nanoparticulates have been reported to show better biodistribution and pharmacokinetics of therapeutic agents and therefore reduce the toxicity by preferentially accumulating at the targeted site (Martinho et al., 2011). Solid lipid nanoparticles have been explored for cancer targeting (Wang et al., 2015) with an added advantage of stealth shielding by polyethylene glycol or any other suitable agent, which prevents their uptake by the reticuloendothelial system (Yang et al., 2007).

Linkage between siRNA and diseases considerably drives the scope of delivering nanocarrier-mediated siRNA therapeutics for treating several disorders (Tekade, 2015). Nanocarrier-mediated combination therapy has revolutionized cancer treatment by delivering siRNA/miRNA in amalgamation with other drugs to the particular target (Youngren et al., 2013a, 2013b; Gandhi et al., 2014). Further innovation in nanotechnology guided the development of phospholipids-based vesicular drug delivery systems like liposomes, niosomes, transferosomes, ethosomes, virosomes, etc., with the capacity to encapsulate hydrophilic as well as lipophilic drugs. These vesicles, especially liposomes, are advantageous concerning drug compatibility, enhanced permeation, and tuned pharmacokinetic profile (Tiwari et al., 2012). However, liposomes are associated with stability issues and low entrapment efficiency and controlled release, which restricts their pharmaceutical use. Liposomes made of nonionic surfactants have been reported, which are called as niosomes. These are considered as one of the carriers for targeted drug delivery with enhanced stability (Kazi et al., 2010). Liposomes have also been modified into ethosomes (Maheshwari et al., 2012) and transferosomes with the addition of surfactants and ethanol to enhance their flexibility and permeation properties (Bragagni et al., 2012). On the other hand, liposomes conjugated with viral proteins have been developed and are called as virosomes. Several studies have demonstrated their potential for immunization, (Schwendener, 2014) cancer treatment, and drug targeting (Liu et al., 2015). These specialized liposomes can be fabricated to increase the uptake of drugs through receptor-mediated routes (Babar et al., 2013). A recent advancement in liposomal technology for drug targeting involves liposomes with internal emulsion droplets called as e-liposomes, which have been designed to target the cancer tissue with enhanced permeability and retention. Functions of e-liposomes such as location and time of release have been controlled with the application of ultrasound (Lattin et al., 2012).

Many technological developments have been achieved in the field of site-specific drug targeting during the last few decades (Park, 2014; Tibbitt et al., 2016). Drug targeting to the specific sites aroused the interest of the scientists toward the novel carriers, which led to the development of new macromolecules that allow their extensive application in the delivery of drugs, genes, and proteins as well as for the designing and manufacturing of devices for the delivery of drugs (Coelho et al., 2010). Dendrimers have been used as one of the novel carriers for delivering drugs and other therapeutics to the selective targets (Frechet and Tomalia, 2001; Newkome et al., 2001). Dendrimers are the new group of well-defined, immensely branched, nanoscopic macromolecules with excellent biological compatibility and structural uniformity.

In 1978, dendritic structures were first designed and successfully synthesized (Buhleier et al., 1978). Another breakthrough occured when highly functionalized macromolecules referred as "Stardust Polymer" were synthesized by chemical bridging of a new type of oligomer, which was coined as "dendrimer." These were the first synthesized polyamidoamine dendrimers (PAMAM) that were unique and different from classical oligomers/monomers by their astonishing symmetry, hyper-branching, and maximized terminal functionality (Tomalia et al., 1985). Similar type of mono-cascade macromolecules were synthesized and called as "Arborols" (Newkome et al., 1985).

Dendritic architecture possesses unique characteristics such as monodispersity (Mody et al., 2014), modifiable surface functionality (Patri et al., 2005), membrane transport efficiency, high drug payload (Satija et al., 2007), biocompatibility, and well-defined molecular structure and composition (Jain and Asthana, 2007), which explain its enormous potential in drug delivery over the other carrier systems. Several dendrimer formulations have been developed and explored with the rationale of sustaining the drug delivery (Gajbhiye et al., 2009b) and increasing the solubilization of drugs (Prajapati et al., 2009; Jain and Tekade, 2013). Dendrimer-based delivery of siRNA therapeutics has also been extensively explored (Tekade et al., 2015a). These novel architects have been comprehensively studied as carriers for drug targeting to specific sites. Usually, dendrimer architecture consists of three topological domains: (a) A central core comprising an atom or a molecule with a minimum of two identical functional groups; (b) Branches with several interior repeating units structured geometrically in a sequence of radically aligned layers known as "generations"; (c) Terminal groups that determine the surface properties of dendritic structure (Kesharwani et al., 2014a, 2015a; Martinho et al., 2014) (Figure 7.1).

Dendrimers possess better control of dispersity as well as lesser dependence on the polymerization conditions in comparison with polydisperse polymers. Moreover, compared with the linear polymers, the solubility of the dendrimers has been reported in a range of solvents, with reasonably lesser viscosity along with controlled hydrophilic, lipophilic, or amphiphilic characteristics. Dense outward (and inward) architecture of dendrimers provides specific sites for the attachment of drug molecules (Duncan et al.,

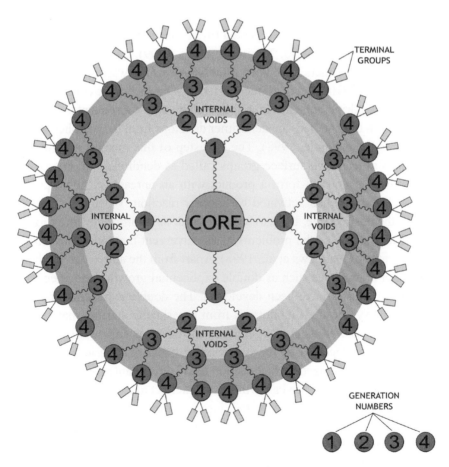

FIGURE 7.1 Structured diagram of dendrimer presenting its major domains: (a) core, (b) generations, (c) terminal groups, and (d) internal voids.

2006). Low-generation dendrimers can encapsulate hydrophobic drugs into their internal cavities. Dendrimers exhibit an exceptional attribute of actual mimicry of globular proteins as their sizes and contours matched with some proteins. They are also termed as "artificial proteins," because of their systematic, electrophoretic, biomimetic, and dimensional length scaling properties (Esfand and Tomalia, 2001).

Two major approaches have been developed to synthesize dendrimers since 1979. One approach is called as "divergent method," which was introduced by Tomalia in which the dendrimer develops from a multifunctional core to the periphery. This method involves the reaction of the core with the reagent (2 moles or more) having a minimum of two branching sites. The reactive sites thus liberated results in the formation of first-generation dendrimers. The procedure is repeated till the desired size of the dendrimer is achieved (Tomalia, 1996; Nanjwade et al., 2009). Another strategy for synthesizing dendrimers have been established, which was termed as "convergent method" (Hawker and Frechet, 1990). This technique constitutes the building of dendritic branches first, followed by their subsequent attachment to the core. In this method, the dendrimer develops from the periphery to the core (Archut and Vogtle, 1998). The first step of this method involves the reaction of at least two surface groups called as dendrons with one monomer (branching unit) to form a product with an unreactive "focal point." The focal point is then activated by another reaction. When the dendrons grow to the desired size, they are connected to an appropriate core to form a dendrimer. Some of the problems of the divergent method are overcome by this strategy (Matthews et al., 1998). Apart from the above two methods, some other approaches such as double exponential growth, lego chemistry, and click chemistry have been developed. The double exponential method involves the synthesis of monomers from a particular starting material for the divergent as well as convergent growth, which is somewhat related to a rapid growth process used for synthesizing linear polymeric materials. A trimer is then synthesized by the reaction the two resultant products, that can be utilized to get the desired growth (Kawaguchi et al., 1995).

Lego chemistry was discovered for dendrimer preparation with the aim to simplify the synthetic procedure concerning duration and cost. This method involves the synthesis of phosphorus dendrimers from highly functional cores and branched monomeric units. After some modifications to the general synthetic scheme, a new procedure is developed in which the terminal functional groups can be multiplied from "48 to 250' in a single step

(Maraval et al., 2003). Click chemistry involves an accelerated growth strategy that permits dendritic growth without the activation or deprotection of the peripheral end groups. High chemical yield, simple reactions, and high purity of the resulting dendrimers are the major advantages of this approach.

An outburst of interest in the dendrimer-based formulations for site-specific drug delivery has led to this technology as a promising, safe, and rapidly expanding in the the area of modern drug-delivery system. Some important patents have been issued in the field of dendrimer-based drug targeting as shown in Table 7.1. In this chapter, an attempt has been made to describe strategies for targeted drug delivery based on dendrimers. A brief introduction to the various targets along with an emphasis on the synthesis and evaluation of delivery systems has been focused. Molecular inclusion and binding interactions between guest molecules and dendritic molecules have also been highlighted.

7.2 DENDRIMERS-BASED TARGETED DRUG DELIVERY TO THE BRAIN

Drug targeting to the particular sites of the brain always represents a herculean challenge for the scientist community and pharmaceutical industry. The existence of highly specific biological barriers in the brain makes it the least accessible organ for delivering therapeutic agents (Scherrmann, 2002). The blood–brain barrier (BBB) is considered to be complex, highly selective, anatomical, and physiological barrier that protects the brain against invading organisms and noxious circulating substances. The BBB consists of tight junction capillary endothelial cells that restrict the entry of harmful blood-borne substances like xenobiotics and endogenous molecules into the brain to maintain homeostasis of neuronal microenvironment for brain function (Stamatovic et al., 2008). The highly selective BBB create obstacles in the treatment of neurological disorders (Laksitorini et al., 2014) with significant consequences such as Huntington's disease, gliomas, Parkinson's disease, Alzheimer's disease etc. Several neurological disorders remain untreated by available therapies because therapeutic agents fail to cross the selective BBB or other specialized CNS barriers (Chen and Liu, 2012). The endothelial cells of BBB linked via tight junctions, along with other restricting factors that include enzymes, transporters, receptors, and efflux pumps that control and minimize the entry of about 98% of therapeutic agents into the brain (Aryal et al., 2014).

TABLE 7.1 List of Important Patents Claimed in the Field of Dendrimer-based Drug Targeting

Title	Patent Number	Year	Inventor
Dendrimers for use in targeted delivery	US20050019923 A1	2005	Ijeoma Uchegbu, Avril Munro, Andreas Schatzlein, Alexander Gray, Bernd Zinselmeyer,
Targeted dendrimer-drug conjugates	US9345781 B2	2016	Mohamed E. H. El-Sayed, William Ensminger, Donna Shewach
Psma-targeted dendrimers	US20130336888 A1	2013	John W. Babich, John L. Joyal, Craig Zimmerman
Flexible, polyvalent antiviral dendritic conjugates for the treatment of HIV/aids and enveloped viral infection	WO2009038605 A3	2009	Sriram Subramaniam, Adam Bennett
Bioactive and/or targeted dendrimer conjugates	US5714166 A	1998	Donald A. Tomalia, James R. Baker, Roberta C. Cheng, Anna U. Bielinska, Michael J. Fazio, David M. Hedstrand, Jennifer A. Johnson, Donald A. Kaplan, deceased, Scott L. Klakamp, William J. Kruper, Jr., Jolanta Kukowska-Latallo, Bartley D. Maxon, Lars T. Piehler, Ian A. Tomlinson, Larry R. Wilson,Rui Yin, M. Brothers II Herbert
Dendrimer based compositions and methods of using the same	US20090053139 A1	2009	Xiangyang Shi, Suhe Wang, James R. Baker Jr.
Methods for tumor treatment using dendrimer conjugates	US20060204443 A1	2006	Hisataka Kobayashi, Peter Choyke

Some other specialized neural barriers that require attention includes the blood-cerebrospinal fluid barrier, blood–nerve barrier, blood–retinal barrier, blood–labyrinth barriers, etc., which adds further complexities for drug molecules to enter the brain (Neuwelt et al., 2008; Banks, 2012). Despite the complex nature of the BBB, the expression of several receptors and transporters on the brain endothelial cells can mediate the transport particular molecules to the brain (Omidi and Barar, 2012). Three main transport mechanisms across the BBB includes: carrier-assisted transport, receptor-based transport, and active efflux transport. Carrier-assisted transport of includes transporters like glucose transporter (GLUT1), large neutral amino acid transporter (LAT1), cationic amino acid transporter (CAT1), mono-carboxylic acid transporter (MCT1), concentrative nucleoside transporter (CNT2), nucleobase transporter (NBT), and choline transporter (CHT) (Tsuji and Tamai, 1999; Guo et al., 2012; Gao, 2016). These transporters are expressed on the endothelial cells, which can bind exclusively and transport selective molecules into the brain (Gabathuler, 2010).

Some receptors are also overexpressed on the BBB that can bind with corresponding ligands and mediate transport of specific molecules across the BBB and trigger subsequent internalization into the brain cells. Receptor-based transport includes receptors like transferrin receptor (TfR), insulin receptor, leptin receptor, nicotinic acetylcholine receptor, low-density lipo-protein receptor-related proteins 1 and 2 (LPR1 and LPR2), etc. (Jones and Shusta, 2007; Wang et al., 2009; Lajoie and Shusta, 2015). In the past few decades, considerable attention has also been paid on active efflux transport at the BBB. ATP-binding cassette (ABC) transporter like P-glycoprotein, breast cancer resistance protein (BCRP), and multidrug resistance protein (MRP) family were thoroughly investigated for exploring the possibilities of targeting drugs to the brain to generate new therapeutic alternatives for vari-ous CNS disorders (Kusuhara and Sugiyama, 2005; Loscher and Potschka, 2005; Pardridge, 2007).

Several strategies were developed for transporting the therapeutic agents across the brain in a site-specific manner. Among all the approaches, den-drimers have emerged as promising vectors with a great potential of deliver-ing the drugs into the CNS. In an investigation, PAMAM dendrimers have been tested to deliver the therapeutics to the targeted cell in the brain in an animal model. Significant uptake and selective localization of dendritic carriers were observed in the brain of the injured animal. Two dendrimer conjugates were prepared, one with N-acetyl cysteine and the other with val-proic acid. Both the conjugates were investigated in a combination therapy in a model of the large animal. Significant activity of the conjugates was observed in comparison with the high-dose conventional therapy with val-proic acid and N-acetyl cysteine (Mishra et al., 2014).

In another study, a water-insoluble antipsychotic drug haloperidol was targeted to the brain via intranasal and intraperitoneal routes by using

PAMAM dendrimer-based formulation of the drug in a rat model. Considerably increased distribution of haloperidol was observed in the brain as well as plasma in comparison with the conventional haloperidol formulation. Lower doses of the dendrimer conjugate produce comparable behavioral responses to that of conventional haloperidol formulations (Katare et al., 2015). Various dendrimer-based drug-delivery systems for targeting specific transporters like GLUT1, TfR, LPR1, LPR2, etc. have been successfully formulated and evaluated.

7.2.1 DENDRIMER-BASED TARGETING OF TRANSFERRIN RECEPTOR

A transmembrane glycoprotein TfR is mainly overexpressed on the brain endothelial cells, but it is also found on hepatocytes, erythrocytes, and proliferating cells. TfR primarily mediates iron transport to the brain, which is bound to transferrin by receptor-mediated endocytosis (Roberts et al., 1992; Moos and Morgan, 2000; Visser et al., 2004). In the last 20 years, extensive research has been done on Tf (Transferrin) as a ligand for delivering drug molecules to the brain (Dufes et al., 2013). Dendrimer-based targeting of transferrin receptor was thoroughly studied for target specific delivery of genes and therapeutic agents (Figure 7.2).

Diversified transferrin receptor ligands were discovered due to their capabilities to assist transport across BBB, which includes Lactoferrin (Lf), (Huang et al., 2008) HAIYPRH- T7 (a specific peptide for transferrin receptor), and OX26 (a monoclonal antibody) (Gosk et al., 2004). Lf was explored as a ligand for targeting the brain in the formulation consisting of PAMAM dendrimer conjugated to Lf through PEG spacer. The conjugate exhibits concentration-dependent cellular uptake. In comparison with PAMAM conjugated with Tf, the brain uptake of PAMAM conjugated with Lf (*in vivo*) was higher (about 2.2 times). The conjugation of Lf significantly increases the gene expression in endothelial cells of the brain (Huang et al., 2008).

In an investigation, a dendrimer-based novel brain-targeting gene vector was reported, in which polyamidoamine (PAMAM) dendrimer was conjugated with Tf via PEG (polyethylene glycol) spacer, which is a hydrophilic polymer. PEG was incorporated to enhance the biocompatibility of the gene vector as an additional function. The vector was further conjugated

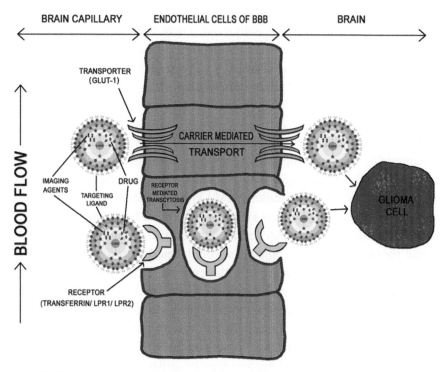

FIGURE 7.2 Dendrimer-based targeting of transferrin receptor and GLUT1 transporter.

to the plasmid DNA. The biodistribution, transfection efficiency in the brain endothelial cells, and potency of the targeting gene vector were successfully evaluated. The PAMAM-PEG-Tf vector exhibits concentration-dependent cellular uptake. The uptake of this vector in the brain was about 2.5 times higher than that of PAMAM as well as PAMAM-PEG conjugates *in vivo*. The quality analysis suggests the distribution of gene expression for PAMAM-based formulated vector in many areas of the brain (Huang et al., 2007). Lectins like wheat germ agglutinin (WGA) were also revealed as the ligand for brain-targeting and shows marked affinity for the brain capillary endothelial cells (Smith and Gumbleton, 2006) and cancer cells (Mo and Lim, 2005).

Synthesis of generation 4 PEGylated PAMAM dendrimer-based dual-targeting nano-constructs conjugated with WGA and Tf on the surface and encapsulated with doxorubicin (DOX) within the void space of the dendrimer was reported. The rate of growth of C6 cancer cells was considerably inhibited by the novel carrier, and the cytotoxicity of DOX was found to be reduced considerably in the healthy cells. There was a significant increase in DOX accumulation in the tumor cells. An increase in transport ratio of dual-targeting nano-construct has also been reported in comparison with free DOX- and PAMAM-based conjugates tailored with WGA and Tf alone. The formulation was also reported to demonstrate its inhibitory effect on C6 cells (He et al., 2011).

In another investigation, generation 4.5 PAMAM-based dendritic nanoparticles were developed for targeting opioid peptide DPDPE into the brain. DPDPE was conjugated to PEG-modified generation 4.5 PAMAM dendrimer or with PEG-modified PAMAM dendrimer tailored with monoclonal antibody OX26, which is a transferrin receptor ligand. In this research, buccal mucosa was investigated as an optional site for administering therapeutic nanoparticles. The conjugates were labeled with 5-aminoacetamido fluorescein (AAF) or fluorescein isothiocyanate (FITC) for the evaluation of their transport across buccal mucosa quantitatively. Administration of sodium glycodeoxycholate along with the formulation or with the application of gelatin/PEG-based mucoadhesive system the permeability of dendritic nanoparticles was increased by several folds. The findings suggest that transport through the buccal mucosa may be utilized as a possible route for delivering dendritic formulation to the CNS (Yuan et al., 2011).

Tamoxifen (TAM), an anti-estrogen, (Pearson et al., 1982) directly inhibits transmembrane glycoprotein (P-170) pump, and it has been reported to reverse multidrug resistance in the cells expressing P-glycoprotein (Callaghan and Higgins, 1995). Based on the concept, dual-targeting system was synthesized by conjugating the fourth generation PAMAM dendrimer with transferrin on the surface and encapsulating TAM within the voids. TAM and Tf were used as targeting groups to improve the transport through the BBB as well as to enhance the drug accumulation in glioma cells. DOX was attached via the acyl hydrazone bond to the amino group of the PAMAM dendrimer. The results of the study revealed higher transportation ratios in an *in vitro* BBB model for the dual-targeting system and the considerable decrease in viability of C6 glioma cells (Li et al., 2012).

In a research, generation 3-polypropylenimine dendrimer modified with transferrin has been synthesized for transporting plasmid DNA across the BBB. On intravenous administration of the transferrin-bearing dendrimer conjugate, the gene expression was found to be more than twice in comparison with the unmodified conjugate. Threefold higher gene expression has been reported in the brain as compared to the other organs tested (Somani et al., 2014).

7.2.2 DENDRIMER-BASED TARGETING OF GLUCOSE TRANSPORTER (GLUT-1)

Overexpression of glucose transporter GLUT-1 is found in BBB (Pardridge et al., 1990) as well as on the tumors cells (Luciani et al., 2004). GLUT-1 was discovered as one of the transporters that could be efficiently utilized for targeting therapeutic agents to the brain (Figure 7.2). In a study, D-glucosamine-modified polyether-copolyester (PEPE) dendrimers conjugated with methotrexate (MTX) were synthesized and investigated for the targeting of gliomas. D-glucosamine was incorporated as a ligand to potentiate the transportation across the BBB and specific targeting of tumor cells. The transport of MTX across the BBB after loading in dendrimers was three to five times, which was further increased by glucosylation. It was also found that the distribution of dendrimers conjugated in the tumor cells takes place within 6 h in comparison with nonglycosylated dendrimers in 12 h. Accumulation of high amounts of MTX in tumor spheroids and enhanced permeability across the BBB suggest the effectiveness of glucosylated PEPE dendrimers (Dhanikula et al., 2008).

7.2.3 DENDRIMER-BASED TARGETING OF LOW-DENSITY LIPOPROTEIN RECEPTOR-RELATED PROTEIN 1 AND 2 (LRP1 AND LRP2)

Expression of LRP1 and LRP2 is found in brain endothelial cells (Uchida et al., 2011). They have been found to bind with strong affinity to various ligands like lactoferrin, proteases, and protease/inhibitor complexes, lipoprotein lipase, intracellular proteins like calreticulin, etc. (Lillis et al., 2008). Angiopep-2 was shown to have an exceptional characteristic to cross the BBB, and it is a ligand for LRP1 (Demeule et al., 2008). Angiopep was investigated for targeting gene to the brain, and based on the concept, PAMAM-based nanoconstruct was synthesized and engineered with angiopep via PEG and loaded with DNA. The resulting nanoparticles were reported to exhibit higher expression of the gene in the brain and better transport efficiency across BBB in comparison with unmodified nanoparticles (Ke et al., 2009).

In another study, PAMAM-based dual targeting nanoparticles were synthesized by utilizing Angiopep-2 as a ligand that was conjugated via bifunctional polyethylene glycol (PEG) and further linked with DNA. The resulting nanoparticles showed higher *in vivo* biodistribution in the brain and targeted tumor cells in comparison with PAMAM-PEG/DNA nanoparticles. The findings suggest that Angiopep-2 can be utilized as a ligand for brain targeting (Huang et al., 2011). In another investigation, poly-l-lysine dendrigraft modified with angiopep via PEG and linked to a plasmid DNA has been prepared and evaluated for its neuroprotective effects in a Parkinson's disease. The reported nanoparticles reveal higher gene expression and uptake in cells of the brain than unmodified nanoparticles. Improvement of the locomotor activity and apparent revival of the dopaminergic neurons were also reported (Huang et al., 2013).

7.3 DENDRIMER-BASED TARGETING OF THE PULMONARY SYSTEM

The pulmonary route has been investigated as a potential route for delivering therapeutic agents into the pulmonary system for the management of several diseases like asthma, chronic bronchitis, cystic fibrosis, chronic obstructive pulmonary disorders (COPD), etc. (Labiris and Dolovich, 2003). Aerosolized formulations have been used for many years for drug administration to the pulmonary system. Several particulate drug carriers have been explored for delivering therapeutic and diagnostic agents as aerosolized formulations. The short duration of clinical effects and high dosing frequency are the associated drawbacks with aerosol systems.

An important consideration in pulmonary drug targeting involves the development of functional carriers having the ability to bind to the receptors for targeted drug delivery (Kurmi et al., 2010). Even though, pulmonary delivery of therapeutic agents is considered as controversial especially after the withdrawal of a marketed product: Exubera, (Heinemann, 2008). Several attempts have been made to explore the pulmonary route for delivering macromolecules like insulin, calcitonin, growth hormones, somatostatin, etc., in the systemic circulation. The route offers some exceptional advantages of the sufficiently large surface area, considerable vascularization, good capacity for exchange of solutes, and thin alveolar epithelium (Patton, 1996; Agu et al., 2001; Patton and Byron, 2007). Over the past decades, increasing attention has been focused on nanocarriers like nanoparticles, nanoemulsions, liposomes, or dendrimers for site-specific pulmonary drug delivery (Mansour et al., 2009). Nanoparticulates deposited in the respiratory tract, primarily by the inhalation of the aerosolized formulation, reaches the targeted organs by rapid translocation and their subsequent uptake may occur into the systemic circulation (Kayat et al., 2011). Dendrimers were thoroughly explored as systemic delivery carriers for the therapeutic agents to the pulmonary route.

Gene therapy may be considered as a viable strategy for the treating various disorders of lungs like cystic fibrosis (CF), acute respiratory distress syndrome (ARDS), cancer, etc. (Griesenbach et al., 2004). In an investigation, branched polyethyleneimine (PEI) and fractured dendrimers were investigated for *in vivo* delivery of gene to the lungs. The results suggested the superiority of PEI for delivering the gene to the airways over fractured dendrimers, as the gene transfer efficiency mediated by PEI is about twice in magnitude in comparison with fractured dendrimers (Rudolph et al., 2000). In another study, gene therapy was emphasized as Conti et al. designed generation four PAMAM dendrimer complexed with siRNA for effectively delivering siRNA to the epithelial cells of the lungs. The result shows that epithelial cells (A549), as well as silence genes, were efficiently targeted by the siRNA dendriplexes (Conti et al., 2014).

Deep vein thrombosis refers to the development of blood clots in the deep veins (Kesieme et al., 2011). Enoxaparin is an anticoagulant drug classified as low-molecular-weight heparin (LMWH) (Ibbotson and Goa, 2002) and is currently used for treating deep vein thrombosis (Mismetti et al., 2005). In a study, enoxaparin was complexed with PAMAM dendrimers for effective delivery to the pulmonary system. The dendrimer-based formulations were evaluated in animal models of deep vein thrombosis. The bioavailability of enoxaparin increased when conjugated with positively charged dendrimers, while no effect was observed with negatively charged dendrimer. Enoxaparin linked with PAMAM dendrimer (generations 2 and 3) was found effective in a rodent model of deep vein thrombosis. The findings suggest that the pulmonary absorption of enoxaparin was enhanced by the PAMAM dendrimers (Bai et al., 2007). In another study, PEGylated dendrimers were synthesized and utilized as nanocarriers for the pulmonary delivery of LMWH. About 40% entrapment efficiency was reported for LMWH. Pulmonary absorption of LMWH entrapped in PEG-dendrimer was found to increase significantly, and the effectiveness of the formulation in reducing the weight of thrombus was found to be similar to that of subcutaneously administered LMWH in a rodent model. The half-life of the nano-formulation was found to be 2.4 times greater than that of saline control formulation of LMWH (Bai and Ahsan, 2009).

Ryan et al. in 2013 investigated PEG-modified polylysine dendrimers as carriers for the pulmonary system. A variety of polylysine dendrimers complexed with PEG of different chain lengths were evaluated for pharmacokinetic behavior and pulmonary stability in rat models. The findings interpret that chain length of PEG along with the overall molecular weight of the construct influence the absorption of the formulation from the lungs as well as the stability of formulation in rat models. Comparatively small dendrimers exhibit better absorption kinetics with enhanced bioavailabilities (Ryan et al., 2013).

Beclometasone dipropionate (BDP) is a poorly water soluble corticosteroid that is frequently used against asthma via inhalation (Ahmadiafshar et al., 2010). In a study, aerosol-based PAMAM dendrimers were investigated for the pulmonary delivery of BDP. Characterization of PAMAM-BDP complexes was done to evaluate solubility, complexation efficiency, and aerosolization properties and release profile of the drug. It was revealed by the *in vitro* release studies that complexation of BDP with PAMAM results in the sustained release of the drug. It was found that the aerosol characteristics are highly affected by the design of nebulizer rather than the generation of a dendrimer. The work suggests the use of PAMAM dendrimers as novel carriers for poorly soluble corticosteroids such as BDP in pulmonary drug delivery (Nasr et al., 2014).

Tuberculosis (TB) is primarily a pulmonary disease, and it is considered as one of the major and oldest public health problem caused by *Mycobacterium tuberculosis*, which principally affects the lungs (Smith, 2003; Harries and Dye, 2006; Sasindran and Torrelles, 2011). Rifampicin is a potent first-line anti-tuberculosis drug. However, it has been associated with serious limitations like poor water solubility, variable bioavailability, instability in the gastrointestinal tract, and requirement of high dose. These limitations are particularly associated with conventional dosage forms (Henwood et al., 2001; Shishoo et al., 2001; Singh et al., 2001). In a study, dried rifampicin was co-sprayed with PAMAM dendrimers (generations 1, 2, and 3) to prepare microspheres to deliver rifampicin into lung tissues. Desirable *in vitro* aerodynamic behavior was exhibited by the microspheres. Evaluation of various pharmacokinetics parameters was done, and it was revealed that the t_{max} value of the formulations was decreased while the C_{max} value was increased. The lower generation PAMAM dendrimers significantly affect the pharmacokinetics and bioavailability of rifampicin, which makes them suitable carriers for pulmonary drug delivery (Rajabnezhad et al., 2016).

7.4 DENDRIMERS-BASED OCULAR DRUG DELIVERY

Delivering therapeutic agents to the ocular tissue represent a challenging task for scientists owing to its complex structure and resistance to drugs and foreign substances. Conventional drug delivery systems encounter several issues like lacrimal drainage, tear dilution, low corneal permeability, a transient corneal residence time, etc. Despite the associated drawbacks, conventional formulations like eye drops, eye ointments, and eye suspensions account for about 90% of the marketed products (Lang, 1995). Improvements have been made to lengthen the residence time in the cornea and to enhance the bioavailability of ophthalmic preparations by adopting different strategies like enhancing the viscosity of the formulation, adding various permeation enhancers, incorporation of cyclodextrins, etc. But significant improvement has not been observed by such approaches (Gaudana et al., 2009).

Delivering drugs to the posterior eye segment is considered to be a challenging task in ophthalmology. Many ophthalmic disorders like glaucoma, macular degeneration, different types of retinitis pigmentosa and diabetic retinopathy are considered to be responsible for the damage of posterior ocular segment, which may cause diminished vision and blindness in several cases (Ranta et al., 2010). Over the last few decades, remarkable transformations have been observed in the expanding area of ophthalmic drug targeting with the application of novel drug-delivery carriers like dendrimers, nanoparticles, micelles, liposomes, and gene delivery systems. Use of ocular implants, iontophoresis, and ultrasound has also gained momentum in assisting drug delivery to the eye. Unprecedented advancement of the novel drug-delivery devices has revolutionized this field with their usage in a range of ophthalmological purposes and ocular drug targeting (Del Amo and Urtti, 2008; Eljarrat-Binstock et al., 2010; Maheshwari et al., 2015). Recent years have witnessed several types of research and investigations for the development of dendrimeric systems to provide effective delivery of several therapeutic agents to intraocular targets by noninvasive delivery techniques.

Dendrimers offer several advantages like better water solubility, high permeability, enhanced bioavailability, and biocompatibility as potential carriers for delivering drugs to the ocular tissue.

7.4.1 DENDRIMER-BASED TARGETING OF THE ANTERIOR EYE SEGMENT

Associated drawbacks of the conventional formulation account for the evolution of innovative formulations strategies for targeting the anterior segment of the eye.

7.4.1.1 Application of Dendrimers in Corneal Scar Tissue Formation

Glaucoma infiltration and cataract surgeries sometimes result in scar tissue formation and inflammation. In an investigation, two different dendrimeric formulations were designed and investigated in an animal model which involves glaucoma filtration surgery for the formation of scar tissue in rabbits. One formulation was prepared by conjugating generation 3.5 PAMAM dendrimer with D(+)-glucosamine and other by conjugating D(+)-glucosamine 6-sulfate. Both the formulations were found to increase the success rate of surgery. Lipopolysaccharide (LPS)-induced pro-inflammatory chemokines and cytokines in human dendritic cells and macrophages were successfully inhibited by the conjugates. Proinflammatory mediator synthesis was reported to be inhibited dendrimer glucosamine conjugate. The dendrimer conjugated with glucosamine 6-sulfate was found to inhibit fibroblast growth factor-controlled proliferation and neo-angiogenesis of human umbilical vein endothelial cells. Furthermore, any signs of tissue inflammation, systemic microbial infections and toxicity have not been observed with these conjugates throughout the experimental period (Shaunak et al., 2004).

7.4.1.2 Application of Dendrimers in Open-Angle Glaucoma

Open-angle glaucoma is considered as one of the primary reasons of blindness affecting almost 66.8 million individuals of the world's population. Destruction of the optic nerve takes place in glaucoma, which is the cause of blindness (Beidoe and Mousa, 2012). Permeability issues and lacrimal drainage associated with conventional pilocarpine formulations accounts for the investigation of novel delivery systems that can deliver the therapeutic agent precisely to the targeted location. In a research, PAMAM dendrimers were studied as carriers for delivering tropicamide and pilocarpine to the ocular tissue. Miotic and mydriatic activities were evaluated on albino rabbits, and the results suggest that the bioavailability of pilocarpine and tropicamide was considerably enhanced by the PAMAM dendrimer solutions. Prolonged residence time was also reported with the solutions of dendrimers having hydroxyl and terminal carboxylic groups. The physicochemical characteristics of dendrimer formulations like refractive index, surface tension, pH, etc., were comparable to that of marketed Carbopol eye drops (Vandamme and Brobeck, 2005).

In another study, Spataro and co-workers designed phosphorous-containing dendrimers as novel carriers for delivering an anti-hypertensive drug carteolol to the eye to treat glaucoma. No irritation was reported after the instillation of dendrimers-carteolol conjugates in the eyes of the rabbit. The penetration of generation zero dendrimer-carteolol conjugates inside the aqueous humor was comparable to that of carteolol alone. However in the case of second generation dendrimer-carteolol conjugate, the penetration inside the eye was considerably higher than that of carteolol alone, suggesting the enhanced bioavailability (Spataro et al., 2010).

Conventional eye drops formulations of various antiglaucoma agents like timolol, brimonidine, latanoprost, travoprost, etc., have low bioavailability due to the tear drainage and low permeability of corneal epithelium. Their frequent instillation leads to patient noncompliance. In an investigation, PAMAM and poly(lactic-co-glycolic acid)-based nano-formulation was reported for delivering timolol maleate and brimonidine. The formulation was evaluated in adult male rabbits. Significantly higher concentrations of both the drugs were observed in the aqueous humor and cornea without any signs of inflammation or discomfort. The findings suggest that the formulation causes sustained and effective reduction of intraocular pressure. It has been demonstrated that the formulation increases mucoadhesion and enhance the bioavailability of conjugated drugs without producing cytotoxicity (Yang et al., 2012). In another investigation, photocurable dendrimer derivatives were synthesized in which PAMAM dendrimers were linked with PEG and photoreactive groups like acrylate group. Crosslinking of the reactive acrylate groups was accomplished by the exposure dendrimer derivatives to UV light resulting in the formation of dendrimer hydrogel (DH). The formed DH was investigated for delivering brimonidine and timolol maleate, which are antiglaucoma drugs, to the ocular tissue. *In vitro* release studies conclude the sustained release of both drugs for 56–72 hours. The formulation was also reported to be nontoxic and mucoadhesive to human corneal epithelial cells (HCET). Marked rise in intracellular uptake of both the drugs by HCET cells was observed. The transport of drugs through the bovine corneal endothelium was also increased by the application of DH formulation, as compared to conventional formulation. With an application of the gel formulation, appreciably increased concentration of timolol maleate was observed in the corneal endothelium, epithelium, and stroma according to *ex vivo* bovine eye studies (Holden et al., 2012).

7.4.1.3 Application of Dendrimers in Corneal Infections

Many of the viruses, bacteria, parasites, and fungi are capable of attacking the surface or interior of the eye, leading to corneal infections that may result in corneal opacity and blindness if not identified quickly and managed appropriately (Upadhyay et al., 2009). Fungal keratitis, one of the infections of the cornea, is considered to be the main eye blinding disease in China (Xie et al., 2001). The bioavailability of less than 5% of the applied dosage of topically instilled eye drops presents a challenging task in curing corneal infections. Sufficient penetration of the antimicrobial drugs into the eye through systemic administration is also hindered due to blood, retinal, and aqueous barriers (Velpandian, 2009).

Dendrimer-based drug delivery of antimicrobials in corneal infections offers a favorable option for improved delivery of therapeutic agents and tissue targeting. A new category of dendrimers with trioyl branching and terminal guanidine groups was reported, which are termed as dendrimeric polyguanidilyated translocators (DPTs) (Durairaj, 2009). In a study, a DPT vehicle for ocular drug delivery of gatifloxacin (GFX) was developed. The solubility of GFX was reported to be increased four times in DPT-GFX formulation. Dendrimer formulation was found to be significantly permeable across human corneal epithelial cells as well as enhance the transport of GFX by 40% across isolated bovine sclera and sclera-choroid-RPE. *In vivo* studies suggest that the formulation improves the corneal and conjunctival delivery of GFX when compared to GFX alone, potentially allowing decreased frequency of administration. The formulation also demonstrates the faster rate of killing *Staphylococcus aureus* (methicillin-resistant) in comparison with GFX solution (Durairaj et al., 2010).

7.4.2 DENDRIMER-BASED TARGETING OF THE POSTERIOR EYE SEGMENT

Drug targeting to the posterior eye segment necessitated the drug molecule to pass through the fat-water-fat structure of the corneal barrier with minimum wastage through tear washout and ensured systemic absorption. This represents a challenging task for the researchers to effectively deliver the therapeutic agents to the posterior segment of the eye. Several investigations were performed to design novel formulations, which can cross the ocular barriers for the efficient treatment of the disorders of posterior eye segment like uveitis, cytomegalovirus retinitis, retinoblastoma, uveal melanomas, neurodegenerative, and neovascular retinal diseases, etc. (Shah et al., 2010). Repeated injection of drugs into the vitreous cavity may result in problems such as vitreous hemorrhage, endophthalmitis, patient noncompliance, retinal injury, etc. (Sheardown and Saltzman, 2006). Dendrimer-based systems have been explored for the efficient delivery of drugs to the ocular tissue, which can provide prolonged release of drugs with high specificity for targets while reducing the frequency of administration as well as associated side effects.

7.4.2.1 Application of Dendrimers in Retinoblastoma and Uveal Melanomas

Primary cancers of the eye like uveal melanoma and retinal neoplasm retinoblastoma present a high risk of complications worldwide (Eagle, 2013). In an investigation, a subconjunctival injection of dendrimeric nanoparticles loaded with carboplatin was evaluated in transgenic murine retinoblastoma in mice. Results suggest the enhanced effectiveness of sub-conjunctival nanoparticle formulation in comparison with carboplatin in aqueous solution. The formulation has been reported to cross the sclera to reach the contralateral eye through the local vasculature and sustained the release of the drug by retaining in the tumor vasculature for a longer period. Dendrimer-carboplatin conjugate was reported to decrease the ocular mean tumor burden considerably in comparison with the control and conventional carboplatin group (Kang et al., 2009).

Photodynamic therapy (PDT) is one of the latest techniques used alternatively in the treatment of retinoblastoma owing to its therapeutic effectiveness and lesser adverse effects than chemotherapy and radiotherapy (Stephan et al., 2008; Aerts et al., 2010). The efficacy of PDT depends on photosensitizers' intracellular accumulation in the tumor tissues along with their exposure to light. Porphyrin-based compounds have been explored as photosensitizers in PDT. On the other hand, mannose and lectin-like receptors are found on the tumor cell's surface, and glycopolymers were reported to interact with surface mannose and lectins. This interaction can be utilized for tumor targeting. Both the above concepts were utilized to develop porphyrin-based glycodendrimers with Concanavalin A as mannose-specific ligand protein, to target retinoblastoma. In this case, mannosylated dendrimers have been developed as photosensitizers for PDT, which exhibits preferential accumulation in malignant ocular tissue (Makky et al., 2011). In another study, Wang and co-workers synthesized glycodendrimeric phenylporphyrins, which demonstrated better photo efficiency, increased cellular uptake, and significant phototoxicity in retinoblastoma cells. The dendrimeric conjugate was reported to enhance the plasma proteins binding affinity and proved to be highly sensitive to PDT treatment of retinoblastoma (Wang et al., 2012).

7.4.2.2 Application of Dendrimers in Neurodegenerative and Neovascular Retinal Diseases

Several ocular complications like retinitis pigmentosa, macular degeneration, diabetic retinopathy, etc., may cause neurodegeneration of the retinal ganglion cells (Schmidt et al., 2008). Advancement in treatments for the retinal neurodegenerative disorders has led to the development and design of therapeutic dendrimer-drug conjugates. Retinal neurodegenerative disorders are associated with neuroinflammation. In research, Iezzi and co-workers investigated PAMAM dendrimers (generation 4) with terminal hydroxyl groups conjugated with fluocinolone acetonide for treating neuroinflammation of retina. The conjugates were reported to accumulate preferentially into activated microglia and photoreceptors upon intravitreal administration. The investigation revealed that the conjugates provided significant neuroprotection and reduced neuroinflammation significantly (Iezzi et al., 2012). In an investigation, PAMAM dendrimers were explored for the delivering neuroprotective genes to the retina. In another study, PAMAM-dexamethasone conjugates were developed as efficient carriers for genes along with the drug molecules. Dexamethasone plays a dual role of therapeutic agent and assists gene translocation into the nucleus by dilating the nuclear pore (Choi et al., 2006).

7.5 DENDRIMER-BASED CARDIOVASCULAR DRUG DELIVERY

Cardiovascular disorders (CVDs) are one of the main reasons of morbidity and death worldwide. Owing to changes in the lifestyle characterized with more stressful sedentary working hours, consumption of fat-rich fast foods, lack of proper exercise and chronic alcoholism and use of tobacco, CVDs are now inflicting more fatalities than earlier. CVDs are the group of diseases associated with pathological changes in structural, functional, and metabolic faculties of heart and blood vessels. Rheumatic heart disease; coronary heart disease; congenital heart disease; disease of blood vessels supplying to heart muscles, brain, arms and legs; cerebrovascular disease; peripheral arterial disease; damages to the blood vessels and heart valves due to rheumatic fever; abnormality of heart structure at birth; deep venous thrombosis and embolism; blood clots in blood vessels; diseases associated with fluid imbalance such as hypertension; changes in lipid metabolism leading to atherosclerosis are the main CVDs (Gothwal et al., 2015). Many types of research are being carried out in the CVDs sector to alleviate the existing conditions. Dendrimer-based delivery of cardiovascular drugs is one of the hot topics for researchers today to explore the optimized pharmacokinetic and pharmacodynamic profile of drugs, best suited to benefit the recipients (Sharma et al., 2015).

Simvastatin (SMV) is a most frequently prescribed drug to manage lipid disorders in cardiac patients. It belongs to statin category of drugs that inhibit 3-hydroxy-3-methylglutaryl coenzyme reductase (HMG-Co-A reductase), which regulates lipid metabolism (Kulhari et al., 2013). They reduce LDL (low-density lipoproteins) cholesterol level and tend to increase the HDL (high-density lipoproteins) cholesterol level (Crouse et al., 1999). *In vivo* investigations revealed that PAMAM dendrimer-based SMV formulation significantly lowered the increased levels of triglycerides and cholesterol in the experimental model of animals. Furthermore, kinetic profiles such as absorption, residence time, peak plasma SMV concentration, elimination time were modulated to demonstrate a controlled release pattern of the drug in comparison to plain SMV formulation (Kulhari et al., 2013). In another investigation, Qi et al. demonstrated that generation 5 PAMAM dendrimer-based SMV formulation showed better *in vitro* transepithelial transport in Caco-2 monolayer cell culture and improved oral absorption *in vivo* (Qi et al., 2015). Enhanced oral absorption is one of the key advantages with a dendrimer-based formulation of SMV, seen in experiments as it is less than 5% available through the oral route.

Probucol is another lipid lowering drug frequently used in CVDs. It is structurally different to lipid-lowering drugs. Probucol decrease the levels of cholesterol and very low-density lipoprotein cholesterol (VLDL-c) in plasma. It also decreases atheromatous plaques and improves restenosis. This atypical drug works on the antioxidant mechanism. Poor solubility and low rate of oral absorption limit the use of the drug probucol (Sawayama et al., 2002; Wakeyama et al., 2003; Yamamoto et al., 2011). In a study, it was demonstrated that PEGylated G5 PAMAM dendrimer conjugate considerably improved the rate of oral drug absorption and consequently enhanced pharmacodynamic properties of probucol (Ma et al., 2015).

Nifedipine belongs to the class of calcium-channel blocker (CCB), which was developed mainly for the treatment of angina symptoms, but later, it was used as an anti-hypertensive drug (Lundy et al., 2009). Devarakonda et al. found that Nifedipine conjugated with both amines as well as ester terminated G0-G3 PAMAM dendrimers showed enhanced solubility (Devarakonda et al., 2004). However, conjugation with G3 and G5 PAMAM dendrimers exhibited higher release profile of Nifedipine (Devarakonda et al., 2005). Several findings have revealed the usefulness of β blocker drug Propranolol. Propranolol is a substrate for the efflux transporter P-glycoprotein and it is poorly soluble (Kulhari et al., 2013). *In vitro* studies on cell lines have shown that PAMAM dendrimers (generation 3) increases the solubility of Propranolol and the conjugate was reported to circumvent efflux transporters, consequently enhancing the bioavailability of the drug. Furthermore, cytotoxicity of Propranolol was also lowered after conjugating with dendrimer (D'Emanuele et al., 2004). Successful efforts have also been made to deliver anti arrhythmic drug Quinidine. Quinidine consists of asterically hindered hydroxyl group that makes it almost nonreactive with traditional polymeric molecules. Because of the presence of the few functional groups, the reaction between quinidine and conventional polymers becomes quite difficult resulting in poor drug loading efficiency. In an investigation anionic (generation 2.5) and cationic (generation 3.0), PAMAM dendrimers covalently attached to quinidine via glycine spacer has been reported. The findings suggest enhanced drug loading efficiency and drug delivery by the dendrimer conjugate (Yang and Lopina, 2007).

Cardio-protective effects of various adenosine receptor agonists delivered through dendrimers have been extensively studied. Conjugations of different adenosine receptor agonists with generation-3 and generation-5 PAMAM dendrimers yielded diverse action such as platelet antiaggregatory action and cardioprotection in ischemia perfusion (Ivanov and Jacobson, 2008; Kim et al., 2008, 2009; Keene et al., 2010; Wan et al., 2011).

7.6   DENDRIMERS-BASED CANCER TARGETING

Targeted delivery of anti-cancer drugs has been a major challenge before the medical and scientific community. The acute toxicities of anti-cancer drugs on non-cancerous tissue and organs of recipients lead to various complexities. Bone marrow suppression, hair loss, hepatic and renal toxicity, ototoxicity, teratogenic effects, metabolic abnormalities, sexual dysfunction and infertility are major problems associated with anti-cancer therapy (Remesha, 2012; Schover et al., 2014). To achieve targeted delivery, scientists are working to enhance the availability of therapeutic agents on the tumor site for efficient uptake by the tumor cells. To accomplish this task, several strategies have been adopted such as by sustaining or delaying the drug release from the formulations for prolong actions; using drugs entrapped in lipid-based carrier like liposomes, solid lipid nanoparticles etc. for reduced toxicity and prolonged effect; designing safe and effective prodrugs for activation at the tumor site; and delivering the drugs anchored with targeting ligands, vectors and specific carriers to target the tumor (Zee-Cheng and Cheng, 1989). Several hybrid nanoformulations have been successfully developed and thoroughly investigated (Tekade and Chougule, 2013; Tekade et al., 2014). However, after the development of dendrimer-based systems, the targeted delivery of anti-cancer drugs has gained good momentum. In an investigation, methotrexate and all-trans retinoic acid-loaded dendrimer formulation was evaluated for cytotoxicity in HeLa cell lines, where the formulation was found more effective than conventional drug combination (Tekade et al., 2009a). In another study, Tekade et al. developed paclitaxel loaded dendrimer nanoparticles for cancer targeting. The formulation significantly inhibits the growth of cancer cells in various cell lines and shows increased cellular uptake as compared to conventional paclitaxel formulation. The formulation was also reported to be biocompatible and stable at physiological pH (Tekade et al., 2015b). In another research, doxorubicin-loaded PPI dendrimers and multi-walled carbon nanotubes conjugated with glycyrrhizin were developed. Both the formulations were found to inhibit the growth of cancer cells in MTT [3-(4,5-dimethylthiazol-2-Yl)-2,5-diphenyltetrazolium Bromide] cytotoxicity studies on HepG2 cells. Hemolytic toxicity of doxorubicin was reported to decrease significantly due to glycyrrhizin conjugation (Chopdey et al., 2015). Recently, dendrimers have been developed that allow the conjugation of drugs as well as targeting ligands like folic acid, monoclonal antibodies, and peptides for target specific drug delivery (Wolinsky and Grinstaff, 2008).

7.6.1 FOLIC ACID–DENDRIMER CONJUGATE-BASED CANCER TARGETING

Folic acid is a key intermediate product in purine metabolism and target of folate metabolic inhibitors, used as anti-cancer drugs. Several studies suggest that the high-affinity folate receptors are selectively overexpressed in many tumors (Elnakat and Ratnam, 2004). Folic acid has been explored as the efficient anchor in tumor targeting and has been continually used for targeting folate receptors found on the cancer cells. The idea was employed to investigate polyether dendritic compounds containing folate residues as drug carriers with specificity to the tumor cells (Kono et al., 1999).

PEGylated dendrimers attached with targeting moieties were also explored for cancer targeting. In a study, PEG-conjugated PAMAM or PEI dendritic cores were synthesized with branched poly(L-glutamic acid) chains. The dendritic construct was modified with folic acid as a targeting moiety and further linked with indocyanine green as diagnostic agent. The resulting formulation bound effectively and specifically to the tumor cells (Gajbhiye et al., 2007). In another research, folate-conjugated poly(propyleneimine) (PPI) dendrimer was designed to target the delivery of Melphalan. Dendrimers of various generations have been used, and it has been observed that in comparison with third and fifth generation PPI dendrimers, fourth generation dendrimer was proven to be a better carrier for folic acid-based targeting of the tumor (Kesharwani et al., 2015b). Folate-modified poly (propyleneimine)dendrimers bearing ethylenediamine core conjugated with MTX and all-trans retinoic acid have also been explored, which demonstrates enhance targeting, receptor upregulation, and lower IC50 values on HeLa cell lines (Tekade et al., 2008).

A targeted drug-delivery system using generation 5 PAMAM dendrimer for targeting MTX was also formulated. The terminal functionality of the dendrimer was modified by folic acid as a targeting ligand. The synthesized conjugates include covalently conjugated MTX with dendrimer as well as MTX dendrimer inclusion complex. Cytotoxicity and drug release characteristics of both the conjugates were compared. The findings conclude enhanced the solubility of the inclusion complex, but the suitability of dendrimer conjugate linked covalently for tumor targeting. Cytotoxicity tests revealed that dendrimer-based inclusion complex had similar activity as that of free drug *in vitro*. On the other hand, modified dendrimer conjugate exclusively killed KB cells (cell line with expression of folate receptors) by receptor-mediated endocytosis of the drug (Patri et al., 2005).

In another research, fifth generation PAMAM dendrimer having ethylenediamine core was covalently attached to fluorescein (for imaging), folic acid (for targeting), and MTX. The findings suggest the selective uptake and internalization of the formulation in KB cells. The dendrimer-based nano device enhanced the cytotoxic activity of methotrexate (Quintana et al., 2002). Similar targets were explored by designing a generation five dendrimer-based nano-device containing folic acid, MTX, and fluorescein. This conjugate has been reported to bind to the KB cells with subsequent internalization. The targeted conjugate inhibited the growth of KB cells in comparison with the non-targeted conjugate (fabricated without targeting agent), which, on the other hand, was not able to inhibit the cellular growth (Thomas et al., 2005). In another study, modified folic acid anchored PAMAM dendritic polymers were conjugated with MTX/tritium and 6-carboxytetramethylrhodamine/fluorescein. The different nanoconstructs thus formulated were investigated in immunodeficient mice bearing human KB tumors by i.v. administration. Folate-conjugated nanoparticles exhibit specific uptake by the tumor and liver tissue in comparison with the nontargeted system. Selective internalization of the targeted conjugate was reported in the tumor cells, which correspond to an increase in anticancer activity of MTX (Kukowska-Latallo et al., 2005). In a study, PPI dendrimers were investigated for their generation-dependent cancer-targeting ability of Melphalan. Findings suggest that generation four PPI dendrimer was more suitable as nanocarriers for anticancer drug delivery (Kesharwani et al., 2014b. In another investigation, folate-modified PPI dendrimer loaded with docetaxel was evaluated in all trans retinoic acid and dexamethasone pre-treated cell lines. In the presence of these folate receptor upregulators, the formulation shows significant reduction in IC50 values as compared to conventional docetaxel and non-functionalized dendrimer formulation (Dhakad et al., 2013).

7.6.2 ANTIBODY–DENDRIMER CONJUGATE-BASED CANCER TARGETING

High specificity of antigen-antibody interaction serves as the tool in achieving multiple targets for clinical purposes. Tumors are found to express various receptors that make them susceptible to the specific antibody and induce the cascade of reaction including immune-mediated destruction (Tekade et al., 2009b). Monoclonal antibodies are suitable agents to deliver anti-cancer drugs to cancerous tissue or organ expressing specific antigens. Identification of an exclusively expressed antigen on the tumor cells is of vital significance for targeting tumors, such that the targeting should be site specific while sparing normal tissue (Patri et al., 2004; Chacko et al., 2013).

In a study, alexaFluor 488-labeled generation five PAMAM dendrimer conjugated to anti-growth factor receptor-2 monoclonal antibody (anti-HER2 mAb) has been synthesized. The study is based on tumor targeting potential and specific affinity of the conjugate for human HER2. In comparison with free antibody, the conjugate confirmed specific uptake and internalization in cells expressing HER2. The conjugate was also reported to target tumors owing to the affinity of monoclonal antibodies for HER2 in animal models (Shukla et al., 2006). In another study, cetuximab-bearing generation 5- PAMAM dendrimer has been covalently conjugated with methotrexate. Cetuximab (a monoclonal antibody) shows binding affinity for epidermal growth factor receptor (EGFR). The bioconjugate has been reported to exhibit binding characteristics and cytotoxicity against rat glioma cell line with an overexpression of EGFR (Wu et al., 2006).

7.6.3 CANCER TARGETING BASED ON GLYCODENDRIMERS

Dendrimers containing sugar moieties like lactose, glucose, mannose, or some disaccharide in their structure are termed as glycodendrimers (Woller and Cloninger, 2001; Roy and Baek, 2002). Glycodendrimers were investigated for targeting anti-mitotic agent colchicines to tumor cells. Glycosylated dendrimers containing colchicines were reported to inhibit the proliferation of cancer HeLa cells (Lagnoux et al., 2005). PAMAM and polylysine glycodendrimers were investigated for the eradication of tumor using rat B16 murine melanoma and colorectal carcinoma models (Tekade et al., 2009a).

7.7 CONCLUSION

Unique architectural properties, biocompatibility, drug entrapment efficiency, and adjustable surface properties elucidate the extraordinary potential of dendrimers in delivering therapeutic agents to the desired site. Dendrimers offer many advantages over conventional drug-delivery methods such as increased solubility, higher loading capacity, increased bioavailability, site-specific targeting, and uptake, enhanced efficiency and reduced toxicity. Dendrimers are successfully utilized as potential vectors for site-specific delivery of genes and drugs to cancer and brain cells. Folate-conjugated dendrimers; glycosylated dendrimers, PEGylated hybrid dendrimers, the monoclonal antibody conjugated dendrimer, and peptide conjugated dendrimers are extensively being researched for targeting chemotherapeutic agents to the specific sites to minimize the hazardous side effects. Dendrimers are being extensively explored for delivering/targeting therapeutic agents to the brain by utilizing the specific receptors and transport systems present on the endothelial cells of the blood–brain barrier. Dendrimers have offered many advantages in the ocular delivery of drugs such as enhanced corneal residence time, target retinal inflammation, sustained neuroprotection in retinal degeneration and deliver drugs to the retina upon systemic administration. This chapter clearly illustrates the possibilities in the form of several investigations successfully done on dendrimers as novel carriers of therapeutic agents, despite of the associated drawbacks. This chapter also illuminates the vast utilitarian potential of dendritic carriers in drug targeting with advanced level of sophistication, specificity, and selectivity.

7.8 FUTURE PROSPECTS

Despite all these promising features and potentials as drug-delivery system, dendrimers suffer from some drawbacks. Due to the lack of more persuasive information on the safety profile as well as toxicity, most of the dendrimers are yet not approved for the clinical purpose. The range of diversified release mechanisms and kinetics of dendritic drug-delivery systems is another bottle-neck. Therapeutic agents encapsulated in dendritic carriers sometimes show premature burst release profile, before reaching to the target. The release of chemotherapeutic agents from dendrimer systems mainly depends on the chemical linkage and type of interaction between drug and dendrimer. Cyto-toxicity and biocompatibility issues are the major challenges in pulmonary delivery of dendrimers. As most of the CNS disorders require chronic treat-ment and drug administration, there is an urgent requirement of more thor-ough toxicological investigations of brain targeting dendritic formulations. The major targets of dendrimer-based medicine for CNS disorders include receptors like transferrin and insulin, which were also reported to be pres-ent on other tissues; this suggest the identification of more specific targets to assist the development of safer dendrimer-based drug delivery systems.

With all their pros and cons, dendrimers are full of potentials that yet to be explored. Hybrid dendrimers, including PEGylated dendrimers with tailored function, were reported to have comparatively reduced toxicity and immunogenicity along with enhanced biocompatibility and biodegradation. Research in the structural domain of dendrimers to create wonder carriers that could carry the drug, a ligand for specific receptor and diagnostic or imaging tool at the same time yet to be realized. Several aspects like bio-availability, biocompatibility, pharmacokinetics, and targeted delivery of therapeutics to specific sites need to be refined to make dendrimers, ideal agents for drug delivery. Although several investigations have been reported in the past, some more specific and detailed experimental protocols are required to precisely explain the *in vitro/in vivo* co-relationship of the perfor-mance of dendrimer formulations to produce more persuasive information to set up these novel constructs for clinical settings.

KEYWORDS

- antibody conjugated dendrimers
- biomaterials
- cancer therapy
- ligands
- PAMAM dendrimers
- targeting to organs

REFERENCES

Aerts, I., Leuraud, P., Blais, J., Pouliquen, A. L., Maillard, P., Houdayer, C., Couturier, J., Sastre-Garau, X., Grierson, D., Doz, F., & Poupon, M. F., (2010). *In vivo* efficacy of photodynamic therapy in three new xenograft models of human retinoblastoma. *Photodiagnosis. Photodyn. Ther.*, *7*(4), 275–283.

Agu, R. U., Ugwoke, M. I., Armand, M., Kinget, R., & Verbeke, N., (2001). The lung as a route for systemic delivery of therapeutic proteins and peptides. *Respir. Res.*, *2*(4), 198–209.

Ahmadiafshar, A., Mogimi, H. M., & Rezaei, N., (2010). Comparison of effectiveness between beclomethasone dipropionate and fluticasone propionate in treatment of children with moderate asthma. *World Allergy. Organ. J.*, *3*(10), 250–252.

Archut, A., & Vogtle, F., (1998). Functional cascade molecules. *Chem. Soc. Rev.*, *27*, 233–240.

Aryal, M., Arvanitis, C. D., Alexander, P. M., & McDannold, N., (2014). Ultrasound-mediated blood-brain barrier disruption for targeted drug delivery in the central nervous system. *Adv. Drug Deliv. Rev.*, *72*, 94–109.

Babar, M. M., Zaidi, N. S., Kazi, A. G., & Rehman, A., (2013). Virosomes hybrid drug delivery systems. *J. Antivir. Antiretrovir.*, *5*, 166–172.

Bai, S., & Ahsan, F., (2009). Synthesis and evaluation of pegylated dendrimeric nanocarrier for pulmonary delivery of low molecular weight heparin. *Pharm. Res.*, *26*(3), 539–548.

Bai, S., Thomas, C., & Ahsan, F., (2007). Dendrimers as a carrier for pulmonary delivery of enoxaparin, a low-molecular weight heparin. *J. Pharm. Sci.*, *96*(8), 2090–2106.

Banks, W. A., (2012). Drug delivery to the brain in Alzheimer's disease, consideration of the blood-brain barrier. *Adv. Drug Deliv. Rev.*, *64*(7), 629–639.

Beidoe, G., & Mousa, S. A., (2012). Current primary open-angle glaucoma treatments and future directions. *Clin. Ophthalmol.*, *6*, 1699–1707.

Bragagni, M., Mennini, N., Maestrelli, F., Cirri, M., & Mura, P., (2012). Comparative study of liposomes, transfersomes and ethosomes as carriers for improving topical delivery of celecoxib. *Drug. Deliv.*, *19*(7), 354–361.

Buhleier, E., Wehner, W., & Vogtle, F., (1978). Cascade and nonskid-chain-like synthesis of molecular cavity topologies. *Synthesis*, *1978*(2), 155–158.

Callaghan, R., & Higgins, C. F., (1995). Interaction of tamoxifen with the multidrug resistance P-glycoprotein. *Br. J. Cancer*, *71*(2), 294–299.

Chacko, A. M., Li, C., Pryma, D. A., Brem, S., Coukos, G., & Muzykantov, V., (2013). Targeted delivery of antibody-based therapeutic and imaging agents to CNS tumors: crossing the blood-brain barrier divide. *Expert. Opin. Drug. Deliv.*, *10*(7), 907–926.

Chen, Y., & Liu, L., (2012). Modern methods for delivery of drugs across the blood-brain barrier. *Adv. Drug Deliv. Rev.*, *64*(7), 640–665.

Choi, J. S., Ko, K. S., Park, J. S., Kim, Y. H., Kim, S. W., & Lee, M., (2006). Dexamethasone conjugated poly(amidoamine) dendrimer as a gene carrier for efficient nuclear translocation. *Int. J. Pharm.*, *320*(1–2), 171–198.

Chopdey, P. K., Tekade, R. K., Mehra, N. K., Mody, N., & Jain, N. K., (2015). Glycyrrhizin conjugated dendrimer and multi-walled carbon nanotubes for liver specific delivery of doxorubicin. *J. Nanosci. Nanotechnol.*, *15*(2), 1088–1100.

Coelho, J. F., Ferreira, P. C., Alves, P., Cordeiro, R., Fonseca, A. C., Gois, J. R., & Gil, M. H., (2010). Drug delivery systems, advanced technologies potentially applicable in personalized treatments. *EPMA J.*, *1*, 164–209.

Conti, D. S., Brewer, D., Grashik, J., Avasarala, S., Da Rocha, S. R., (2014). Poly(amidoamine) dendrimer nanocarriers and their aerosol formulations for siRNA delivery to the lung epithelium. *Mol. Pharm.*, *11*(6), 1808–1822.

Crouse, J. R., Frohlich, J., Ose, L., Mercuri, M., & Tobert, J., (1999). Effects of high doses of simvastatin and atorvastatin on high-density lipoprotein cholesterol and apolipoprotein A-1. *Am. J. Cardiol.*, *83*, 1476–1477.

D'Emanuele, A., Jevprasesphant, R., Penny, J., & Attwood, D., (2004). The use of a dendrimer-propranolol prodrug to bypass efflux transporters and enhance oral bioavailability. *J. Control. Release*, *95*, 447–453.

Del Amo, E. M., & Urtti, A., (2008). Current and future ophthalmic drug delivery systems. A shift to the posterior segment. *Drug Discov. Today*, *13*(3–4), 135–143.

Demeule, M., Currie, J. C., Bertrand, Y., Ché, C., Nguyen, T., Régina, A., Gabathuler, R., Castaigne, J. P., & Béliveau, R., (2008). Involvement of the low-density lipoprotein receptor-related protein in the transcytosis of the brain delivery vector angiopep-2. *J. Neurochem.*, *106*(4), 1534–1544.

Devarakonda, B., Hill, R. A., & De Villiers, M. M., (2004). The effect of PAMAM dendrimer generation size and surface functional group on the aqueous solubility of nifedipine. *Int. J. Pharm.*, *284*, 133–140.

Devarakonda, B., Li, N., & De Villiers, M. M., (2005). Effect of polyamidoamine (PAMAM) dendrimers on the *in vitro* release of water soluble nifedipine from aqueous gel. *AAPS Pharm. Sci. Tech.*, *6*(3), Article 63.

Dhakad, R. S., Tekade, R. K., & Jain, N. K., (2013). Cancer targeting Potential of folate targeted nanocarrier under comparative influence of tretinoin and dexamethasone. *Curr. Drug. Deliv.*, *10*(4), 477–491.

Dhanikula, R. S., Argaw, A., Bouchard, J. F., & Hildgen, P., (2008). Methotrexate loaded polyether-copolyester dendrimers for the treatment of gliomas, enhanced efficacy and intratumoral transport capability. *Mol. Pharm.*, *5*(1), 105–116.

Dufes, C., Al Robaian, M., & Somani, S., (2013). Transferrin and the transferrin receptor for the targeted delivery of therapeutic agents to the brain and cancer cells. *Ther. Deliv.*, *4*(5), 629–640.

Duncan, R., Ringsdorf, H., & Satchi-Fainaro, R., (2006). Polymer therapeutics—polymer as drugs, drug and protein conjugates and gene delivery systems, past, present and future opportunities. *J. Drug Targeting*, *14*, 337–341.

Durairaj, C. K. U., (2009). Dendritic polyguanidilyated translocators for ocular drug delivery. *Drug. Deliv. Technol.*, *9*, 36–43.

Durairaj, C., Kadam, R. S., Chandler, J. W., Hutcherson, S. L., & Kompella, U. B., (2010). Nanosized dendritic polyguanidilyated translocators for enhanced solubility, permeability, and delivery of gatifloxacin. *Invest. Ophthalmol. Vis. Sci.*, *51*(11), 5804–5816.

Eagle, R. C. Jr., (2013). The pathology of ocular cancer. *Eye (Lond)*, *27*(2), 128–136.

Eljarrat-Binstock, E., Pe'er, J., & Domb, A. J., (2010). New techniques for drug delivery to the posterior eye segment. *Pharm. Res.*, *27*(4), 530–543.

Elnakat, H., & Ratnam, M., (2004). Distribution, functionality and gene regulation of folate receptor isoforms: Implications in targeted therapy. *Adv. Drug Deliv. Rev.*, *29*, *56*(8), 1067–1084.

Esfand, R., & Tomalia, D. A., (2001). Poly(amidoamine) (PAMAM) dendrimers, from biomimicry to drug delivery and biomedical applications. *Drug. Discov. Today.*, *6*(8), 427–436.

Frechet, J. M. J., & Tomalia, D. A., (2001). (Eds.), *Dendrimers and Other Dendritic Polymers*, Wiley, West Sussex, UK.

Gabathuler, R., (2010). Approaches to transport therapeutic drugs across the blood-brain barrier to treat brain diseases. *Neurobiol. Dis.*, *37*(1), 48–57.

Gajbhiye, V., Kumar, P. J., Tekade, R. K., & Jain, N. K., (2007). Pharmaceutical and biomedical potential of PEGylated dendrimers. *Curr. Pharm. Des.*, *13*(4), 415–429.

Gajbhiye, V., Vijayaraj, K. P., Tekade, R. K., & Jain, N. K., (2009b). PEGylated PPI dendritic architectures for sustained delivery of H2 receptor antagonist. *Eur. J. Med. Chem.*, *44*(3), 1155–1166.

Gandhi, N. S., Tekade, R. K., & Chougule, M. B., (2014). Nanocarrier mediated delivery of siRNA/miRNA in combination with chemotherapeutic agents for cancer therapy: current progress and advances. *J. Control. Release*, *194*, 238–256.

Gao, H., (2016). Progress and perspectives on targeting nanoparticles for brain drug delivery. *Acta. Pharmaceutica. Sinica B.*, *6*(4), 268–286.

Gaudana, R., Jwala, J, Boddu, S. H., & Mitra, A. K., (2009). Recent perspectives in ocular drug delivery. *Pharm. Res.*, *26*(5), 1197–216.

Gosk, S., Vermehren, C., Storm, G., & Moos, T., (2004). Targeting anti-transferrin receptor antibody (OX26) and OX26- conjugated liposomes to brain capillary endothelial cells using in situ perfusion. *J. Cereb. Blood Flow Metab.*, *24*(11), 1193–1204.

Gothwal, A., Kesharwani, P., Gupta, U., Khan, I., Cairul, M., Amin, I. M., Banerjee, S., & Iyer, A. K., (2015). Dendrimers as an effective nanocarrier in cardiovascular disease. *Curr. Pharm. Desi.*, *21*, 4519–4526.

Griesenbach, U., Geddes, D. M., & Alton, E. W., (2004). Gene therapy for cystic fibrosis, an example for lung gene therapy. *Gene Ther.*, *11(Suppl 1),* 43–50.

Guo, L., Ren, J., & Jiang, X., (2012). Perspectives on brain-targeting drug delivery systems. *Curr. Pharm. Biotechnol.*, *13*(12), 2310–2318.

Harries, A. D., & Dye, C., (2006). Tuberculosis. *Ann. Trop. Med. Parasitol.*, *100*(5–6), 415–431.

Hawker, C. J., & Fréchet, J. M. J., (1990). Preparation of polymers with controlled molecular architecture. A new convergent approach to dendritic macromolecules. *J. Am. Chem. Soc.*, *112*(21), 7638–7647.

He, H., Li, Y., Jia, X. R., Du, J., Ying, X., Lu, W. L., Lou, J. N., & Wei, Y., (2011). PEGylated poly(amidoamine) dendrimer-based dual-targeting carrier for treating brain tumors. *Biomaterials*, *32*(2), 478–487.

Heinemann, L., (2008). The failure of exubera, are we beating a dead horse? *J. Diabetes Sci. Technol.*, *2*(3), 518–529.

Henwood, S. Q., Liebenberg, W., Tiedt, L. R., Lötter, A. P., & De Villiers, M. M., (2001). Characterization of the solubility and dissolution properties of several new rifampicin polymorphs, solvates, and hydrates. *Drug Dev. Ind. Pharm.*, *27*(10), 1017–1030.

Holden, C. A., Tyagi, P., Thakur, A., Kadam, R., Jadhav, G., Kompella, U. B., & Yang, H., (2012). Polyamidoamine dendrimer hydrogel for enhanced delivery of antiglaucoma drugs. *Nanomedicine*, *8*(5), 776–783.

Huang, R. Q., Qu, Y. H., Ke, W. L., Zhu, J. H., Pei, Y. Y., & Jiang, C., (2007). Efficient gene delivery targeted to the brain using a transferrin-conjugated polyethyleneglycol-modified polyamidoamine dendrimer. *FASEB J.*, *21*(4), 1117–1125.

Huang, R., Ke, W., Liu, Y., Jiang, C., & Pei, Y., (2008). The use of lactoferrin as a ligand for targeting the polyamidoamine-based gene delivery system to the brain. *Biomaterials*, *29*(2), 238–246.

Huang, R., Ma, H., Guo, Y., Liu, S., Kuang, Y., Shao, K., Li, J., Liu, Y., Han, L., Huang, S., An, S., Ye, L., Lou, J., & Jiang, C., (2013). Angiopep-conjugated nanoparticles for targeted long-term gene therapy of Parkinson's disease. *Pharm. Res., 30*(10), 2549–2559.

Huang, S., Li, J., Han, L., Liu, S., Ma, H, Huang, R., & Jiang, C., (2011). Dual targeting effect of Angiopep-2-modified, DNA-loaded nanoparticles for glioma. *Biomaterials, 32*(28), 6832–6838.

Ibbotson, T., & Goa, K. L., (2002). Enoxaparin, an update of its clinical use in the management of acute coronary syndromes. *Drugs, 62*(9), 1407–1430.

Iezzi, R., Guru, B. R., Glybina, I. V., Mishra, M. K., Kennedy, A., & Kannan, R. M., (2012). Dendrimer-based targeted intravitreal therapy for sustained attenuation of neuroinflammation in retinal degeneration. *Biomaterials, 33*(3), 979–988.

Ivanov, A. A., & Jacobson, K. A., (2008). Molecular modeling of a PAMAMCGS21680 dendrimer bound to an A2A adenosine receptor homodimer. *Bioorg. Med. Chem. Lett., 18*, 4312–4315.

Jain, N. K., & Asthana, A., (2007). Dendritic systems in drug delivery applications. *Expert. Opin. Drug Deliv., 4*(5), 495–512.

Jain, N. K., & Tekade, R. K., (2013). *Dendrimers for Enhanced Drug Solubilization*. In: *Drug Delivery Strategies for Poorly Water-Soluble Drugs*, Douroumis, D., Fahr, A., (ed.), John Wiley & Sons Ltd., Oxford, UK, p 373–409.

Jain, S., & Jain, N. K., (1997). Engineered erythrocytes as a drug delivery system. *Indian J. Pharm. Sci., 59*(6), 275–281.

Jones, A. R., & Shusta, E. V., (2007). Blood-brain barrier transport of therapeutics via receptor-mediation. *Pharm. Res., 24*(9), 1759–1771.

Kang, S. J., Durairaj, C., Kompella, U. B., O'Brien, J. M., & Grossniklaus, H. E., (2009). Subconjunctival nanoparticle carboplatin in the treatment of murine retinoblastoma. *Arch. Ophthalmol., 127*(8), 1043–1047.

Katare, Y. K., Daya, R. P., Sookram, G. C., Luckham, R. E., Bhandari, J., Chauhan, A. S., & Mishra, R. K., (2015). Brain targeting of a water insoluble antipsychotic drug haloperidol via the intranasal route using PAMAM dendrimer. *Mol. Pharm., 12*(9), 3380–3388.

Kawaguchi, T., Walker, K. L., Wilkins, C. L., & Moore, J. S., (1995). Double exponential dendrimer growth. *J. Am. Chem. Soc., 117*(8), 2159–2165.

Kayat, J., Gajbhiye, V., Tekade, R. K., & Jain, N. K., (2011). Pulmonary toxicity of carbon nanotubes: A systematic report. *Nanomedicine, 7*(1), 40–9.

Kazi, K. M., Mandal, A. S., Biswas, N., Guha, A., Chatterjee, S., Behera, M., & Kuotsu, K., (2010). Niosome: A future of targeted drug delivery systems. *J. Adv. Pharm. Technol. Res., 1*(4), 374–380.

Ke, W., Shao, K., Huang, R., Han, L., Liu, Y., Li, J., Kuang, Y., Ye, L., Lou, J., & Jiang, C., (2009). Gene delivery targeted to the brain using an Angiopep-conjugated polyethyleneglycol-modified polyamidoamine dendrimer. *Biomaterials, 30*(36), 6976–6985.

Keene, A. M., Balasubramanian, R., Lloyd, J., Shainberg, A., & Jacobson, K. A., (2010). Multivalent dendrimeric and monomeric adenosine agonists attenuate cell death in HL-1 mouse cardiomyocytes expressing the A3receptor. *Biochem. Pharmacol., 80*, 188–196.

Kesharwani, P., Jain, K., & Jain, N. K., (2014). Dendrimer as nanocarrier for drug delivery. *Prog. Polym. Sci., 39*, 268–307.

Kesharwani, P., Tekade, R. K., & Jain, N. K., (2014). Generation dependent cancer targeting potential of poly (propyleneimine) dendrimer. *Biomaterials, 35*(21), 5539–5548.

Kesharwani, P., Tekade, R. K., & Jain, N. K., (2015). Dendrimer generational nomenclature: the need to harmonize. *Drug Discov. Today, 20*(5), 497–499.

Kesharwani, P., Tekade, R. K., & Jain, N. K., (2015). Generation dependent safety and efficacy of folic acid conjugated dendrimer based anticancer drug formulations. *Pharm. Res.*, *32*(4), 1438–1450.

Kesieme, E., Kesieme, C., Jebbin, N., Irekpita, E., & Dongo, A., (2011). Deep vein thrombosis, a clinical review. *J. Blood Med.*, *2*, 59–69.

Kim, Y., Hechler, B., Klutz, A. M., Gachet, C., & Jacobson, K. A., (2008). Towardmultivalent signaling across G protein coupled receptors from poly(amidoamine) dendrimers. *Bioconjugate Chem.*, *19*, 406–411.

Kim, Y., Klutz, A. M., Hechler, B., Gao, Z. G., Gachet, C., & Jacobson, K. A., (2009). Application of the functionalized congener approach to dendrimerbased signaling agents acting through A2A adenosine receptor. *Purinergic Signal*, *5*, 39–50.

Kono, K., Liu, M., & Frechet, J. M., (1999). Design of dendritic macromolecules containing folate or methotrexate residues. *Bioconjug. Chem.*, *10*(6), 1115–1121.

Kukowska-Latallo, J. F., Candido, K. A., Cao, Z., Nigavekar, S. S., Majoros, I. J., Thomas, T. P., Balogh, L. P., Khan, M. K., & Baker, J. R. Jr., (2005). Nanoparticle targeting of anticancer drug improves therapeutic response in animal model of human epithelial cancer. *Cancer Res.*, *65*(12), 5317–5324.

Kulhari, H., Kulhari D. P., Prajapati, S. K., & Chauhan, A. S., (2013). Pharmacokinetic and pharmacodynamic studies of poly(amidoamine) dendrimer based simvastatin oral formulations for the treatment of hypercholesterolemia. *Mol. Pharmaceutics.*, *10*, 2528–2533.

Kurmi, B. D., Kayat, J., Gajbhiye, V., Tekade, R. K., & Jain, N. K., (2010). Micro- and nanocarrier-mediated lung targeting. *Expert. Opin. Drug. Deliv.*, *7*(7), 781–94.

Kusuhara, H., & Sugiyama, Y., (2005). Active efflux across the blood-brain barrier, role of the solute carrier family. *Neuro. Rx.*, *2*(1), 73–85.

Labiris, N. R., & Dolovich, M. B., (2003). Pulmonary drug delivery. Part I, physiological factors affecting therapeutic effectiveness of aerosolized medications. *Br. J. Clin. Pharmacol.*, *56*(6), 588–599.

Lagnoux, D., Darbre, T., Schmitz, M. L., & Reymond, J. L., (2005). Inhibition of mitosis by glycopeptide dendrimer conjugates of colchicines. *Chemistry*, *11*, 3941–3950.

Lajoie, J. M., & Shusta, E. V., (2015). Targeting receptor-mediated transport for delivery of biologics across the blood-brain barrier. *Annu. Rev. Pharmacol. Toxicol.*, *55*, 613–631.

Laksitorini, M., Prasasty, V. D., Kiptoo, P. K., & Siahaan, T. J., (2014). Pathways and progress in improving drug delivery through the intestinal mucosa and blood-brain barriers. *Ther. Deliv.*, *5*(10), 1143–1163.

Lang, J. C., (1995). Ocular drug delivery conventional ocular formulations. *Adv. Drug Deliv. Rev.*, *16*(1), 39–43.

Lattin, J. R., Belnap, D. M., & Pitt, W. G., (2012). Formation of eliposomes as a drug delivery vehicle. *Colloids. Surf. B. Biointerfaces.*, *89*, 93–100.

Li, Y., He, H., Jia, X., Lu, W. L., Lou, J., & Wei, Y., (2012). A dual-targeting nanocarrier based on poly(amidoamine) dendrimers conjugated with transferrin and tamoxifen for treating brain gliomas. *Biomaterials*, *33*(15), 3899–3908.

Lillis, A. P., Van Duyn, L. B., Murphy-Ullrich, J. E., & Strickland, D. K., (2008). LDL receptor-related protein 1, unique tissue-specific functions revealed by selective gene knockout studies. *Physiol. Rev.*, *88*(3), 887–918.

Liu, H, Tu, Z, Feng, F, Shi, H, Chen, K, Xu, X., (2015). Virosome, a hybrid vehicle for efficient and safe drug delivery and its emerging application in cancer treatment. *Acta. Pharm.*, *65*(2), 105–116.

Loscher, W., & Potschka, H., (2005). Blood-brain barrier active efflux transporters, ATP-binding cassette gene family. *Neuro Rx.*, *2*(1), 86–98.

Luciani, A., Olivier, J. C., Clement, O., Siauve, N., Brillet, P. Y., Bessoud, B., Gazeau, F., Uchegbu, L. F., Kahn, E., Frija, G., & Cuenod, C. A., (2004). Glucose-receptor MR imaging of tumors, Study in mice with PEGylated paramagnetic niosomes. *Radiology*, *231*, 135–142.

Lundy, A., Lutfi, N., Beckey C., (2009). Review of nifedipine GITS in the treatment of high risk patients with coronary artery disease and hypertension. *Vasc. Health Risk Manag.*, *5*(1), 429–440.

Ma, Q., Han, Y., Chen, C., Cao, Yi., Wang, Si., Shen, We., Zhang, H., Li, Y., Van Dongen, M. A., He, B., Yu, M., Xu, L., Holl, M. M. B., Liu, G., Zhang, Q., & Qi, R., (2015). Oral absorption enhancement of probucol by PEGylated G5 PAMAM dendrimer modified nanoliposomes. *Mol. Pharmaceutics.*, *12*, 665‒674.

Maheshwari, R. G, Thakur, S., Singhal, S., Patel, R. P., Tekade, M., Tekade, R. K., (2015). Chitosan encrusted nonionic surfactant based vesicular formulation for topical administration of ofloxacin. *Sci. Adv. Mater.*, *7*(6), 1163–1176.

Maheshwari, R. G., Tekade, R. K., Sharma, P. A., Darwhekar, G., Tyagi, A., Patel, R. P., & Jain, D. K., (2012). Ethosomes and ultradeformable liposomes for transdermal delivery of clotrimazole: A comparative assessment. *Saudi. Pharm. J.*, *20*(2), 161–170.

Makky, A., Michel, J. P., Maillard, P., & Rosilio, V., (2011). Biomimetic liposomes and planar supported bilayers for the assessment of glycodendrimeric porphyrins interaction with an immobilized lectin. *Biochim. Biophys. Acta.*, *1808*(3), 656–666.

Mansour, H. M., Rhee, Y. S., & Wu, X., (2009). Nanomedicine in pulmonary delivery. *Int. J. Nanomedicine.*, *4*, 299–319.

Maraval, V., Pyzowski, J., Caminade, A. M., & Majoral, J. P., (2003). "Lego" chemistry for the straightforward synthesis of dendrimers. *J. Org. Chem.*, *68*(15), 6043–6046.

Martinho, N., Damge, C., & Reis, C. P., (2011). Recent advances in drug delivery systems. *J. Biomater. Nanobiotechnol.*, *2*, 510–526.

Martinho, N., Florindo, H., Silva, L., Brocchini, S., Zloh, M., & Barata, T., (2014). Molecular modeling to study dendrimers for biomedical applications. *Molecules*, *19*(12), 20424–20467.

Matthews, O. A., Shipway, A. N., & Stoddart, J. F., (1998). Dendrimers-branching out from curiosities into new technologies. *Prog. Polym. Sci.*, *23*(1), 1–56.

Mishra, M. K., Beaty, C. A., Lesniak, W. G., Kambhampati, S. P., Zhang, F., Wilson, M. A., Blue, M. E., Troncoso, J. C., Kannan, S., Johnston, M. V., Baumgartner, W. A., & Kannan, R. M., (2014). Dendrimer brain uptake and targeted therapy for brain injury in a large animal model of hypothermic circulatory arrest. *ACS. Nano.*, *8*(3), 2134–2147.

Mismetti, P., Quenet, S., Levine, M., Merli, G., Decousus, H., Derobert, E., & Laporte, S., (2005). Enoxaparin in the treatment of deep vein thrombosis with or without pulmonary embolism, an individual patient data meta-analysis. *Chest.*, *128*(4), 2203–2210.

Mo, Y., & Lim, L., (2005). Paclitaxel-loaded PLGA nanoparticles, potentiation of anticancer activity by surface conjugation. *J. Control. Release.*, *108*(2–3), 244–262.

Mody, N., Tekade, R. K., Mehra, N. K., Chopdey, P., & Jain, N. K., (2014). Dendrimer, liposomes, carbon nanotubes and PLGA nanoparticles: one platform assessment of drug delivery potential. *AAPS. Pharm. Sci. Tech.*, *15*(2), 388–399.

Moos, T., & Morgan, E. H., (2000). Transferrin and transferrin receptor function in brain barrier systems. *Cell. Mol. Neurobiol.*, *20*(1), 77–95.

Nanjwade, B. K., Bechra, H. M., Derkar, G. K., Manvi, F. V., & Nanjwade, V. K., (2009). Dendrimers, emerging polymers for drug-delivery systems. *Eur. J. Pharm. Sci.*, *38*(3), 185–196.

Nasr, M., Najlah, M., D'Emanuele, A., & Elhissi, A., (2014). PAMAM dendrimers as aerosol drug nanocarriers for pulmonary delivery via nebulization. *Int. J. Pharm.*, *461*(1–2), 242–250.

Neuwelt, E., Abbott, N. J., Abrey, L., Banks, W. A., Blakley, B., Davis, T., Engelhardt, B., Grammas, P., Nedergaard, M., Nutt, J., Pardridge, W., Rosenberg, G. A., Smith, Q., & Drewes, L. R., (2008). Strategies to advance translational research into brain barriers. *Lancet. Neurol.*, *7*(1), 84–96.

Newkome, G. R., Moorefield, C. N., & Vogtle, F., (2001). *Dendrimers and Dendrons, Concepts, Syntheses, Applications*, Wiley-VCH, Weinheim, Germany.

Newkome, G. R., Yao, Z. Q., Baker, G. R., & Gupta, V. K., (1985). Cascade molecules, A new approach to micelles. *J. Org. Chem.*, *50*(11), 2003–2006.

Omidi, Y., & Barar, J., (2012). Impacts of blood-brain barrier in drug delivery and targeting of brain tumors. *Bioimpacts.*, *2*(1), 5–22.

Pardridge, W. M., (2007). Blood-brain barrier delivery. *Drug. Discov. Today*, *12*(1–2), 54–61.

Pardridge, W., Boado, R., & Farrell, C., (1990). Brain-type glucose transporter (GLUT-1) is selectively localized to the blood-brain barrier. Studies with quantitative western blotting and in situ hybridization. *J. Biol. Chem.*, *265*, 18035–18040.

Park, K., (2014). Controlled drug delivery systems, Past forward and future back. *J. Control. Release*, *190*, 3–8.

Patri, A. K., Kukowska-Latallo, J. F., & Baker, Jr. J. R., (2005). Targeted drug delivery with dendrimers, Comparison of the release kinetics of covalently conjugated drug and non-covalent drug inclusion complex. *Adv. Drug Deliv. Rev.*, *57*, 2203–2214.

Patri, A. K., Myc, A., Beals, J., Thomas, T. P., Bander, N. H., & Baker, J. R. Jr., (2004). Synthesis and *in vitro* testing of J591 antibody-dendrimer conjugates for targeted prostate cancer therapy. *Bioconjug. Chem.*, *15*(6), 1174–1181.

Patton, J. S., (1996). Mechanisms of macromolecule absorption by the lungs. *Adv. Drug Deliv. Rev.*, *19*(1), 3−36.

Patton, J. S., & Byron, P. R., (2007). Inhaling medicines, delivering drugs to the body through the lungs. *Nat. Rev. Drug Discov.*, *6*(1), 67–74.

Pearson, O. H., Manni, A., & Arafah, B. M., (1982). Antiestrogen treatment of breast cancer, an overview. *Cancer Res.*, *42*(8 Suppl), 3424–3429.

Prajapati, R. N., Tekade, R. K., Gupta, U., Gajbhiye, V., & Jain, N. K., (2009). Dendrimer-mediated solubilization, formulation development and *in vitro-in vivo* assessment of piroxicam. *Mol. Pharm.*, *6*(3), 940–950.

Qi, R., Zhang, H., Xu, L, Shen, W., Chen, C., Wang, C., Cao, Y., Wang Y., Van Dongen, M. A., He, B., Wang, Si., Liu, Ge., Holl, M. M. B., & Zhang, Q., (2015). G5 PAMAM dendrimer versus liposome: A comparison study on the *in vitro* transepithelial transport and *in vivo* oral absorption of simvastatin. *Nanomedicine*, *11*(5), 1141–1151.

Quintana, A., Raczka, E., Piehler, L., Lee, I., Myc, A., Majoros, I., Patri, A. K., Thomas, T., Mule, J., & Baker, Jr. J. R., (2002). Design and function of a dendrimerbased therapeutic nanodevice targeted to tumor cells through the folate receptor. *Pharm. Res., 19*(9), 1310–1316.

Rajabnezhad, S., Casettari, l., Lamc, J. K. W., Nomani, A., Torkamani, M. R., Palmieri, G. F., Rajabnejad, M. R., & Darbandi, M. A., (2016). Pulmonary delivery of rifampicin microspheres using lower generation polyamidoamine dendrimers as a carrier. *Powder Technol., 291*, 366–374.

Ranta, V. P., Mannermaa, E., Lummepuro, K, Subrizi, A., Laukkanen, A., Antopolsky, M., Murtomäki, L., Hornof, M., & Urtti, A., (2010). Barrier analysis of periocular drug delivery to the posterior segment. *J. Control. Release, 148*(1), 42–48.

Remesha, A., (2012). Toxicities of anticancer drugs and its management. *I. J. Basic & Clin. Pharmacol., 1*(1), 1–12.

Roberts, R., Sandra, A., Siek, G. C., Lucas, J. J., & Fine, R. E., (1992). Studies of the mechanism of iron transport across the blood-brain barrier. *Ann. Neurol., 32*(Suppl), 43–50.

Roy, R., & Baek, M. G., (2002). Glycodendrimers: novel glycotope isosteres unmasking sugar coding. case study with T-antigen markers from breast cancer MUC1 glycoprotein. *J. Biotechnol., 90*, 291–309.

Rudolph, C., Lausier, J., Naundorf, S., Müller, R. H., & Rosenecker, J., (2000). *In vivo* gene delivery to the lung using polyethylenimine and fractured polyamidoamine dendrimers. *J. Gene Med., 2*(4), 269–278.

Ryan, G. M., Kaminskas, L. M., Kelly, B. D., Owen, D. J., McIntosh, M. P., & Porter, C. J., (2013). Pulmonary administration of PEGylated polylysine dendrimers, absorption from the lung versus retention within the lung is highly size-dependent. *Mol. Pharm., 10*(8), 2986–2995.

Safari, J., & Zarnegar, Z., (2014). Advanced drug delivery systems, nanotechnology of health design a review. *J. Saudi. Chem. Soc., 18*(2), 85–99.

Sasindran, S. J., & Torrelles, J. B., (2011). Mycobacterium tuberculosis infection and inflammation, what is beneficial for the host and for the bacterium? *Front. Microbiol., 2*, 2.

Satija, J., Gupta, U., & Jain, N. K., (2007). Pharmaceutical and biomedical potential of surface engineered dendrimers. *Crit. Rev. Ther. Drug Carrier. Syst., 24*(3), 257–306.

Sawayama, Y., Shimizu, C., Maeda, N., Tatsukawa, M., Kinukawa, N., Koyanagi, S., Kashiwagi, S., & Hayashi, J., (2002). Effects of probucol and pravastatin on common carotid atherosclerosis in patients with asymptomatic hypercholesterolemia. Fukuoka Atherosclerosis Trial (FAST). *J. Am. Coll. Cardiol., 39*, 610–616.

Scherrmann, J. M., (2002). Drug delivery to brain via the blood-brain barrier. *Vascul. Pharmacol., 38*(6), 349–354.

Schmidt, K. G., Bergert, H., & Funk, R. H., (2008). Neurodegenerative diseases of the retina and potential for protection and recovery. *Curr. Neuropharmacol., 6*(2), 164–178.

Schover, L. R., Van der Kaaij, M., Van Dorst, E., Creutzberg, C., Huyghe, E., & Kiserud, C. E., (2014). Sexual dysfunction and infertility as late effects of cancer treatment. *EJC. Supplements, 12*(1), 4 1–5 3.

Schwendener, R. A., (2014). Liposomes as vaccine delivery systems: a review of the recent advances. *Ther. Adv. Vaccines., 2*(6), 159–182.

Shah, S. S., Denham, L. V., Elison, J. R., Bhattacharjee, P. S., Clement, C., Huq, T., & Hill, J. M., (2010). Drug delivery to the posterior segment of the eye for pharmacologic therapy. *Expert. Rev. Ophthalmol., 5*(1), 75–93.

Sharma, P. A., Maheshwari, R., Tekade, M., & Tekade, R. K., (2015). Nanomaterial based approaches for the diagnosis and therapy of cardiovascular diseases. *Curr. Pharm. Des.*, *21*(30), 4465–4478.

Shaunak, S., Thomas, S., Gianasi, E., Godwin, A., Jones, E., Teo, I., Mireskandari, K., Luthert, P., Duncan, R., Patterson, S., Khaw, P., & Brocchini, S., (2004). Polyvalent dendrimer glucosamine conjugates prevent scar tissue formation. *Nat. Biotechnol.*, *22*(8), 977–984.

Sheardown, H., Saltzman W. M., (2006). *Novel Drug Delivery Systems for Posterior Segment Ocular Disease.* In *Ocular Angiogenesis*, Trombrain- Tink, J., Barnstable, C. J., (ed.), Humana Press Inc., Totowa, N. J., pp. 393–408.

Shishoo, C. J., Shah, S. A., Rathod, I. S., Savale, S. S., & Vora, M. J., (2001). Impaired bio-availability of rifampicin in presence of isoniazid from fixed dose combination (FDC) formulation. *Int. J. Pharm.*, *228*(1–2), 53–67.

Shukla, R., Thomas, T. P., Peters, J. L., Desai, A. M., Kukowska-Latallo, J., Patri, A. K., Kotlyar, A., & Baker, J. R. Jr., (2006). HER2 specific tumor targeting with dendrimer conjugated anti-HER2 mAb. *Bioconjugate Chem.*, *17*(5), 1109–15.

Singh, M. N., Hemant, K. S., Ram, M., & Shivakumar, H. G., (2010). Microencapsulation: A promising technique for controlled drug delivery. *Res. Pharm. Sci.*, *5*(2), 65–77.

Singh, S., Mariappan, T. T., Shankar, R., Sarda, N., & Singh, B., (2001). A critical review of the probable reasons for the poor variable bioavailability of rifampicin from anti-tuber-cular fixed-dose combination (FDC) products, and the likely solutions to the problem. *Int. J. Pharm.*, *228*(1–2), 5–17.

Smith, I., (2003). Mycobacterium tuberculosis pathogenesis and molecular determinants of virulence. *Clin. Microbiol. Rev.*, *16*(3), 463–496.

Smith, M. W., & Gumbleton, M., (2006). Endocytosis at the blood-brain barrier, from basic understanding to drug delivery strategies. *J. Drug Target*, *14*(4), 191–214.

Somani, S., Blatchford, D. R., Millington, O., Stevenson, M. L., & Dufès, C., (2014). Trans-ferrin-bearing polypropylenimine dendrimer for targeted gene delivery to the brain. *J. Control. Release.*, *188*, 78–86.

Spataro, G., Malecaze, F., Turrin, C. O., Soler, V., Duhayon, C., Elena, P. P, Majoral, J. P., & Caminade, A. M., (2010). Designing dendrimers for ocular drug delivery. *Eur. J. Med. Chem.*, *45*(1), 326–334.

Stamatovic, S. M., Keep, R. F., & Andjelkovic, A. V., (2008). Brain endothelial cell-cell junc-tions, how to "open" the blood brain barrier. *Curr. Neuropharmacol.*, *6*(3), 179–192.

Stephan, H., Boeloeni, R., Eggert, A., Bornfeld, N., & Schueler, A., (2008). Photodynamic therapy in retinoblastoma, effects of verteporfin on retinoblastoma cell lines. *Invest. Ophthalmol. Vis. Sci.*, *49*(7), 3158–3163.

Tekade, R. K., (2015). Editorial: Contemporary siRNA therapeutics and the current state-of-art. *Curr. Pharm. Des.*, *21*(31), 4527–4528.

Tekade, R. K., & Chougule, M. B., (2013). Formulation development and evaluation of hy-brid nanocarrier for cancer therapy: Taguchi orthogonal array based design. *Biomed. Res. Int.*, *2013*, 712678.

Tekade, R. K., Dutta, T., Gajbhiye, V., & Jain, N. K., (2009). Exploring dendrimer towards dual drug delivery: pH responsive simultaneous drug-release kinetics. *J. Microencap-sul.*, *26*(4), 287–296.

Tekade, R. K., Dutta, T., Tyagi, A., Bharti, A. C., Das, B. C., & Jain, N. K., (2008). Surface-engineered dendrimers for dual drug delivery: a receptor up-regulation and enhanced cancer targeting strategy. *J. Drug. Target*, *16*(10), 758–772.

Tekade, R. K., Kumar, P. V., & Jain, N. K., (2009). Dendrimers in oncology: an expanding horizon. *Chem. Rev.*, *109*(1), 49–87.

Tekade, R. K., Maheshwari, R. G., Sharma, P. A., Tekade, M., & Chauhan, A. S., (2015a). siRNA therapy, challenges and underlying perspectives of dendrimer as delivery vector. *Curr. Pharm. Des.*, *21*(31), 4614–4636.

Tekade, R. K., Tekade, M., Kumar, M., & Chauhan, A. S., (2015b). Dendrimer-stabilized smart-nanoparticle (DSSN) platform for targeted delivery of hydrophobic antitumor therapeutics. *Pharm. Res.*, *32*(3), 910–928.

Tekade, R. K., Youngren-Ortiz, S. R., Yang, H., Haware, R., & Chougule, M. B., (2014). Designing hybrid onconase nanocarriers for mesothelioma therapy: a Taguchi orthogonal array and multivariate component driven analysis. *Mol. Pharm.*, *11*(10), 3671–3683.

Thomas, T. P., Majoros, I. J., Kotlyar, A., Kukowska-Latallo, J. F., Bielinska, A., Myc, A., & Baker, J. R. Jr., (2005). Targeting and inhibition of cell growth by an engineered dendritic nanodevice. *J. Med. Chem.*, *48*(11), 3729–35.

Tibbitt, M. W., Dahlman, J. E., & Langer, R., (2016). Emerging frontiers in drug delivery. *J. Am. Chem. Soc.*, *138*(3), 704–717.

Tiwari, G., Tiwari, R., Sriwastawa, B., Bhati, L., Pandey, S., Pandey, P., & Bannerjee, S. K., (2012). Drug delivery systems: An updated review. *Int. J. Pharm. Investig.*, *2*(1), 2–11.

Tomalia, D. A., (1996). Starburst dendrimers-nanoscopic macromolecules according to dendritic rules and principles. *Macromol. Symp.*, *101*(1), 243–255.

Tomalia, D. A., Baker, H., Dewald, J., Hall, M., Kallos, G., Martin, S., Roeck, J., Ryder, J., & Smith, P., (1985). A new class of polymers, starburst dendritic macromolecules. *J. Polym.*, *17*(1), 117–132.

Tsuji, A., & Tamai, I. I., (1999). Carrier-mediated or specialized transport of drugs across the blood-brain barrier. *Adv. Drug Deliv. Rev.*, *36*(2–3), 277–290.

Uchida, Y., Ohtsuki, S., Katsukura, Y., Ikeda, C., Suzuki, T., Kamiie, J., & Terasaki, T., (2011). Quantitative targeted absolute proteomics of human blood-brain barrier transporters and receptors. *J. Neurochem.*, *117*(2), 333–345.

Upadhyay, M. P., Srinivasan, M., & Whitcher, J. P., (2009). Managing corneal disease, focus on suppurative keratitis. *Community Eye Health*, *22*(71), 39–41.

Vandamme, T. F., & Brobeck, L., (2005). Poly(amidoamine) dendrimers as ophthalmic vehicles for ocular delivery of pilocarpine nitrate and tropicamide. *J. Control. Release*, *102*(1), 23–38.

Velpandian, T., (2009). Intraocular penetration of antimicrobial agents in ophthalmic infections and drug delivery strategies. *Expert. Opin. Drug Deliv.*, *6*(3), 255–270.

Visser, C. C., Voorwinden, L. H., Crommelin, D. J., Danhof, M., & De Boer, A. G., (2004). Characterization and modulation of the transferrin receptor on brain capillary endothelial cells. *Pharm. Res.*, 21(5), 761–769.

Wakeyama, T., Ogawa, H., Iida, H., Takaki, A., Iwami, T., Mochizuki, M., & Tanaka, T., (2003). Effects of candesartan and probucol on restenosis after coronary stenting: results of insight of stent intimal hyperplasia inhibition by new angiotensin II receptor antagonist (ISHIN) trial. *J. Circ.*, *67*, 519–524.

Wan, T. C., Tosh, D. K., Du, L., Gizewski, E., T Jacobson, K. A., & Auchampach, J. A., (2011). Polyamidoamine (PAMAM) dendrimer conjugate specifically activates the A3 adenosinereceptor to improve postischimic/ reperfusion function in isolated mouse hearts. *BMC. Pharmacol.*, *11*, 11.

Wang, C., Sun, X., Wang, K., Wang, Y., Yang, F., & Wang, H., (2015). Breast cancer targeted chemotherapy based on doxorubicin-loaded bombesin peptide modified nanocarriers. *Drug Deliv.*, *27*, 1–6.

Wang, Y. Y., Lui, P. C., & Li, J. Y., (2009). Receptor-mediated therapeutic transport across the blood-brain barrier. *Immunotherapy*, *1*(6), 983–993.

Wang, Z. J., Chauvin, B., Maillard, P., Hammerer, F., Carez, D., Croisy, A., Sandré, C., Chollet-Martin, S., Prognon, P., Paul, J. L., Blais, J., & Kasselouri, A., (2012). Glycodendrimeric phenylporphyrins as new candidates for retinoblastoma PDT, blood carriers and photodynamic activity in cells. *J. Photochem. Photobiol. B.*, *115*, 16–24.

Wolinsky, J. B., & Grinstaff, M. W., (2008). Therapeutic and diagnostic applications of dendrimers for cancer treatment. *Adv. Drug Deliv. Rev.*, *60*, 1037–1055.

Woller, E. K., & Cloninger, M. J., (2001). Mannose functionalization of a sixth generation dendrimer. *Biomacromolecules*, *2*, 1052–1054.

Wu, G., Barth, R. F., Yang, W., Kawabata, S., Zhang, L., & Green-Church, K., (2006). Targeted delivery of methotrexate to epidermal growth factor receptorpositive brain tumors by means of cetuximab (IMC-C225) dendrimer bioconjugates. *Mol. Cancer Ther.*, *5*, 52–59.

Xie, L., Dong, X., & Shi, W., (2001). Treatment of fungal keratitis by penetrating keratoplasty. *Br. J. Ophthalmol.*, *85*(9), 1070–1074.

Yamamoto, S., Tanigawa, H., Li, X., Komaru, Y., Billheimer, J. T., & Rader, D. J., (2011). Pharmacologic suppression of hepatic ATP-binding cassette transporter 1 activity in mice reduces high-density lipoprotein cholesterol levels but promotes reverse cholesterol transport. *Circulation*, *124*, 1382–1390.

Yang, H., & Lopina, S. T., (2007). Stealth dendrimers for antiarrhythmic quinidine delivery. *J. Mater. Sci: Mater. Med.*, *18*, 2061–2065.

Yang, H., Tyagi, P., Kadam, R. S., Holden, C. A., & Kompella, U. B., (2012). Hybrid dendrimer hydrogel/PLGA nanoparticle platform sustains drug delivery for one week and antiglaucoma effects for four days following one-time topical administration. *ACS Nano.*, *6*(9), 7595–7606.

Yang, T., Cui, F. D., Choi, M. K., Cho, J. W., Chung, S. J., Shim, C. K., & Kim, D. D., (2007). Enhanced solubility and stability of PEGylated liposomal paclitaxel: *in vitro* and *in vivo* evaluation. *Int. J. Pharm.*, *338*(1–2), 317–326.

Youngren, S. R., Tekade, R. K., Gustilo, B., Hoffmann, P. R., & Chougule, M. B., (2013a). STAT6 siRNA matrix-loaded gelatin nanocarriers: formulation, characterization, and ex vivo proof of concept using adenocarcinoma cells. *Biomed. Res. Int.*, 1–13.

Youngren, S. R., Tekade, R. K., Hoffmann, P. R., & Chougule, M. B., (2013b). Biocompatible nanocarrier mediated delivery of STAT-6 siRNA to cancer cells. Proceedings of the 104th Annual Meeting of the American Association for Cancer Research, Washington, DC. Apr. 6–10, *Cancer Research*, *73*(8), Abstract no. 3313.

Yuan, Q., Fu, Y., Kao, W. J., Janigro, D., & Yang, H., (2011). Transbuccal delivery of CNS therapeutic nanoparticles, synthesis, characterization, and *in vitro* permeation studies. *Acschem. Neurosci.*, *2*(11), 676–683.

Zee-Cheng, R. K., & Cheng, C. C., (1989). Delivery of anticancer drugs. *Methods. Find. Exp. Clin. Pharmacol.*, *11*(7–8), 439–529.

CHAPTER 8

DENDRIMERS IN ORAL DRUG DELIVERY

SOUGATA JANA,[1] JOYITA ROY,[1] NILOFER JASMIN,[1] BIBEK LAHA,[1] and SUDIPTA DAS[2]

[1]Department of Pharmaceutics, Gupta College of Technological Sciences, Ashram More, GT Road, Asansol–713301, West Bengal, India, Mobile: (+91) 9434896683, E-mail: janapharmacy@rediffmail.com

[2]Department of Pharmaceutics, Netaji Subhas Chandra Bose Institute of Pharmacy Nadia–741222, West Bengal, India

CONTENTS

ABSTRACT

Oral route (OR) of drug delivery is most important for cost-effectiveness, easy administration, on-invasive nature, and patient compliance. OR is the most common for systemically acting drug and more emphasis to gastro-intestinal (GI) absorption of drug. Drugs those are rapidly absorbed from the gastrointestinal tract (GIT) with short half-life ($t_{1/2}$) are quickly eliminated from the systemic circulation; this leads to incomplete absorption from the small intestine. Currently, nanostructure dendrimers are an important polymeric drug-delivery systems utilized worldwide. Dendrimers are three dimensional (3-D), nanometric size range, globular shape, and large number of surface groups. It is also called versatile biomaterial due to delivery of hydrophobic and hydrophilic drugs. It is biocompatible, monodispersity, and nontoxic nature. Various types of drugs are conjugated with dendrimer and deliver to the target sites for prolong periods. In this chapter, we focused on the especially oral delivery systems of dendrimer and its versatile applications.

8.1   INTRODUCTION

Oral route (OR) of drug delivery is most important for cost-effectiveness, easy administration, on-invasive nature, and patient compliance. Presently, oral dosage forms are modified from immediate release (IR) to controlled release (CR) and site-specific delivery (SSD) (Dhole et al., 2011). OR is the most common for systemically acting drug and more emphasis to gastrointestinal (GI) absorption of drug. Drugs are rapidly absorbed from the gastrointestinal tract (GIT) with short half-life ($t_{1/2}$) and are quickly eliminated from the systemic circulation; this leads to incomplete absorption from the small intestine. In that case, multiple dosing is essential for desired therapeutic effect. Oral sustained release (OSR) formulations are fabricated to deliver the drug for prolong period in the GIT, and thus maintain plasma drug concentration for longer duration. So, the design and fabrication of effective controlled release drug delivery systems (CRDDS) for favorable absorption of drugs from the GIT is a complex process. Because of extent of drug absorption through GIT is related to the residence time of the delivery system in the small intestinal mucosa. Small intestine is the major site for most drug absorption, long intestinal transit time is desirable for complete drug absorption.GI residence time of formulation depends of the GI motility and drug concentration. Generally, intestinal transit time is 5 minutes in the duodenum, 2 h in the jejunum, 3–6 h in the ileum, 0.5–1 h in the cecum, and 6–12 h in the colon. Delayed intestinal transit time is essential for a) sustained release (SR) formulations (drugs that release/dissolve slowly from their dosage form); b) intestinal dissolve drug; and c) drugs that absorb from specific sites in the intestine. So, intestinal transit time is a key parameter for poorly absorbed drug. Mucoadhesion mechanism has been adapted for the development of GI retention and bioavailability of solid oral dosage form (Deshpande et al., 1996; Cuña et al., 2001; Suk et al., 2009; Tang et al., 2009; Li et al., 2015; Luo et al., 2015). However, drug-loaded polymeric system has been developed for the stability of GI formulation for oral drug absorption. Currently, dendrimers are an important drug-delivery systems utilized worldwide.

8.2 STRUCTURE

The term dendrimer is derived from a Greek word "dendron" that means "tree," which is logical in view of their typical structure having number of branching units. Dendrimers are defined as three-dimensional (3D), nanometric size range, globular shape and large number of surface groups. It is synthetic macromolecules that are characterized by high branching points. From the literature, it also called as "dendritic molecules," "nano-metric architectures," "arborols," and "cascade molecules" (Newkome et al., 1985; Tomalia et al., 1990). Dendrimers are polymeric macromolecular structure that is composed of a series of branches around an inner core (Figure 8.1) (Basavaraj et al., 2009). Where dendritic cores can act in a "host" and drug molecules act as "guests," drugs are released from the dendritic core in a controlled manner. The components of dendrimers are (i) an initiator core; (ii) an interior layer (generations), that made of repeating units, radially attached to the initiator core; and (iii) exterior part (terminal functionality) that attached to the outermost interior generation (Tomalia et al., 1985; Stevelmens et al., 1996).

It has various advantages such as 1–100 nm size, less susceptible to uptake by the RS (reticuloendothelial system), and easily passes through biological barriers (vascular endothelial tissues) compared to other branched polymers. It is also called versatile biomaterial due to delivery of hydrophobic and hydrophilic drugs (Tajarobi et al., 2001; El-Sayed et al., 2002; Jevprasesphant et al., 2004). Generally, dendrimer is synthesis by two methods, i) divergent (starting from the central core) and ii) convergent (starting from the outermost residues that are top-down approach). Drug-loaded polymer matrix increases biological half-life, increases the drug concentration at the site of action, and decreases the non-specific toxicity (Duncan, 2003, 2006).

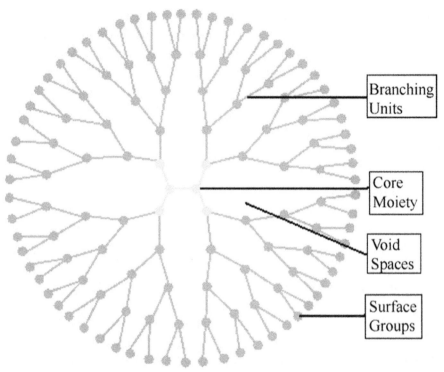

FIGURE 8.1   Structure of dendrimer (Reprinted with permission from Nanjwade, B. K., Bechra, H. M., Derkar, G. K., Manvi, F. V., & Nanjwade, V. K., (2009). Dendrimers: emerging polymers for drug-delivery systems. Eur. J. Pharm. Sci., 38, 185–196. © 2009 Elsevier.).

8.3   CHARACTERISTIC OF DENDRIMERS

Dendrimers showed significantly improved physicochemical properties as compared to the general linear polymers. Versatile characteristics properties of dendrimer are given below:

- Monodispersity
- Branched structure
- Large number of surface groups
- Surface charge
- Nanosize and shape
- Biocompatibility
- Nontoxic
- Nonimmunogenic (except for vaccines)

- To cross intestinal bio-barriers (biopermeable), blood–tissue barriers, and cell membranes, etc.
- To stay in circulation for the time needed to have a clinical effect
- To target to specific organ/tissues.
- Increases the biological half-life of drugs.

8.3.1   TRANSEPITHELIAL PERMEABILITY

The transepithelial permeability is also important for oral drug delivery. Jevprasesphant et al. (2003) investigated PAMAM dendrimers and surface-modified PAMAM dendrimers permeation across the Caco-2 cells. The result revealed that PAMAM dendrimers and surface-modified dendrimers with lauroyl groups exhibited better permeation by transcellular and paracellular pathways. In another study, D'Emanuele et al. (2004) developed propranolol-loaded PAMAM dendrimer permeation through Caco-2cell monolayers, and result showed that reduction in the effect of P-glycoprotein on intestinal absorption of propranolol. From this study, it is revealed that the dendrimer can by-pass P-glycoprotein efflux transporter and increases the oral bioavailability. Sweet et al. (2009) described the impact on the transepithelial transport on PEGlytion of the PAMAM (poly amidoamine) dendrimer. Two generation (G3.5 and G4.5) dendrimers were selected for the uptake, cytotoxicity, and transport study across the Caco-2 cells. The cell viability (>90%) was maintained by PEGylation of the dendrimers. The carboxylic acid terminated dendrimer was conjugated with methoxy polyethylene glycol (PEG-750 D_a). PEGylation of anionic dendrimer did not show any effect on the cytotoxicity activity even at the concentration of 0.1mM compared to the control. There was a decrease of uptake of the cells and transepithelial transport upon PEGylation of the G3.5 dendrimer after 60 m (with single strand PEG); on the other hand, there was an increase in the cellular uptake for the PEGylated G4.5 dendrimer but decrease in the transport through the cells. Through the confocal microscopy techniques, it was proved that PEGylated dendrimer causes modulation to the tight junction in epithelial cells (Caco-2). The change in the epithelial wall with the modulation of the dendrimer suggest that the paracellular route is being commonly used for the transport of the anionic dendrimers.

8.3.2 BIODISTRIBUTION

Sakthivel et al. (1999) investigated the biodistribution of lipidic peptide dendrimer in female Sprague–Dawley rat which was 180 g in weight and ~ 9-week old; administration of the dendrimer was done through OR. Blood was collected and different organs (stomach, kidney, spleen, small intestine, liver, and large intestine) were removed, and at different interval of time (3, 6, and 24 h), the % of dendrimer in them was determined. For blood, it was found to be 3% in 6 h. The concentration in stomach after 3 h was found to be approximately 2% of the drug administered, while in the intestine, it was 15% of drug administered and more than 20% of the administered (14 mg/kg) was obtained in the stomach. The dendrimer distribution in GIT was found to be maximum (26.4% in 6 h) than other organs like the liver (1.2%), spleen and kidney (1%). Further extension of the study was done in the lymphoid tissue and the administered dose was increased to 28mg/kg. It found that there was likely more uptake of dendrimer from the small intestine lymphoid tissue (1% in 3 h) but no absorption took place in the large intestine. The non-lymphoid absorption in small intestine (3.8, 2.5, and 0.3% in 3, 6, and 12 h, respectively) and in large intestine (1.06%–3.8%) was observed to increase with time. The absorption from the lymphoid tissue was found to be more than that from the non-lymphoid tissue.

8.4 ORAL DELIVERY

Dendrimer acts as promising biomaterials for delivery of therapeutics at targeted sites due to nanostructures (Menjoge et al., 2010). Presently, dendrimer drug conjugates have been developed to deliver the therapeutic agent to specific tissue or organ that leads to increase the efficacy of the drug and decrease the side effect. The nature of drug release from the drug-loaded dendrimer depends on the chemistry of the dendrimer structure (Florence et al., 2000; Navath et al., 2008). Dendritic polymer can tailor and permeate the drug through epithelial barrier of gut wall due to high surface charge density and compact structure. Poly(amido amine) (PAMAM) and poly(lysine) have been widely use to oral drug delivery (Wiwattanapatapee et al., 2000; El-Sayed et al., 2003; D'Emanuele et al., 2004; Jevprasesphant et al., 2004; Kolhatkar et al., 2007; Ke et al., 2008; Pisal et al., 2008; Lin et al., 2011).

Various dendrimer systems are applied in the OR that are summarized in Table 8.1.

8.4.1 PAMAM DENDRIMERS

Tomalia et al. (1985) developed hyperbranched polymers of PAMAM dendrimers. The polydispersity index (PDI) of 5.0–10.0 G PAMAM dendrimers is <1.08, which means that the particle size distribution is very uniform for each generation (Cowie, 1991). Ethylene diamine core and amido amine branching structure of the PAMAM lead alternatively to amine-terminated full generation or carboxyl-terminated half-generation dendrimers (Kolhatkar et al., 2008). A full-generation PAMAM dendrimer showed tertiary amine groups within the core (pKa = 3.86) and primary amine groups on the surface (pKa = 6.85) (Tomalia et al., 1990). Due to high surface charge, density revealed the versatile application in drug delivery and biomedical fields (Figure 8.2) (Sadekar et al., 2012).

Wiwattanapatapee et al. (2003) described the designing of water-soluble PAMAM dendrimer in conjugation with 5-amino salicylic acid (5-ASA) for colon delivery. The conjugation of the dendrimer with the drug was done by using azo-bond, p-aminobenzoic acid (PABA) and p-aminohippuric acid (PAH) containing different spacer molecules. The PABA and PAH spacers containing dendrimer along with rat cecal contents release the drug (5-ASA) after 24 h was 45.6% and 57% of the dose at 37 °C. The release of 5-ASA from the prodrug (sulphasalazine) was found to be faster than the dendrimer conjugates, nearly 80.2% of the dose in 6 h. No evidence of the drug was found from the dendrimer conjugate incubation with homogenate stomach (pH 1.2) or phosphate buffer (pH 6.8). Na et al. (2006) investigated PAMAM dendrimer as a potent carrier of hydrophobic drugs. Ketoprofen was the drug of choice for the experiment, which work by inhibiting the prostaglandin synthetase and has some side effects such as local and systemic disturbances in GIT. It has been used in the management of diseases like ankylosing spondylistis, mild-to-moderate pain and rheumatoid arthritis and dysmenorrheal. The *in vitro* release study was done in a dialysis bag with distilled water as the media and the release after two (2 h) was showed that 66% and 33% for pure and dendrimer complex of ketoprofen and after ten (10 h) it was showed that 76% and 57%, respectively. Thus, this gave evidence that the dendrimer complex has slower releasing profile.

TABLE 8.1 Various Application of Conjugated Dendrimer in Oral Delivery

Dendrimer–drug conjugate	Therapeutic agent	Therapeutic approaches	References
PPI dendrimer	Albendazole	Anthelmintic drug delivery	Mansuri et al., 2016
PAMAM dendrimers	Ketoprofen	Anti-inflammatory drug delivery	Na et al., 2006
Polyamidoamine (PAMAM) dendrimer	5-aminosalicylic acid	Anti-inflammatory drug delivery	Wiwattanapatapee et al., 2003
PAMAM dendrimers	Naproxen	Anti-inflammatory drug delivery	Najlah et al., 2007
Poly(amidoamine) dendrimer	Resveratrol	Permeability	Chauhan 2015
Poly(amido amine) (PAMAM) dendrimers	Mannitol	Permeability	Hubbard et al., 2015
Lauroyl-G3 PAMAMdendrimers	Propranolol	Anti-cancer drug delivery	D'Emanuele et al. 2004
PAMAM dendrimers	Camptothecin	Anti-cancer drug delivery	Sadekar et al., 2013
Polyamidoamine (PAMAM) dendrimers graft	5-fluorouracil	Anti-cancer drug delivery	Tripathi et al., 2002
Hyaluronic acid-conjugated polyamidoamine dendrimers	3,4-difluorobenzylidene curcumin	Anti-cancer drug delivery	Kesharwani et al., 2015
3.5 GPAMAM dendrimers	7-ethyl-10-hydroxy-campothecin (SN38)	Anti-cancer drug delivery	Goldberg et al., 2011
Polyamidoamine (PAMAM)	Short hairpin RNA (shRNA)	Anti-cancer drug delivery	Liu et al. 2011
Hydroxylated and internally quaternised poly(propylene imine) dendrimers	Norfloxacin	Anti-bacterial drug delivery	Murugan et al., 2013
poly (amidoamine) (PAMAM) dendrimers	Mannitol	Permeability	El-Sayed et al., 2002
anionic PAMAM dendrimers	Serosal fluid	Permeability	Wiwattanapatapee et al., 2000

TABLE 5.1 (Continued)

Dendrimer–drug conjugate	Therapeutic agent	Therapeutic approaches	References
Polyamidoamine (PAMAM)	Silybin	Enhanced aqueous solubility and oral bioavailability	Huang et al., 2011
PAMAM dendrimer	Simvastatin	Lipid-lowering drug delivery	Kulhari et.al (2011)
amine terminated PAMAM) dendrime	Simvastatin	Lipid-lowering drug delivery	Qi et al. (2015a)
G5-PEG PAMAM dendrimer (PEGylated poly(amidoamine) dendrimer)	Probucol	Lipid-lowering drug delivery	Qi et al. (2015b)
Poly (amidoamine) dendrimer	Ramipril and hydrochloro-thiazide	Antihypertensive and diuretics	Singh et al., 2017
Thiolated dendrimer	Acyclovir	Ant-viral delivery	Yandrapu et al., 2013

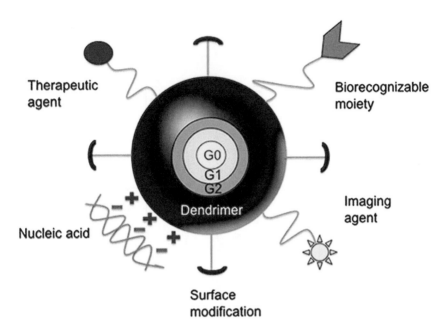

FIGURE 8.2 A cartoon representing a dendrimer-based delivery system functionalized with therapeutic agent, biorecognizable moiety, imaging agent, nucleic acid and surface-modifying groups (Reprinted with permission from Sadekar, S., & Ghandehari, H., (2012). Transepithelial transport and toxicity of PAMAM dendrimers: Implications for oral drug delivery. Adv. Drug Deliv. Rev., 64, 571–588. © 2012 Elsevier.).

The antinociceptive study was done using acetic acid induced writhing mice model. The average writhing of control was found to be 27.4 and the writhing for the pure drug showed a significant decrease in the number of writhing with max response shown at 0.5 h which last for 2 h, whereas the mice treated with the drug complex showed the maximum anti-nociceptive effect within 3 h and lasted for 6 h, thus it revealed that a prolong action of the ketoprofen drug complex. *In vivo* investigation in mice model showed that a favorable effect. The maximum plasma concentration (C_{max}) attained by the pure drug was 48.31μ/mL/hrs and $[AUC]_{1-12}$ was 137.23 μ/mL/h, whereas the C_{max} attained by the dendrimer complex was found to be 51.53μ/mL/hrs and $[AUC]_{0-12}$ was 160.96 μ/mL/h. Najlah et al. (2007) prepared naproxen-PAMAM dendrimer conjugates for oral delivery of enhancement of poor water soluble drug. The poor water solubility of naproxen hinders its oral bioavailability. Human plasma (80%) and rat liver homogenate (50%) were used for the evaluation of the trans-epithelial permeability of naproxen-PAMAM dendrimer. Two different linkers (lactate acid and diethylene glycol) were used to conjugate between naproxen and PAMAM dendrimer and these linkages stated good stability of formulated dendrimer. PAMAM-naproxen-lactate dendrimer were more stable and slowly hydrolysis (($t_{1/2}$= 180 min) in liver homogenate (50%) and human plasma (80%), whereas PAMAM-naproxen-diethylene glycol dendrimer showed the first release of naproxen and hydrolyzed more rapidly in human plasma (80%) (i.e., $t_{1/2}$ is 51 m) and was rapidly cleaved in liver homogenate (50%) ($t_{1/2}$ is 4.7 m). Conjugate dendrimer showed non-toxic nature when exposed to the caco-2 cells for 3 h. Finally, authors revealed that naproxen-loaded conjugated PAMAM dendrimer could enhance the oral bioavailability and conjugate PAMAM–naproxen-diethylene glycol dendrimer may be utilized controlled delivery of naproxen. Recently, Murgan et.al (2014) describe the quaternized poly(propylene imine) dendrimer (QPPI-G3) as a carrier of poor water soluble anti-inflammatory drug, nimesulide (NMD). The surface amine groups of poly(propylene amine) dendrimer were treated with glycidytrimethyl ammonium chloride to form QPPI(G3). Different techniques such as FTIR, 1H, and 13C NMR as well as MALDI–TOF mass spectral techniques are used for the characterization of QPPI dendrimer. The *in vitro* drug release, drug solubility, and cytotoxic studies are done to determine the drug carrying potential of the dendrimer I(QPPI). Studies showed that the solubility and sustain release character of the drug was increased in the presence of QPPI (G3) dendrimer; even the complexation gave the same results. This was proved by UV, DSC, and DLS, NMR (1H and 2D) techniques. There was an increase of bioavailability and tolerance concentration of the drug (NMD) when formulated with dendrimer, which was revealed by the cytotoxicity study by MTT assay on the vero and HBL-100 cellies. Though the drug loading for the commercial PPI (G3) is 108% but in 300 m the drug release is only 90% with higher toxicity, whereas the quaternized dendrimer (G3) has low drug loading (72%) and drug release (35%) compared to the commercial dendrimer, however, it has low toxicity. The dendrimer was an effective carrier for the poor aqueous soluble drug (NMD).

8.4.2   DELIVERY OF ANTHELMINTIC DRUG

Mucoadhesive formulation has been important for drug delivery for prolong period due to high intimate contact and surface charge. Mansuri et al. (2016)

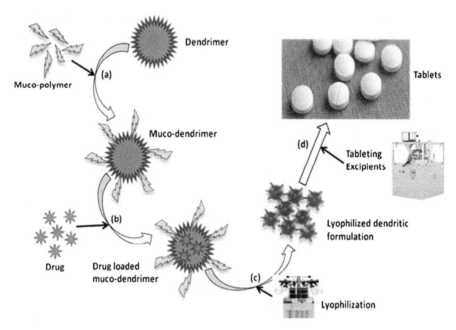

FIGURE 8.3 Illustration showing formulation of mucoadhesive dendrimer containing tablet (Reprinted with permission from Mansuri, S., Kesharwani, P., Tekade, R. K., & Jain, N. K., (2016). Lyophilized mucoadhesive-dendrimer enclosed matrix tablet for extended oral delivery of albendazole. Eur. J. Pharm. Biopharm., 102, 202–213. © 2016 Elsevier.).

fabricated mucodendrimer tablet of anthelmintic drug albendazole (ABZ) (Figure 8.3).

Fifth generation (5.0 G) PPI dendrimer (muco-PPI) was attached with chitosan to produce mucoadhesive complex which was further characterized by FTIR, electron microscopy, UV, and 1H NMR methods. ABZ entrapment efficiency was performed in phosphate buffer Saline pH 7.4. *In vitro* ABZ release from the matrix tablet was evaluated in pH 1.2 (simulated gastric fluid; SGF), pH 7.4 PBS and pH 6.8 (simulated intestinal fluid; SIF) at 37 ± 0.5 °C. *In vitro* release of ABZ from F4 (muco-PPI-ABZ matrix tablet) was found to be slower (~72% at intestinal pH up to 48 h) in respect to F1(ABZ tablet) ~92% at physiological pH up to 6 h, F2(convectional ABZ matrix tablet) ~96% at pH 6.8 up to20 h and ~98% release from plain ABZ at pH 1.2 up to1 h (Figure 8.4).

The *in vivo* pharmacokinetic parameters were observed in rabbit model. Plasma C_{max} of the muco-PPI dendrimer showed ~2 μg/mL, which was higher than the free ABZ (0.19 μg/mL) and 0.20 μg/mL of conventional tablet. The

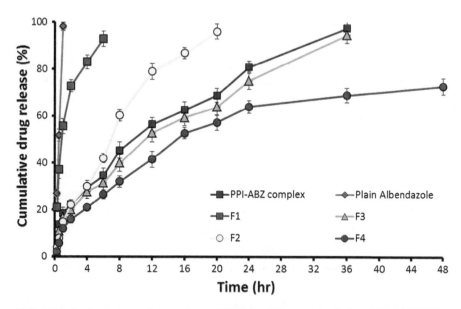

FIGURE 8.4 *In vitro* cumulative release of ABZ for different formulations. Where, F1(ABZ tablet), F2(convectional ABZ matrix tablet), F3(PPI-ABZ matrix tablet) and F4(muco-PPI-ABZ matrix tablet) Results are represented as mean ± SD from three distinct studies (n = 3) (Reprinted with permission from Mansuri, S., Kesharwani, P., Tekade, R. K., & Jain, N. K., (2016). Lyophilized mucoadhesive-dendrimer enclosed matrix tablet for extended oral delivery of albendazole. Eur. J. Pharm. Biopharm., 102, 202–213. © 2016 Elsevier.).

F4 formulation showed increased MRT (mean retention time) and biological half-life ($t_{1/2}$) two-fold as compared to other formulation.

8.4.3   DELIVERY OF LIPID-LOWERING DRUG

Qi et al. (2015a) observed, transepithelial transport and oral bioavailability of amine terminated PAMAM dendrimer (G5-NH$_2$) containing simvastatin (SMV). SMV is an inhibitor of 3-hydroxy-3-methyl coenzyme A (HMG-CoA) reductase, decreases blood cholesterol level. SMV is poorly water soluble and irregular intestinal absorption that leads to decreases oral bioavailability (<5%). *In vitro* transepithelial transport of formulated dendrimer was evaluated in caco-2 cells, and the result revealed significant increase the solubility and transepithelial transport. *In vivo* experiments were performed in Male Sprague Dawley rats, and the result showed that better oral bioavailability (2.5 times higher for SMV/G5-NH$_2$ complexes and 3.7 times higher for the SMV-liposomes, as compared to free SMV).

Further Qi et al. (2015b) investigated for the increasing of oral bioavailability and plasma lipid lowering effect of probucol (PB) by nanostructure of lipid carriers (NLCs) and PEGylated poly (amidoamine) dendrimer (PEG-PAMAM). Probucol-combined drug delivery system (PB-CDDS), PB-NLCs/G5-PEG, and PB-NLCs/G7-PEG were formed by incorporating PEG –PAMAM dendrimer (G5/G7) in PB-NILCs. Particle size, zeta-potential and PDI of PB-NLCs evaluated in Malvern Zetasizer Nano-ZS. Cellular uptakes of PB were performed in caco-2 cells and result showed that greater uptake of NLCs. *In vitro* PB releases from the formulation were evaluated in dialysis bag method SIF pH 6.8 (phosphate buffer). *In vivo* cholesterol lowering effects performed in mice model. In another study, Kulhari et al. (2011) described the performance of the PAMAM dendrimer as oral drug delivery with the help of three varieties of surface groups. Different characterization of dendrimer such as solubility, dissolution, stability and *in vitro* release studies were performed. The PEGylated dendrimer was found to have maximum solubility enhancement property, that is, 33 times than amine dendrimer, which was 23 times followed by hydroxyl dendrimer, that is, 17.5 times. The dissolution of the drug complex was done in USP apparatus with USP dissolution media (pH7) and SGF(pH1.2), it was noted that the dissolution of simvastatin from dendrimer complex was more compared to that of pure simvastatin, and the dissolution was found more in the SGF media (pH-1-2). The *in vitro* release study was done by dialysis tube diffusion technique using phosphate buffer saline (PBS-pH7.4) as the media. There was linear correlation found between solubility and concentration when solubility profile of simvastatin-dendrimer complex (SMV–dendrimer) was studied, which indicates it follows Higuchi AL-type diagram. The FTIR characterization of the drug dendrimer complexes (SMV-dendrimer) showed the bonding between the functional groups of drug, that is, -OH and -COOH and dendrimer, that is, NH_2 and OH. Controlled release of simvastatin was also noted in case of drug dendrimer complex (PEGylated-SMV-dendrimer), where the drug dendrimer complex releases the drug for 5 days on the other hand the free simvastation was released within 5 h. Other dendrimers showed the release of simvastatin for 24 h. Stability study stated that storage condition of PEGylated –SMV-dendrimer, which need to be stored at room temperature and in dark (amber color vials), whereas for amine dendrimer it was dark and at 0°C temperature. Further Kulhari et al. (2013) evaluated the pharmacokinetic and pharmacodynamic parameters of SMV-loaded PAMAM dendrimer for oral preparation. The formulated nano-dendrimer characterized by Fourier-transformed infrared (FTIR) spectroscopy. SMV loaded dendrimer showed better pharmacokinetic characteristics than pure SMV suspension. The C_{max} of SMV concentration is higher (3.8 μg/mL) in dendrimer formulation than pure SMV (2.3 μg/mL). The formulated dendrimer showed higher mean residence time (3 to 5 times) than pure SMV and SMV elimination rate and absorption were decreased.

8.4.4 DELIVERY OF ANTI-CANCER DRUG

Sadekar etal. (2013) investigated PAMAM dendrimer as a promising intestinal penetration enhancer, increased drug solubility and as carrier for the poor soluble drug *in vitro* as well as in situ. Camptothecin, which is the model drug for the experiment, has been found to have poor bioavailability and GI toxicity, which limits its use as oral delivery even though it has therapeutic value. There were very limited tests for the evaluation of the *in vivo* characterization of PAMAM dendrimer through oral delivery. In the following experiment, cationic dendrimer, anionic dendrimer, carboxyl-terminated PAMAM dendrimer generation 3.5 (G3.5), amine terminated PAMAM dendrimer of generation 4 (G4) was delivered along with the camptothecin (dose 5mg/kg) in a 8 to 6 weeks female mice model through oral gavage. The dose of PAMAM G4-NH$_2$ and PAMAM G3.5-COOH was found to be 100 and 300 mg/kg and 300 to 1000 mg/kg, respectively. It was noted that camptothecin has greater affinity for PAMAM dendrimer (G-4) than PAMAM G-3.5 in the formulation, due to the presence of surface electrostatic charge on PAMAMA G4. After the time interval of 2 h (t_{max}), it was observed that both the generation (G-4/G-3.5) PAMAM dendrimer increased the camptothecin solubility and oral absorption in SGF, which was nearly 2 to 3 fold. The PAMAM dendrimer does not influence the transport of mannitol, indicating that the absorption of the drug was not due to opening of the tight junction of the transepithelial. The dose of the drug or along with dendrimer do not show any disruption on the microvilli; sometimes some significant changes are observed when the drug was given alone, but no such disruption was observed with dendrimer (drug + dendrimer). This gives evidence that the dendrimer provides a protective coating to the potential of the drug (free). Goldberg et al. (2011)

fabricated that complexes of anticancer drug 7-ethyl-10-hydroxy-campothecin (SN38) with 3.5 GPAMAM dendrimers revealed good oral bioavailability with reduced toxicity. In another experiment, Kesharwani et al. (2015) describe the formulation of 3,4 diflurobenylidene curcumin (CDF) as a targeted delivery and evaluated it in CD44 cells in pancreatic cancer. The targeted PAMAM dendrimer of CDF for CD44 was formulated with the help of using amine-terminated poly(amidoamine) (PAMAM –G4) as nanocarrier and hyaluronic acid as the ligand. The nano-system dendrimer has 9.3 nm and −7.02 mv, particle size and surface charge, respectively. A dose-dependent toxicity was observed when over expressing CD44 receptor, MiaPaCa-2 and AsPC-1 in human pancreatic cancer cells were treated with HA-PAMAM-CDF (dendrimer-G4). When the CD44 receptor on the MiaPaCa-2 cells was blocked with free excess HA(soluble) before the treatment with the nano-formulation (HA-PAMAM-CDF), there was 1.71 times increase in the IC50 value in comparison with non-targeted formulation (PAMAM CDF); thus, it confirms the target specificity of the nano-system formulation (HA-PAMAM-CDF). The comparison between the HA-PAMAM-CDF and to PAMAM-CDF formulation was done by fluorescence studies and it revealed better cellular uptake of HA-PAMAM-CDF to the cancer cells. It showed site-specific targeting of CDF via CD44 receptor with therapeutic and safety margin. Another study by Liu et al. (2011) described the dependence of therapeutic application of RNA interference (RNAi) on the efficient delivery system. The important targets for oral cancer are human telomerase reverse transcriptase (hTERT), the catalytic subunit of telomerase complex. The anticancer effect of polyamidoamine (PAMAM) dendrimer-mediated short hairpin RNA (shRNA) against hTERT in oral cancer was characterized in *in vitro* the dendrimer-mediated shPNA silences the hTERT gene, which results in inhibition of cell growth and leads to apoptosis. Treatment of the tumor cell on xenograft model was done with shRNA dendriplex; it showed that the silencing of gene with the help of dendrimer complex (along with RNAi-mediated hTERT gene silencing) may result in promising approach, where hTERT has been abundantly expressed. For the preventing of the development of resistance, Tekade et al. (2009) developed the combination systems for leukemia. The work include the loading of two anti-leukemia drugs (Methotrexate and All-trans Retinoic acid, [ATRA]), simultaneous within the PAMAM dendrimer to develop an novel technology. One molecule of PAMAM dendrimer was found to entrap approximately 27.02 and 8.00 molecules of drugs (Methotrexate and all-trans Retinoic acid) at optimized PH and dialysis time. At different pH 4, 7.4, and 10, the *in vitro* release profiling of both the drugs were done in a dialysis bag with PBS pH 7.4. The analysis was done with the HPLC method. At pH 7.4% cumulative release of drug (MTX and ATRA) was found to be approximately 21.92, 56.95 at 2h and 88.43, 93.35 at 8h, respectively. The cumulative release of MTX and ATRA was found to be 46%, 64% at 2 h and 91%, 94% at 8 h, respectively, at pH 4 and at pH 10, even after 8 h a hampered release of 33.84% and 50.12% were observed as opposed to a significantly rapid release of 91.65% and 94.61%, respectively. This effect was due to the bulkiness of the dendrimer which is caused due to de-protonation of the dendrimer in alkaline condition. The release kinetics of the formulation was dependent upon degree of dendrimer protonation and the sustained and controlled release of the drug was found best at pH 7.4. The toxicity of the dendrimer formulation was reduced by modifying the terminal loading dendrimer with less-hemolytic bioactive, ATRA. To determine the cytotoxicity study was performed by MTT assay on the HeLa cell lines, after 72 h the cytotoxicity was observed to be insignificant and the dual drug loading was found to be more efficient, that is, IC50 = 0.5mM then the free combination of drug, that is, IC 50 = 0.75 Mm.

8.4.5 DELIVERY OF ANTI-VIRAL DRUG

Yandrapu et al. (2013) described the development of mucoadhesive drug-delivery system in the form of a novel thiolated dendrimer. It was prepared by conjugating cysteamine Hcl (Cys) of two different molar concentrations (1:30 and 1:60) of DCys1 and DCys2, respectively, with PAMAM dendrimer of generation 3.5. The model drug acyclovir (AC) was encapsulated within the thiolated dendrimer and various experiments were done such as drug release character in PBS buffer (pH 7.4), drug loading was determined by HPLC analysis, thiol content and mucoadhesive characteristics was determined using rat small intestinal. The conjugated dendrimer revealed that the thiol content was approximately 10.56±0.34 and 68.21±1.84 µM/mg of the conjugate (DCys1 and DCys2). The loading of acyclovir was found to be highest in dendrimer conjugate as compared to others (other DCys1Ac and DCys2Ac conjugates). There was higher sustained release of the drug and mucoadhesion with dendrimer conjugation of the drug (acyclovir). The mucoadhesive of DCys1AC and DA was 1.53 and 2.89 times, respectively, lesser than DCys2AC.

8.4.6 DELIVERY OF ANTI-TUBERCULAR DRUG

Bllini et al. (2015) synthesized the conjugation of the anti-tuberculosis drug (rifampicin) with the fourth generation poly (amidoamine) dendrimer by, means of molecular dynamics simulation. At neutral pH, the load capacity of rifampicin (RIF) was found to be 20 RIF per 4-PAMAM (G4). At neutral and acidic different pH, the 20 RIF complex with dendrimer was subjected to 100 ns molecular dynamic simulations. The dendrimer complex (RIF20-PAMAM) was found to be stable neutral pH simulation in comparison with the acidic pH simulation where the drug (RIF) was rapidly expelled out to the media or solvent. The property of being stable at physiological pH and high release of the drug at acidic pH makes it a good targeting drug delivery to the Mycobacterium, which forms colonies at the macrophages. Thus, in view of PH and stability, PAMAM dendrimer could be considered as a suitable drug-delivery system of RIF drug.

8.4.7 DELIVERY OF COMBINED DRUG

Singh et al. (2017) prepared the hybrid PAMAM dendrimer as a drug carrier in combination therapy of anti-hypertensive drug ramipril (RAPL) and diuretic drug hydrochlorothiazide (HCTZ). The complex of dendrimer was formed by phase equilibrium method. The concentration and PH of the dendrimer solution affect the solubility of the both drugs (RAPL and HCTZ). Both the drugs showed a nonlinear relationship with the concentration of dendrimer. There is 4.91 times increase of solubility of RAPL along with amine-terminated dendrimer and 3.72 times increase of solubility of HCTZ with carboxy-terminated, with a concentration of 0.8% (w/v) concentration of dendrimer. The solubility gets enhanced with most with carboxy-terminated dendrimer. The characterization of the complexes was done by Fourier transform infrared (FIR) spectroscopy, nuclear magnetic resonance and high performance liquid chromatography techniques. The *in vitro* dissolution study of pure, dendrimer formulation with individual drug loaded and hybrid formulation of dendrimer was done in USP dissolution apparatus with dissolution medium PH-7 and stimulated gastric fluid (PH 1.2). The dissolution becomes faster with dendrimer than the pure drug RAPL and HCTZ. The hybrid formulation of dendrimer also shows the similar dissolution profile as compared to the dendrimer loaded with individual drug. The hybrid formulation was much more stable in dark and in refrigerator for 5 weeks. The formulation could be used in multitasking in loading and delivering the drug.

8.4.8   MISCELLANEOUS

Lin et al. (2011) described the influence of PAMAM dendrimer of the poorly absorbable drug in intestine with the help of rats using loop method. The drug used in the experiment are calcitonin and insulin of various molecular weight, carboxyflurorescein (CF) and fluorescein isothiocyanate-dextrans (FDs), which were used as poorly water absorbable model drugs. It was observed that in the presence of PAMAM dendrimer, there was a significant increase in absorption of calcitonin, CF and FD4 drugs from the small intestine of the rat. It was noted that the absorption from the small intestine was both generation and concentration dependent, here the absorption of CF drug was enhanced in the presence of 0.5% w/v PAMAM dendrimer G2 (11 times) and G3 than generation G1 and G0 dendrimers. But the absorption of macromolecules was not enhanced by PAMAM dendrimer (G2) which was explained by the loop method where it showed that the absorption of FD4 (M.W-4400) was increased in the presence of 0.5 w/ v (G2) PAMAM dendrimer, whereas there was no enhancement of absorption in case of FD10 (M.W-1900) and insulin from the small intestine as well as in large intestine. There was a decrease in absorption activity as there was increase in molecular weight of drugs. There was an increase of protein release from the membrane of intestine and enhancement of the activity of lactate dehydrogenase (LDH) with the use of PAMAM dendrimer (G2 0.5w/v). The amount of toxic maker and activity was found to be less in the presence of Triton X-100 (3%) which was used as positive control; the toxic maker were also not formed in the presence of G2. PAMAM dendrimer (0.05 and 0.1 w/v). Thus, PAMAM dendrimer could be safe for intestinal absorption enhancer at low concentration. In another study, Thiagarajan et al. (2013) developed anionic generation (G) 6.5 poly-(amido amine) (PAMAM) dendrimers for oral delivery. Prepared dendrimer characterized in mice model. The dendrimers were tagged with ^{125}I and their stability study evaluated. Radioactivity analysis of plasma data concluded that the presence of both small and large molecular weight (MW) compounds. Plasma area under the curve (AUC) analysis stated that an effective 9.4% bioavailability of ^{125}I marker tagged with G6.5 PAMAM dendrimer. In another experiment, Hubbard et al. (2014) investigated FITC-labeled dendrimers (carboxyl-terminated G3.5 and amine-terminated G4 PAMAM) for oral delivery. Transepithelial transport evaluated in isolated rat jejunal mucosae mounted in Ussing chambers. From this experiment, it is concluded that Papp (apparent permeability coefficients) from the apical to the basolateral side were significantly higher for FITC tagged with G3.5 PAMAM dendrimer compared to FITC single. Recently, Chauhan (2015) described the enhancement of pharmacokinetic profile of resveratrol by conjugating it with poly(aminoamine) dendrimer in the form of aqueous soluble nanostructures. The design of this drug delivery provides novel features by applying spherical structure and polyvalence in it. The model drug used here is resveratrol, which has low water solubility and it get readily metabolize. Its photosensitivity and instability in high pH also add to its limitation to formulate in oral drug delivery. The drug has poor absorptive profile and thus give poor bioavailability. The designing of the novel drug delivery not only increases the bioavailability but also its water solubility. There was enhancement of the drug stability and solubility when the PAMAM dendrimer was given in conjugation with the drug (resveratrol), and hence, the drug has good bioavailability and thus can be formulated in oral drug delivery.

8.5   CONCLUSION

Oral SR formulations are fabricated to deliver drug to the GIT, thus to maintain effective concentration of drug/therapeutic agents for a prolonged period. Genes or drugs loaded within the dendrimers matrix through non-covalent linkage such as hydrogen, ionic bond, or conjugated with the peripheral groups of dendrimers via covalent linkage. Dendrimers are versatile drug carriers due to their chemical versatility well-defined nanostructures. It is 3-D, monodispersed, highly branched, macromolecular structures, ~1–10 nm in size. Different functional groups present in the surface, monodispersity, and biocompatibility are potential for pharmaceutical and biomedical applications. The toxicity of the formulated dendrimers can be eliminated by optimizing the generation, surface modification and dose. In this chapter, special emphasis is given on dendrimer on oral drug delivery. In this article, the authors have included various drugs such as anti-inflammatory, anthelmintics, anti-cancer, anti-tubercular, anti-bacterial, and anti-hyperlipidemics conjugated with different functionalized dendrimers. The various research reports were summarized and we conclude that conjugated dendrimers successfully deliver the drug/therapeutics and increases the oral bioavailability, permeability through OR.

KEYWORDS

- **biocompatible**
- **dendrimer**
- **drug delivery**
- **oral route**

REFERENCES

Bellini, R. G., Guimaraes, A. P., Pacheco, M. A. C., Dias, D. M., Furtado, V. R., Alencastro, R. B., & Horta, B. A. C., (2015). Association of the anti-tuberculosis drug rifampicin with a PAMAM Dendrimer. *J. Mol. Graph Mode.*, *60*, 34–42.

Chauhan, A. S., (2015). Dendrimer nanotechnology for enhanced formulation and controlled delivery of resveratrol. *Ann. New York Acad. Sci.*, *1348*, 134–140.

Cowie, J. M. G., (1991). *Polymers: Chemistry & Physics of Modern Materials*. 2nd ed. Glasgow UK, Chapman & Hall., pp. 436.

Cuna, M., Alonso, M. J., & Torres, D., (2001). Preparation and *in vivo* evaluation of mucoadhesive microparticles containing amoxicillin-resin complexes for drug delivery to the gastric mucosa. *Eur. J. Pharm. Biopharm.*, *51*, 199–205.

D'Emanuele, A., Jevprasesphant, R., Penny, J., & Attwood, D., (2004). The use of a dendrimer propranolol prodrug to bypass efflux transporters and enhance oral bioavailability. *J. Control Release*, *95*, 447–453.

D'Emanuele, A., Jevprasesphant, R., Penny, J., & Attwood, D., (2004). The use ofa dendrimer-propranolol prodrug to bypass efflux transportersand enhance oral bioavailability. *J. Control. Release*, *95*, 447–53.

Dallin, H., Hamidreza, G., & David, J. B., (2014). Transepithelial transport of PAMAM dendrimers across isolated Rat Jejunal Mucosae in ussing chambers. *Biomacromol.*, *15*, 2889–2895.

Deshpande, A. A., Rhodes, C. T., Shah, N. H., & Malick, A. W., (1996). Controlled-release drug delivery systems for prolonged gastric residence: an overview, *Drug Dev. Ind. Pharm.*, *22*, 31–39.

Dhole, A., Gaikwad, P., Bankar, V., & Pawar, S., (2011). A review on floating multiparticulate drug delivery system: a novel approach to gastric retention. *Int. J. Pharm. Sci. Rev. Res.*, *6*, 205–211.

Duncan, R., (2003). The dawning era of polymer therapeutics. *Nat. Rev. Drug Discov.*, *2*, 347–360.

Duncan, R., (2006). Polymer conjugates as anticancer nanomedicines. *Nat. Rev. Cancer*, *6*, 688–701.

El-Sayed, M., Ginski, M., Rhodes, C., & Ghandehari, H., (2002). Transepithelial transport of poly (amidoamine) dendrimers across Caco-2 cell monolayers. *J. Control. Release*, *81*, 355–365.

El-Sayed, M., Rhodes, C. A., Ginski, M., & Ghandehari, H., (2003). Transport mechanism (s) of poly (amidoamine) dendrimers across Caco-2 cell monolayers. *Int. J. Pharm.*, *265*, 151–157.

Florence, A. T., Sakthivel, T., & Toth, I., (2000). Oral uptake and translocation of a polylysine dendrimer with a lipid surface. *J. Control. Release*, *65*, 253–259.

Goldberg, D. S., Vijayalakshmi, N., Swaan, P. W., & Ghandehari, H., (2011). G3. 5PAMAM dendrimers enhance transepithelial transport of SN38while minimizing gastrointestinal toxicity. *J. Control. Release*, *150*, 318–325.

Huang, X., Wu, Z., Gao, W., Chen, Q., & Yu, B., (2011). Polyamidoamine dendrimers as potential drug carriers for enhanced aqueous solubility and oral bioavailability of silybin, *Drug. Dev. Ind. Pharm.*, *37*, 419–427.

Hubbard, D., Enda, M., Bond, T., Moghaddam, S. P., Conarton. J., Scaife, C., Volckmann, E., & Ghandehari, H., (2015). Transepithelial transport of PAMAM dendrimers across isolated human intestinal tissue. *Mol. Pharm.*, *12*, 4099–107.

Jevprasesphant, R., Penny, J., Attwood, D., & D'Emanuele, A., (2004). Transport of dendrimer nanocarriers through epithelial cells via the transcellular route. *J. Control. Release*, *97*, 259–267.

Jevprasesphant, R., Penny, J., Attwood, D., & D'Emanuele, A., (2004). Transport of dendrimer through epithelial cells via the transcellular route. *J. Control. Release*, *97*, 259–267.

Jevprasesphant, R., Penny, J., Attwood, D., McKeown, N. B., & D'Emanuele, A., (2003). Engineering of dendrimer surface to enhance transepithe-lial transport and reduce cytotoxicity. *Pharm. Res.*, *20*, 1543–1550.

Ke, W., Zhao, Y., Huang, R., Jiang, C., & Pei, P., (2008). Enhanced oral bioavailability of doxorubicin in a dendrimer drug delivery system. *J. Pharm. Sci.*, *97*, 2208–2216.

Kesharwani, P., Xie, L., Banerjee, S., Mao, G., Padhye, S., Sarkar, F. H., & Iyer, A. K., (2015). Hyaluronic acid-conjugated polyamidoamine dendrimers for targeted delivery of 3, 4-difluorobenzylidene curcumin to CD44 over expressing pancreatic cancer cells. *Colloids. Surf. B. Biointerfaces.*, *136*, 413–423.

Kolhatkar, R. B., Kitchens, K. M., Swaan, P. W., & Ghandehari, H., (2007). Surface acetylation of polyamidoamine (PAMAM) dendrimers decreases cytotoxicity while maintaining membrane permeability. *Bioconjug. Chem.*, *18*, 2054–2060.

Kolhatkar, R., Sweet, D., & Ghandehari, H., (2008). Functionalized dendrimers as nanoscale drug carriers, multifunctional pharmaceutical nanocarriers. *Spring. Sci., 4*, 201–232.

Kulhari, H., Pooja, D., Prajapati, S. K., & Singh Chauhan, A., (2011). Performance evaluation of PAMAM dendrimer based simvastatin formulations. *Int. J. Pharm., 405*, 203–209.

Kulhari, H., Pooja, D., Prajapati, S. K., & Singh Chauhan, A., (2013). Pharmacokinetic and pharmacodynamic studies of poly(amidoamine) dendrimer based simvastatin oral formulations for the treatment of hypercholesterolemia. *Mol. Pharm., 10*, 2528–2533.

Li, C., Liu, Z., Yan, X., Lu, W., & Liu, Y., (2015). Mucin-controlled drug release from mucoadhesive phenylboronic acid-rich nanoparticles. *Int. J. Pharm., 1*, 261–264.

Lin, Y., Fujimori, T., Kawaguchi, N., Tsujimoto, Y., Nishimi, M., Dong, Z., Katsumi, H., Sakane, T., & Yamamoto, A., (2011). Polyamidoamine dendrimers as novel potential absorption enhancers for improving the small intestinal absorption of poorly absorbable drugs in rats. *J. Control Release, 149*, 21–28.

Liu, X., Huang, H., Wang, J., Wang, C., Wang, M., Zhang, B., & Pan, C., (2011). Dendrimers-delivered short hairpin RNA targeting hTERT inhibits oral cancer cell growth *in vitro* and *in vivo*. *Biochem. Pharmacol., 82*, 17–23.

Luo, Y., Teng, Z., Li, Y., & Wang, Q., (2015). Solid lipid nanoparticles for oral drug delivery chitosan coating improves stability, controlled delivery, mucoadhesion and cellular uptake. *Carbohydr. Polym., 20*, 221–229.

Mansuri, S., Kesharwani, P., Tekade, R. K., & Jain, N. K., (2016). Lyophilized mucoadhesive-dendrimer enclosed matrix tablet for extended oral delivery of albendazole. *Eur. J. Pharm. Biopharm., 102*, 202–213.

Menjoge, A. R., Kannan, R. M., & Tomalia, D. A., (2010). Dendrimer-based drug and imaging conjugates: Design considerations for nanomedical applications. *Drug Discov. Today, 15*, 171–185.

Murugan, E., Geetha Rani, D. P., & Yogaraj, V., (2014). Drug delivery investigations of quaternised poly(propylene imine)dendrimer using nimcsulide as a model drug. *Colloids. Surf. B. Biointerfaces., 114*, 121–129.

Murugan, E., Geetha Rani, D. P., Srinivasan, K., & Muthumary, J., (2013). New surface hydroxylated and internally quaternised poly(propylene imine) dendrimers as efficient biocompatible drug carriers of norfloxacin *Expert. Opin. Drug Deliv., 10*, 1319–1334.

Na, M., Yiyun, C., Tongwen, X., Yang, D., Xiaomin, W., Zhenwei, L., Zhichao, C., Guanyi, H., Yunyu, S., & Longping, W., (2006). Dendrimers as potential drug carriers. Part II. Prolonged delivery of ketoprofen by *in vitro* and *in vivo* studies. *Eur. J. Med. Chem., 41*, 670–674.

Najlah, M., Freeman, S., Attwood, D., & D'Emanuele, A., (2007). *In vitro* evaluation of dendrimer prodrugs for oral drug delivery. *Int. J. Pharm., 336*, 183–190.

Nanjwade, B. K. Bechraa, H. M., Derkara, G. K., Manvia, F. M., & Nanjwade, V. K., (2009). Dendrimers: Emerging polymers for drug-delivery systems. *Eur. J. Pharm. Sci., 38*, 185–196.

Navath, R. S., Kurtoglu, Y. E., Wang, B., Kannan, S., Romero, R., & Kannan, R. M., (2008). Dendrimer-drug conjugates for tailored intracellular drug release based on glutathione levels. *Bioconjug. Chem., 19*, 2446–2455.

Newkome, G. R., Yao, Z. Q., Baker, G. R., & Gupta, V. K., (1985). Micelles. Part 1. Cascade molecules: a new approach to micelles, a-arborol. *J. Org. Chem., 50*, 2003–2004.

Pisal, P. S., Yellepeddi, V. K., Kumar, A., & Palakurthi, S., (2008). Transport of surface engineered polyamidoamine (PAMAM) dendrimers across IPEC-J2 cell monolayers. *Drug Deliv., 15*, 515–522.

Qi, R., Li, Y. Z., Chen, C., Cao, Y. N., Yu, M. M., Xu, L., He, B., Jie, X., Shen, W. W., Wang, Y. N., Dongen, M. A. V., Liu, G. Q., Banaszak, H. M. M, Zhang, Q., & Ke, X., (2015b). G5-PEG PAMAM dendrimer incorporating nanostructured lipid carriers enhance oral bioavailability and plasma lipid-lowering effect of probucol. *J. Control. Release, 210*, 160–168.

Qi, R., Zhang, H., Xu, L., Shen, W., Chen, C., Wang, C., Cao, Y., Wang, Y., Mallory, A., Dongen, V., He, B., Wang, S., Liu, G., Mark, M., Holl, B., & Zhang, Q., (2015a). G5PAMAM dendrimer versus liposome: A comparison study on the *in vitro* transepithelial transport and *in vivo* oral absorption of simvastatin. *Nanomedicine, 11*, 1141–1151.

Sadekar, S., & Ghandehari, H., (2012). Transepithelial transport and toxicity of PAMAM dendrimers: Implications for oral drug delivery. *Adv. Drug Deliv. Rev., 64*, 571–588.

Sadekar, S., Thiagarajan, G., Bartlett, K., Hubbard, D., Ray, A., McGill, L. D., & Ghandehari, H., (2013). Poly (amido amine) dendrimers as absorption enhancers for oral delivery of camptothecin. *Int. J. Pharm., 456*, 175–185.

Sakthivel, T., Toth, I., & Florence, A. T., (1999). Distribution of a lipidic 2. 5 nm diameter dendrimer carrier after oral administration. *Int. J. Pharm., 183*, 51–55.

Singh, M. K., Pooja, D., Kulhari, H., Jain, S. K., Sistla, R., & Singh Chauhan, A., (2017). Poly (amidoamine) dendrimer-mediated hybrid formulation for combination therapy of ramipril and hydrochlorothiazide. *Eur. J. Pharm. Sci., 96*, 84–92.

Stevelmens, S., Hest, J. C. M., Jansen, J. F. G., Boxtel, D. A. F., De Bravander Van, Den, B., & Meijer, E. W., (1996). Synthesis, characterization and guest-host properties of inverted unimolecular micelles. *J. Am. Chem. Soc., 118*, 7398–7399.

Suk, J. S., Lai, S. K., Wang, Y. Y., Ensign, L. M., Zeitlin, P. L., Boyle, M. P., & Hanes, J., (2009). The penetration of fresh undiluted sputum expectorated by cystic fibrosis patients by non-adhesive polymer nanoparticles. *Biomaterials, 30*, 2591–2597.

Sweet, D. M., Kolhatkar, R. B., Ray, A., Swaan, P., & Ghandehari, H., (2009). Transepithelial transport of PEGylated anionic poly (amidoamine) dendrimers: Implications for oral drug delivery. *J. Control. Release, 138*, 78–85.

Tajarobi, F., El-Sayed, M., Rege, B. D., Polli, J. E., & Ghandehari, H., (2001). Transport of polyamido amine dendrimers across Madin-Darby canine kidney cells. *Int. J. Pharm., 215*, 263–267.

Tang, B. C., Dawson, M., Lai, S. K., Wang, Y. Y., Suk, J. S., Yang, M., Zeitlin, P., Boyle, M. P., Fu, J., & Hanes, J., (2009). Biodegradable polymer nanoparticles that rapidly penetrate the human mucus barrier. *Proc. Natl. Acad. Sci. USA., 106*, 19268–19273.

Tekade, R. K., Dutta, T., Gajbhiye, V., & Jain, N. K., (2009). Exploring dendrimer towards dual drug delivery: pH responsive simultaneous drug-release kinetics. *J. Microencapsul., 26*, 287–296.

Thiagarajan, G., Sadekar, S., Greish, K., Ray, A., & Ghandehari, H., (2013). Evidence of oral translocation of anionic G6. 5 dendrimers in mice. *Mol. Pharm., 10*, 988–998.

Tomalia, D. A., Baker, H., Dewald, J., Hall, M., Kallos, G., Martin, S., Roeck, J., Ryder, J., & Smith, P., (1985). A new class of polymers: starburst-dendritic macromolecules. *Polymer., 17*, 117–132.

Tomalia, D. A., Naylor, A. M., & Goddard, W. A., (1990). Starburst dendrimers: Molecular level control of size, shape, surface chemistry topology and flexibility from atoms to macroscopic matter. *Angew. Chem. Int. Ed. Engl., 29*, 138–75.

Tripathi, P. K., Khopade, A. J., Nagaich, S., Shrivastava, S., Jain, S., & Jain, N. K., (2002). Dendrimer grafts for delivery of 5-fluorouracil. *Pharmazie., 57*, 261–264.

Wiwattanapatapee, R., Carreno-Gomez, B., Malik, N., & Duncan, R., (2000). Anionic PAMAM dendrimers rapidly cross adult rat intestine *in vitro*: A potential oral delivery system?. *Pharm. Res.*, *17*, 991–998.

Wiwattanapatapee, R., Lomlim, L., & Saramunee, K., (2003). Dendrimers conjugates for colonic delivery of 5-aminosalicylic acid. *J. Control. Release*, *88*, 1–9.

Yandrapu, S. K., Kanujia, P., Chalasani, K. B., Mangamoori, L., Kolapalli, R. V., & Chauhan, A., (2013). Development and optimization of thiolated dendrimer as a viable muco-adhesive excipient for the controlled drug delivery: An acyclovir model formulation. *Nanomedicine*, *9*, 514–522.

Yellepeddi, V. K., & Ghandehari, H., (2016). Poly (amido amine) dendrimers in oral delivery, *Tissue Barriers*, *4*, e1173773, doi: 10.1080/21688370.2016.1173773.

PART III

DENDRIMERS IN SPECIALIZED THERAPEUTICS

CHAPTER 9

DENDRIMERS IN GENE DELIVERY

PIYOOSH A. SHARMA,[1] RAHUL MAHESHWARI,[2]
MUKTIKA TEKADE,[3] SUSHANT K. SHRIVASTAVA,[1]
ABHAY S. CHAUHAN,[4,5] and RAKESH K. TEKADE[2*]

[1]Department of Pharmaceutical Engineering & Technology, Indian
Institute of Technology (Banaras Hindu University), Varanasi–221005,
India

[2]National Institute of Pharmaceutical Education and Research
(NIPER)– Ahmedabad, Opposite Air Force Station Palaj, Gandhinagar,
Gujarat 382355, India

[3]TIT College of Pharmacy, Technocrats Institute of Technology,
Anand Nagar, Bhopal, Madhya Pradesh–462021, India

[4]School of Medicine and Public Health, University of Wisconsin
Madison, Madison, WI 53705, USA

[5]School of Pharmacy, Concordia University Wisconsin, Mequon,
WI 53097, USA
* E-mail: rakeshtekade@gmail.com

CONTENTS

ABSTRACT

In the present scenario, gene therapy is one of the promising approach for treating pathology or altered genetic expressions by making corrections in the host DNA expression. However, gene therapy works through either transfer, repair, or silencing of the investigated/altered gene and could be challenged by extracellular (gene packing, immune stimulation) and intracellular barriers (endosome escape, serum instability, RES recognition, cytoplasmic trafficking, nuclear entry, vector unpackaging). Therefore, the utilization of this novel approach with full potential requires the appropriate delivery of genetic materials to the host cells. The requirement of specialized delivery agent embarks the development of polymer-based nano-dimensional particles in genetic transfer/delivery. In context, highly specialized, branched and surface engineered nanoparticles well established as dendrimers have demonstrated potential binding and transporting properties in gene delivery applications. Dendrimers can be tailored to well match with the genetic material and improve the binding affinity and transport properties of the resulting complexes. In this chapter, we have explored various aspects of gene delivery through dendrimers with special emphasis on dendritic architect and dendriplexes formation. The chapter will also provide the complete applications of dendrimer delivery system for various complex diseases associated with brain, heart, and lungs.

9.1 GENE THERAPY AND GENE DELIVERY

In recent time, gene transfection has been successfully developed as very efficient technique to transfer gene related contents into particular target cell of the patients to cure diseases by altering the defective genes (Maheshwari et al., 2015a; Tekade, 2015). It is a very useful strategy to treat various diseases by replacing or disrupting the defective gene. Apart from its evolvement to treat various gene related diseases such as hemophilia (Gerrard et al., 1993), muscle dystrophy (Harper et al., 2002), thalassemia (Raja et al., 2012), sickle cell anemia (Pawliuk et al., 2001), and cystic fibrosis (Hyde et al., 1993; Zabner et al., 1993), this therapy is also being developed to treat myocardial infarction (Losordo et al., 1998; Li et al., 2001), stroke (Yenari et al., 1998; Yenari et al., 2003), cancer (Gomez et al., 1999; Roth et al., 1999; Xiao et al., 2002; Kesmodel and Spitz, 2003; Baban et al., 2010; Youngren et al., 2013a; Tekade et al., 2016), infectious diseases (Gilboa et al., 1994), neurological disorders (Friedmann, 1994; Kennedy, 1997; Lundberg et al., 2008), and wound healing (Eming et al., 2007). Gene therapy is beneficial in therapeutics by altering specific gene expression or by producing specific proteins that activate certain enzymes. Also, it involves in locating the particular gene responsible for disease condition and transferring of exogenous gene efficiently to the specific cell types. For some of the complex diseases such as cancer, alteration of gene expression for shorter duration could be sufficient, but for most of the chronic genetic diseases, long-term change in gene expression is essential (Youngren et al., 2013b; Gandhi et al., 2014; Tekade et al., 2015a). The major limitation of gene therapy includes effective and controllable transfer of the particular gene to the specific target cells (Verma and Somia, 1997; Pack et al., 2005).

The knowledge of interaction mechanism such as intercellular traffic and targeting between host cells and gene delivery vector is essential for potential gene delivery. Till date, several liposomal and dendrimeric approaches have been identified for gene delivery as tabulated in Table 9.1

9.1.1 METHODS OF GENE DELIVERY

To design an accurate delivery system, there is a necessity to comprehend the specific mechanism by which targeted cell and the delivery system interacted with each other. Also, designed system should have proper indulgent in the intracellular organization of the targeted cell (Prokop and Davidson,

TABLE 9.1 Tabulation Chart Displaying Various Approaches for Gene Delivery

Liposomes	Dendrimers
Polymeric micells	Cationic peptides
Microparticles	Nanoparticles
PEG conjugate	Shear activated nanotherapeutic
Echogenic lipid vesicles	Membrane active agents
Endosomolytic Agent	Ultrasound contrast agents

2007). The delivery of gene to targeted cells could be accomplished by utilizing various gene delivery systems that effectively delivers the particular gene into targeted cells and control the regulation of gene expression.

9.1.1.1   Viral Gene Delivery

Viral vectors are the most recognizable system for delivering exogenous genetic materials to the specific cells, following traverse them to the cell nucleus and thereby activating the genomic expression (Vile et al., 1996; During, 1997). Practically, viruses can be used as a delivery vector mainly because of their characteristic properties. Therapeutic genes can be delivered through various types of viral systems, for example, oncogenic retrovirus, adenovirus, lentivirus, simian virus-40 (SV-40), herpes simplex virus (HSV), adeno-associated virus (AAV), and pox virus. However, the major limitation with viral-based gene delivery system is their safety concern as viruses are naturally immunogenic and that may cause hazardous immune responses. Also, higher cost in manufacturing of viral-based system of gene delivery and its limited specificity for target cells are significant factors in viral-based gene delivery system (Yaron et al., 1997).

9.1.1.1.1   Retroviral Gene Delivery

Retroviruses belong to the family of Retroviridae having two sets of chromosomes. They are mostly single-stranded and circular in shape enveloped into RNA. Its genomic size is 7–11 kb with a diameter of 80–120 nm approximately (Coffin, 1979). Retroviruses have the major role in causing AIDS and cancer. Their role in gene delivery have brought the breakthrough in the treatment therapies (Pages and Danos, 2003).

Retrovirus gene delivery system carries materials for transfer of a particular gene and was first recognized at the start of 1980s. Retrovirus gets integrated with host genomic material for various generation of viral proteins that can be removed during delivery of gene (Elaneed, 2004). The first virus employed for gene delivery was moloney-murine leukemia virus (MMLV) (Muzzonigro, 1999). Retroviral gene delivery mediators have been utilized for efficient transfer of genes into germline stem cells, (Nagano and Shinohara, 2000) as well as into mammalian cells (Kohn et al., 1987). Apart from these, the retroviral gene delivery system has also been used for familial hyperlipidemia gene therapy and tumor vaccination (Grossman et al., 1994; Wakiomoto et al., 1997).

However, several major limitations are associated with retroviral gene delivery system such as their lower effectiveness for *in vivo* experiments, sometimes retroviral vectors can also produce immunogenic reactions to host cells. Retroviral vectors are unable to transduce the cells, which are non-dividing (Lufs et al., 2003; Anson, 2004; Bushman, 2007).

9.1.1.1.2 Adenoviral Gene Delivery

Adenoviruses are relatively bigger in length (26–40 kb) to that of retrovirus and have nonenveloped and linear structure. They are similar to dsDNA type of viruses covered with the icosahedral particle, which is non-integrating in nature (Wickham et al., 1995). As one of their properties, they are not replicated and rapidly causing infection in host cells. Adenoviral delivery vector is also first developed systems and efficient and vital gene delivery vectors due to its characteristics features such as molecular biology; adequate and high capacity to deliver genetic material in many types of host cells (Sullivan, 2003; Campos and Barry, 2007).

Adenoviral vector could be utilized to reprogram human fibroblast into induced pluripotent stem cells (Zhou and Freed, 2009). Adenoviral gene delivery vectors-mediated gene delivery is mainly targeted for human gene diseases such as monogenic inherited diseases (cystic fibrosis, muscular dystrophy, and hemophilia) and infectious diseases like AIDS (Zhang and Zhou, 2016). Recently adenoviral vectors are utilized for delivery of genes in therapy and treatment of malignant glioma (Immonen et al., 2004). Hypertension and reduction in renal failure (Ylaherruala and Martin, 2000), prostate cancer (Eastham et al., 1995; Freytag et al., 2002), and reducing pulmonary hypertension (Mcmurty et al., 2005).

However, there are very significant major limitations associated with adenoviral gene delivery systems. Some of the target cells have very less number of adenoviral receptors and that will need high concentration and dosing of adenoviral vectors to produce its effectiveness. Also, sometimes adenoviral vectors are non-discriminating in nature and can lead to the transfer of genetic material to non-targeted cells (Reynolds et al., 1999). Apart from these, the most significant limitation of adenoviral gene delivery system is to cause marked immunogenicity due to inflammatory reactions and toxin production (Navarro et al., 2008).

9.1.1.1.3   Adeno-Associated Viral Gene Delivery (AAV)

AAV belongs to the family Parvoviridae. AAV is minuscule, replication-defective and non-enveloped. Currently, AAV does not tend to originate any disease in human. This virus is responsible for very mild response to the defense mechanism of the host cells; therefore, this viral delivery system lacks the apparent pathogenicity. Another important characteristic property of AAV is its capability to be integrated into the particular region on chromosome 19 without any noticeable effects. These features of AAV makes it very vital candidate for the delivery vector to deliver the genetic material efficiently. The significant limitations associated with this gene delivery system is complicated production process of vector system and its minuscule transduction capacity to transfer genetic material (Daya and Berns, 2008; Kotterman and Schaffer, 2014; Kotterman et al., 2015).

9.1.1.1.4   Lentivirus Gene Delivery

Lentivirus belongs to the Retroviridae family having single-stranded genomic structure. Lentivirus could be able to transport a substantial quantity of viral RNA into host cell DNA. This virus is having a single ability amongst all retroviruses to naturally integrate non-dividing cells. Therefore, this vector is one of the best and suitable methods of a gene delivery (Quinonez and Sutton, 2002).

Lentivirus gene delivery system majorly utilized for *ex vivo* delivery of exogenous genetic material in the central nervous system without any immunogenicity and unwanted side effects. Lentiviral gene delivery system produces stable expression of a transgene and having the capability to accommodate larger transgene (Federici et al., 2009). Lentiviral vectors have been employed for gene delivery in various CNS-related disorders such as Parkinson, Huntington's disease (Bensadoun et al., 2000; Dealmeida et al., 2001), metabolic disorders related to lysosomal dysfunctions (Haskins, 2009), and spinal injury (Abdellatif et al., 2006).

9.1.1.1.5 Simian Virus-40 (SV-40) Gene Delivery

SV-40 also known as simian vacuolating virus 40 is a type of polyomavirus, which develops significant number of vacuoles in infected green monkey cells. SV-40 is a DNA virus that can produce cancer in animals. It is non-enveloped, double-stranded, circular-shaped virus. SV-40 gene delivery system can infect and transduce all types of mammalian cells in both resting and dividing state to deliver the transgenes efficiently. Also, SV-40 system is nonimmunogenic and does not elicit any pathogenic reactions. The significant limitations of SV-40 gene delivery system include its very low carrying capacity for some transgenes. Also, some of the transgenes cannot be expressed well using this type of vectors due to unknown reasons (Butel and Jarvis, 1986; Breau et al., 1992; Stang et al., 1997).

9.1.1.1.6 Herpes Simplex Virus (HSV) Gene Delivery

HSV has double strands with an envelope-shaped structure. HSV is DNA type of virus belongs to the family of Herpesviridae. HSV is of two types: HSV-1 (mainly produces cold sores) and HSV-2 (responsible for genital herpes). HSV is pathogenic to humans and very useful and latest vector for transfer of genetic material to the nervous system (Liu et al., 2008) and cancerous cells (Goss et al., 2002). During propagation in the host cell, HSV viral particles are released and replicate their genome, but not producing infectious particles (Trobridge, 2009). HSV vector has large carrying capacity and incredible ability to evade the immune system.

9.1.1.1.7 Pox Virus Gene Delivery

Pox virus belongs to the family of Poxviridae and enveloped. Pox virus vary in shape depending upon the type of species. They are oval- or brick-shaped. Pox virus gene delivery vector has vital transduction capacity and efficiently used to deliver genetic material in cancer therapy. The primary limitation associated with Pox virus vectors is their complex structure and biology. Invasion of Pox virus causes lysis of host cells. Therefore, more safety parameters need to be evaluated to reduce the hazards related to cytopathic reactions (Moss, 1996; Gomez et al., 2008).

9.1.1.1.8   Other Viruses for Gene Delivery

Alpha viruses are single-stranded RNA type of virus belongs to the family of Togaviridae. Alpha viruses are found to be useful for gene delivery in cancer therapy due to its high cytotoxicity. Various more similar alpha viruses are also gaining interest amongst researcher to develop gene delivery vectors (Strayer, 1999).

9.1.1.2   Non-Viral Gene Delivery

This type of vector system does not produce rapid and high transfection compared to viral vectors. But viral vectors are hazardous and cause unwanted immunogenic reactions. Other than that, non-viral gene delivery vectors can be more beneficial over viral vectors because of an easy preparation, better host-cell targeting, and lower response to immunogenic reactions. The most limiting factor in using non-viral gene delivery vector is its very low transfection efficacy. Therefore, most of the research is now undertaken to increase transfection efficacy of this type of vectors.

9.1.1.2.1   Naked DNA Transfer

This type of transfer is the simplest of the approach in gene therapy. This type of gene delivery gain success in the gene therapy of DNA vaccination (Herweijer and Wolf, 2003), myocardial ischemia (Wolff and Budker, 2005), angiogenesis (Tsurumi et al., 1996), and cancer (Horn et al., 1995). The major limitation with naked DNA transfer is its hydrophilic property and presence of bulky phosphate groups, which makes the transfer of naked DNA into host cells very challenging. Expression achieved by naked DNA transfer is the stable and long term. Although, naked DNA transfer-mediated gene therapy is simple and safer approach. It achieved very low transfection efficiency, which limits its use for some applications (Mali, 2013).

9.1.1.2.2   Physical Methods

Physical methods facilitate the gene transfer by making short membrane holes using physical forces. For examples, electropermeabilization, bombardment of genetic material by gun, ultrasonic hydrodynamic injectable delivery, or by laser irradiation.

9.1.1.2.2.1   Bombardment by Gene Gun

This is also known as ballistic delivery of genetic material or bombardment of DNA particles. This technique is an ideal alternative approach to naked DNA transfer. The method uses DNA-coated gold, silver, or tungsten spherical microparticles, which was further hastened at very rapid speed to targeted tissue using pressurized inert gas such as helium. This technique is now successfully used for gene therapy into mammalian cells of the skin, mucosa and surgically exposed tissues for DNA vaccination of Alzheimer's disease (Davtyan et al., 2012). This technique was also found to improve gene delivery efficacy of DNA vaccination (Yang et al., 1990). The primary limitation associated with this method is its higher immunogenic response, and it produces short-term gene expression, but in this technique, there is no requirement of the receptor. The larger size of DNA is not an issue and coated DNA particles can be easily prepared (Lin et al., 2000).

9.1.1.2.2.2   Electropermeabilization

Electropermeabilization is the controlled use of electric pulses to increase the permeability of host cells. The concept was first used in the 1960s, and the same experiment has been demonstrated in 1982, in which eukaryotic cells were transfected using electropermeabilization, which is also known as electroporation. This technique transfers foreign genetic material into the cell by the electrical field. In this technique, the permeability of host cell is increased by forming pores on the surface of the membrane that allows DNA to enter the cells more rapidly and efficiently. DNA introduced into host cells depending upon the assembled pore size on the membrane surface. The entry of genetic material and transportation of ions occurred not only through passive diffusion, but also via the electrophoretic and electro-osmotic mechanism. This technique is more reliable and efficient than other methods. Electroporation can be utilized successfully for gene therapy of many types of tissues such as skin, muscle, lung, and cancer therapy (Neumann et al., 1982; Lurquin, 1997; Heller et al., 2005). The major drawback of electroporation method is the difficulty in placing electrodes into the muscles located internally, and the generation of high voltage might damage the tissues and organ (Dean et al., 2003; Mcmahon et al., 2004).

9.1.1.2.2.3   Magnetofection

Magnetofection is a simpler and highly efficient method of gene delivery transfection. It includes chemical and physical transfection systems in one system. In this technique, magnetic fields are generated to concentrate the particles into the targeted host cells. The generated magnetic field allows the rapid and complete entry of genetic material into host cells. By this, total dose of the vector gets in contact with host cells (Plank et al., 2003).

9.1.1.2.2.4   Ultrasound Sonoporation

The increase in uptake of plasmid DNA (pDNA) using microbubble-assisted ultrasonicated gene therapy is called sonoporation. Ultrasound technique makes the nanometric-sized pores into the membranous structure, which can facilitate the host-cell delivery of genetic material. In this procedure, ultrasound is applied on microbubbles, which are modified with pDNA. Recently, it has been demonstrated that positively charged microbubbles provide better gene transfection than microbubbles with no charge. Presently, sonoporation-mediated gene therapy is found to be very effective. The sonoporation technique is very easy and reliable method for gene therapy. Sonoporation technique also produces long-term gene expression system. The primary disadvantage of this approach is its low transfection efficiency (Liange et al., 2004).

9.1.1.2.2.5   Hydrodynamic Delivery

This method uses higher pressure-driven force to deliver the genetic materials to the targeted site. The larger DNA content injected in a shorter period cause changes in permeability of endothelial membrane and generation of temporary pores to allow diffusion of DNA particles. The efficiency of delivery is dependent upon capillary and cell architecture and applied hydrodynamic pressure. Hydrodynamic gene transfection is simpler, safer and highly effective method for gene delivery of any water-soluble compounds and particles into host cells (Liu, 1999).

9.1.1.2.3. Chemical Methods

Chemical methods employed for gene therapy are more common than physical methods. These methods use polycationic nanometric particles such as liposomes or polymers to form complexes with negatively charged nucleic acid materials. These generated compounds shield the DNA and facilitate its cellular uptake and intracellular transport with more stability and long-term gene expression.

9.1.1.2.3.1 Liposomes

Liposomes are the colloidal type of delivery vector formed by the assembly of lipid molecules in vesicular structures and contain aligned hydrophilic head and hydrophobic tail in tail-to-tail style with adjacent ends. Variety of literatures are available depicting liposome as an effective strategy for the transportation of biomolecules (Maheshwari et al., 2012, 2015b; Mody ct al., 2014; Sharma et al., 2015). They can be quickly bound onto surface host-cell surface to render the gene delivery efficiently and with high capacity. The entrapment of DNA inside the liposomes termed as lipoplexes, and the formed structure has protection ability from undesirable degradation of DNA. Liposomes can be cationic, anionic, neutral, or mixture of all. Cationic lipids are most effective as it interacts with negatively charged DNA most efficiently to capture plasmids (Gao and Huang, 1995; Kofler et al., 1998).

Liposomal gene delivery vectors were firstly reported in 1987 (Felgner et al., 1987). Further, cationic liposomal gene delivery system has been investigated very extensively and developed as most universal technique for non-viral gene transfection. Till now, hundreds of liposomal systems have been evolved as gene transfection vectors. The ability to delivery genetic material using liposomal system mainly depends upon the lipids structure (complete geometrical shape, the number of charged ions present, characteristics and nature of lipid and linker properties). The most commonly used lipids in gene transfection are cholesterol and dioleoyl phosphatidylethanolamine (DOPE) for due to their lower charge ratio which in turn reduce the toxicity. Cholesterol-based cationic liposomal systems have better transfection capability *in vivo* because cholesterol stabilizes the liposomal membrane and protect it against the destructive capability of serum components. With comparison to viral and other gene delivery system, liposomal transfection vectors can be formulated more quickly and without any pathogenic and immunogenic adverse effects (Dass et al., 2006). A major limitation associated with liposomal gene delivery system is their short-term gene expression and sometimes this system might produce acute toxicity. In the case of its combination with unmethylated CpG containing pDNA, it may produce inflammatory reactions in host cells. The system can be surface shielded using PEG coating to protect it from inflammatory responses (Harvie et al., 2000).

9.1.1.2.3.2 Polymers

Polymers are made up of many macromolecules which combined to form long chains or complicated structures. Each macromolecule is called monomer. Polymers can be called homopolymer (combination of identical monomers of the same polymer) or heteropolymer (combination of different monomers of the same polymer) or copolymers (combination of two different polymers). Polymers can be obtained naturally (chitosan and dextran) or synthetically (polyvinyl alcohol, polyvinyl acetate, polyethylene glycol, polyacrylic acid, and silicones). Polymers used for packing of DNA can increase its stability and cellular uptake to enter the nucleus of host cells. Polymers can also be categorized as biodegradable or non-biodegradable. Biodegradable polymers are hydrophobic in nature and under the biological condition it degrades chemically or physically. Natural polymers such as chitosan and dextran are biodegradable polymers. Non-biodegradable polymers are not degraded under physiological conditions. Polymer selection for gene delivery is based upon its physicochemical and biochemical characteristics along with extensive preclinical testing. Various methods are used for successful packaging of DNA with polymers such as electrostatic interaction, adsorption, and encapsulation (Pillai and Panchagnula, 2001).

Cationic polymers are studied extensively for gene delivery. Cationic polymers interacted and combined at the anionic site of DNA to form polyplexes. These formed polyplexes are usually more stable than lipoplexes. Polyethyleneimine (PEI) can be termed as most effective cationic polymer for successful delivery of genes and was first reported for gene transfection in 1995. Recently, various polymers have been studied for effective delivery of genes, including poly-L-lysine (PLL), polyornithine, polyarginine, prot-amines, spermidine, spermine, and histones (Midoux et al., 2008).

9.1.1.2.3.3 Dendrimers

Dendrimers are well-defined synthetic polymers with tree-like projections from central core molecule (Gajbhiye et al., 2009a; Kesharwani et al., 2015a). The size of the dendrimers varies from 10 to 200 Å diameter having repetitive branches with functional groups on their surface. These exterior functional groups further utilized as building blocks on which genes can be targeted electrostatically, covalently, or encapsulation. The detailed note on dendrimers as gene delivery vectors is discussed in section 9.3.

9.1.2 PROBLEMS AND CHALLENGES

For the successful gene delivery, the vector must cross through the extracel-lular and intracellular barrier. Viral vectors developed certain functions to challenge these obstacles. In contrast, non-viral vectors lack those features. Therefore, it is important that non-viral gene vectors should be designed to defy such constraints.

9.1.2.1 Extracellular Barrier

Extracellular barriers are obstacles that can come across from the point of entry to the targeted host cell surface. These barriers include binding to pDNA and maintaining polyplex or lipoplex into the solution. Apart from that, there are *in vivo* barriers such as persistence and stability of the delivery system in the blood stream, penetration through blood vessels and tissues and on-target binding (Aldosari et al., 2009).

9.1.2.2 Gene Packing

The pDNA should be packed in polymers or liposomes such as it should be protected from nucleolytic enzymes. Unprotected DNA can be degraded very rapidly by DNase enzyme, although pDNA is protected for several hours with polyplexes or lipoplexes. There is still a need to investigate and understand the process and physicochemical characteristics of these complexes to improve the packaging protocols. The structure and size of ionic charge on lipids or polymers have a significant influence on binding to pDNA. It has been reported that minimum 6–8 cationic charges are needed for high binding and stable complex formation (Schaffer et al., 2000). Also, to improve the binding, cationic moiety should be placed nearer to the backbone of the polymers and keep the least separated charges alongside the polymer chain. It is also important to understand that there should always be balance in binding affinity because tight binding might prevent the transcription of DNA (Zelphati et al., 1996).

9.1.2.3 Immune Stimulation

Systemic administration of gene delivery vectors might release excessive and cytokines and inflammatory mediators, which leads to unwanted side effects and immune system activation. Various other factors are also associated with increase immune system activation such as chemical composition of delivery vectors, physical properties of the prepared gene delivery system, pharmacokinetics, and its routes of administration (Judge et al., 2006).

9.1.2.4 Intracellular Barrier

Endocytosis mediates lipoplexes and polyplexes entry into the cells. The process of endocytosis leads to uptake of the components into vesicles that finally transport all the components to the lysosomal organelle. Therefore, intracellular obstacles include endolysosomal escape, serum stability, RES recognition, cytoplasmic trafficking, nuclear entry, nuclear localization, and vector unpackaging.

9.1.2.4.1 Endosome Escape

Non-viral gene delivery system involves escaping from endosomes before cytoplasmic trafficking. Endosomal escape is the important step to avoid degradation from enzymes. Cationic polymers or liposomes bound to anionic functional groups of the endosomal membrane to make neural ion pairing complexes. This ion pairing will further affect the stability of the membrane and cause a bursting of genetic material into the cytoplasm. The designing mainly depends on pKa, buffer capacity, and pH of ionizable cationic polymers (Hatakeyama et al., 2009).

9.1.2.4.2 Serum Stability

The stable nature of complexes is mainly dependent upon the charge ratio of DNA and polymers/liposomes. Neutral complexes in physiological solution form larger aggregates and lead to cell embolism responsible for cell toxicity. However, positively charged compounds remain in solution and not form aggregates. Nuclease enzyme can rapidly degrade naked RNA in serum. Resistance to the nuclease can be increased by chemically modifying RNAs. The most commonly used modifications are on sugar part and conjugation of fluoro, o-methyl and amine functionality at 2nd position. Such features on RNA increases the nuclease stability and reduces the off-targeting and immunogenicity. This modified RNAs will provide improved delivery, but they lack the site-specific targeting for some tumors and could be excreted rapidly from the systemic circulation; hence, large dosage is required to attain therapeutic effects (Pan et al., 2011).

9.1.2.4.3 RES Recognition

Polycations mainly complexed with nucleic acid components with negative charges. The use of positive ions promotes the off-target interactions. The reticuloendothelial system (RES) recognizes the circulatory nanoparticles with the binding of plasma proteins. The recognition will cause more of the injected dose to concentrate into RES organs (liver, spleen, bone marrow). The nanoparticle surface is shielded with hydrophilic and neutral polymers (PEG). Coating of PEG will hinder the connections of blood components and nanoparticulate surfaces and thereby prevents recognition by RES leads to the circulation of nanoparticles in the circulation for the larger period and reduce the off-target interactions (Li et al., 2010).

9.1.2.4.4 Cytoplasmic Trafficking

After the release from endosomes, the complex of gene delivery vectors will pass through the cytoplasmic components into the nucleus. The cytoplasm consists of proteins and microtubular structures and other components which obstruct the movement of complexes. This action depends on the size of the DNA particles. Larger DNA (>3000 base pairs length) is immobile. The flow of the positively charged complex system might occur by non-specific interactions with anionic microtubular proteins (Suh et al., 2003).

9.1.2.4.5 Nuclear Entry and Localization

The nucleus is the terminus point for gene vectors for expression via transporting through the cytoplasmic components. Transport and entry to the core occur through the pores on nuclear membrane. The nuclear envelope unfolds during mitosis in dividing cells. This unfolding will lead to transfection of genomic material into the nucleus. The quantity of genetic material reached into nucleus is lesser by the enzyme nuclease-based degradation (Lechardeur et al., 1999).

To target the nucleus, specifically many nucleus localizing peptides (NLP) have been developed. These peptides allow the passing of genetic components into the pore of nucleus through active transport mechanism. These peptides are made up of collections of several different amino acids that gets attached to the receptors present in cytoplasm. These peptides attach to the DNA through electrostatic or covalent interactions (Bolhassani et al., 1999).

9.1.2.4.6 Vector Unpackaging

Complexation of DNA with polymers or liposomes protects the enzymatic degradation and protein attachment needed for gene expression. Therefore, complexed DNA should be released during the delivery process. It has been reported that strength of binding between polymers/liposomes and DNA reduces the gene expression (Erbacher et al., 1997). The proposed mechanism for unpackaging of vector system by displacing nucleic acid components utilizing intracellularly present polyanions. The effectiveness of this unpackaging mainly depends on the chemical nature of the gene vectors. The gene delivery system should be designed such that it releases the genes ideally into the nucleus (Chen et al., 2008).

9.2 DENDRIMERS

9.2.1 *STRUCTURAL ARCHITECTURE OF DENDRIMERS*

Dendrimeric architecture represented in Figure 9.1 has a central core atom on which branches of dendrons grow through the chemical reactions. Dendrimers consist of three major domain structures: (a) exterior surface, (b) interior surface, and (c) central core atom, where dendrons are grown. Interior surface along with central atom forms the nano environment, which is protected from the exterior surface (Gajbhiye et al., 2009b; Tekade et al., 2009a). Interior surface is more suitable for interactions and encapsulation with guest molecules. Preparation of dendrimers is controlled throughout the process of its chemical synthesis to form mono dispersible, macromolecular and globular polymeric architecture (Jain et al., 2013; Abbasi et al., 2014; Kesharwani et al., 2014a).

Dendrimeric architectures are prepared using divergent and convergent approaches. In both the approaches, dendrimers are synthesized by step-by-step growth and branching generations. Dendrimers are depicted and classified by their generation numbers. The central core group represented a G0 (generation 0), although the following additions of branching units termed as higher generations of dendrimers (G1, G2, G3, …). This exponential increase with the addition of each group results in sterically crowded dendrimers, which leads to geometrical changes in the dendrimeric structures.

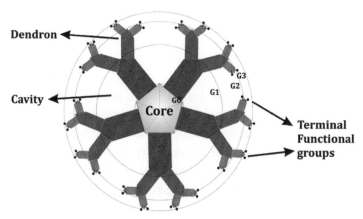

FIGURE 9.1 Representation of dendrimer architecture with different part display.

With the growth in generations, dendrimer takes globular structure due to increasing steric hindrance (Kesharwani et al., 2014b; Kesharwani et al., 2015b). These unique and distinctive characteristics of the dendrimers make them suitable candidate for drug delivery. Dendrimers generate unique biological properties for the bioactive molecules such as high payload capacity, targeting potential, specific biodistribution, improved stability, better internalization, etc.

9.2.2   PROPERTIES OF DENDRIMERS

Dendritic architectures hold significant promise in the research due to its unique characteristic properties. When compared with traditional delivery vectors like polymers and other nanocarriers, dendrimers are well defined and improved physical and chemical characteristics with significantly greater structural diversity.

9.2.2.1   Polyvalent Charges

Dendrimers at their end groups contain positive, negative or neutral charges, which play an important and vital role for dendrimers to be exploited as drug-delivery carrier. Positively charged cation on dendrimers interacts with anionic DNA charges to generate complex structure and with the membrane surface cationic charges for facilitation of intracellular drug delivery.

9.2.2.2   Biocompatibility

Polyvalency of dendrimers may lead to toxic reactions. However, these responses can be overcome by use of surface engineered dendrimers such as glycosylated or PEGylated dendrimers. Biocompatibility and toxicity of dendrimers mainly depend on the peripheral charges present on the terminal end groups. Cationic groups show more toxic and hemolytic reactions than anionic or neutral charged groups (Morgan et al., 2003; Fuchs et al., 2004; Hemmati et al., 2016).

9.2.2.3   Pharmacokinetic Properties

Pharmacokinetic properties of dendrimers mainly depend on physicochemical features of macromolecules and their consensual interactions with biological components. Pharmacokinetics play a significant part in biomedical applications of dendrimers. Mostly, dendrimers are administered through parenteral route to be absorbed across various epithelial barriers. Surface engineered dendrimers exhibit improved pharmacokinetic properties (Nishikawa, 1996; Khan et al., 2005; Szymanski et al., 2011).

9.2.2.4   Covalent Conjugation Properties

To improve the pharmacological properties, the research on covalent conjugation of dendrimers to the guest molecules is under investigation (Gillies et al., 2005; Kolhe et al., 2006). Conjugated molecular complex functions as a prodrug, which liberates the activated drug intracellularly to the target host cell.

9.2.2.5   Electrostatic Interactions

Charged surfaces at the end groups of dendrimers interact electrostatically with oppositely charged DNA and biological membranes of the host cell. An example of electrostatic interactions includes the binding of methylene blue on the surface of dendrimers with the nitroxide cation and copper complexes at EPR probes (Ottaviani et al., 1994).

9.2.2.6   Nanosized Architecture

Dendrimers are uniform and well-defined structural designs, and they have the capability to pass through the biological membranes by effectively interacting with it. The size of dendrimers increases with generation numbers steadily makes them ideal carrier for drug delivery and biomedical applications.

9.2.3   SYNTHESIS OF DENDRIMERS

Dendrimers are synthesized in step-by-step controlled synthesis. Dendrimers are mono-dispersible macromolecular structural architectures and very similar to those structures observed in the biological system. Dendrimeric architectures are synthesized using either divergent or convergent approach (Tomalia et al., 1987) as illustrated in Figure 9.2.

The central core group reacts with the monomeric unit having single active and double inactive groups to produce the G1 dendrimer. Further, newer surface groups get activated to react with additional monomeric units to generate next generations of dendrimers. Dendrimers are synthesized by the cascade of reaction sequences developed by Vogel and coworkers in 1978 either by divergent or convergent approach (Buhleier et al., 1978; Svenson and Tomalia, 2005). The divergent approach employs the synthesis of dendrimers starting from the central core group on which arms of building blocks are attached in a step-by-step controlled manner. In contrast, convergent approach begins with the outermost part of the molecular architecture of dendrimer. Some more methods have been utilized to synthesize the dendrimers such as hyper cored and branched monomers, double exponential approach, lego chemistry, and click chemistry (Liu et al., 2014).

9.3   DENDRIMERS IN GENE DELIVERY

9.3.1   EFFECTS OF DENDRIMER ARCHITECTURE ON GENE DELIVERY

Reports suggested that slight modification in the structure of central dendrimeric atom can have significant outcome on the gene delivery efficiency

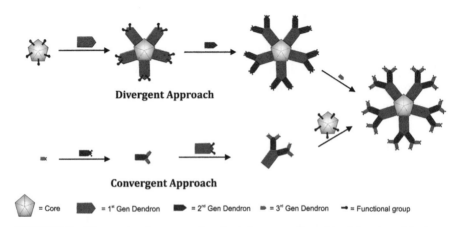

Divergent Approach

Convergent Approach

⬠ = Core   ▬ = 1ˢᵗ Gen Dendron   ▬▶ = 2ⁿᵈ Gen Dendron   ▬ = 3ʳᵈ Gen Dendron   ➝ = Functional group

FIGURE 9.2   Divergent and convergent methods for preparation of dendrimeric architecture.

and biocompatibility (Tekade et al., 2009b; Chang et al., 2014). The mechanism of gene transfection using various types of cores by varying flexibility, lipophilicity, and functionality has been discussed. The study suggested that central core of dendrimer controls the dendrimer conformation and functionality number with size. Apart from these characteristics, central dendrimer core affects the physicochemical properties of the dendrimer like flexibility, hydrophobicity, and functionality. The introduction of the linker to the central core with varying flexibility will modify branching or enhances the central core mobility, which could lead to efficient improvement in overall dendrimer flexibility (Hu et al., 2016). Many novel anticancer formulations utilizing the structural and functional properties of dendrimers have been developed (Tekade et al., 2008; Kesharwani et al., 2011; Tekade et al., 2015b; Singh et al., 2016). The increase in overall dendrimeric flexibility is advantageous for effective delivery of genetic components. The addition of hydrophobic ligands to the central core of the dendrimer can control the balance between hydrophobicity and hydrophilicity. The correct balance in advantageous in increasing cellular uptake and efficiency of gene transfection. Similarly, the addition of functional groups to the central atom can also improve gene delivery of the dendrimers via introduction of new functional characteristics.

In different investigation, the versatility of hyperbranched poly(amidoamine)s (PAMAMs) dendrimer in gene transfection and cytotoxicity. The study compares hyperbranched dendrimers with their linear prototype. Eight different bio-reducible PAMAMs were synthesized using Michael addition reactions. Terminal groups were also modified and studied for their effect on transfection efficacy. Significantly little or no cytotoxicity was seen for hyperbranched PAAs compared to their linear dendrimers (Martello et al., 2012).

9.3.2 DENDRIPLEXES FORMATION

Dendrimer interacts with nucleic acid materials like DNA, RNA, or oligonucleotides to make complexes called as dendriplex. These dendriplexes are fundamentally similar with other cationic polymeric materials in having charged dense material. Dendriplexes formation in mainly based upon the electrostatic interaction of anionic charges present in nucleic acids with that to the cationic charges present on the dendrimeric surface in the form of primary amino groups. Dendriplex protects the nucleic acid materials from degradation. The net neutralization of charge on both leading to alterations in physico-chemical and biological characteristics of the components (Dennig et al., 2002; Nomani et al., 2010).

The nature of the formed complex mainly depends upon the stoichiometry and concentration of the both the components and also on the properties of the solvent in bulk such as its pH, salt concentration, and buffer strength. With the increase in the concentration of sodium chloride, the ionic strength of the solution increases and it also obstructs with the binding mechanism and appears to support in establishing equilibrium (Goula et al., 1998a, 1998b). With the increase in dendrimeric generation, there will be exponential increase in interaction of amine groups with DNA, which will improve the binding of dendrimers with DNA. Therefore, with the higher generation of a dendrimer, improved binding and interactions can be achieved (Ottaviani, 2000). Although some studies also report the better binding of smaller generation dendrimer than larger ones due to the more fluidic structure of smaller dendrimers (Kabanov et al., 2000).

The reports also suggested the difference in the binding interaction of DNA with flexible linear polymers and that of a rigid dendrimeric structure. Flexible linear dendrimer can be able to interact with all the amines available on dendrimer surface as well as inside the dendrimer, whereas rigid dendrimer having double strands only able to interact with the surface amines (Zinselmeyer et al., 2002). These binding interactions would leave some of the charged residues intact without interaction. The morphology and arrangement of dendriplexes were elucidated recently using x-ray diffraction and atomic force microscopy techniques (Bielinska et al., 1997; Tang and Szoka, 1997).

9.3.3 MECHANISTIC ASPECTS OF DENDRIMER-MEDIATED GENE DELIVERY

The dendriplex interacts electrostatically in the form of polycations with the anionic glycoprotein and phospholipids surface. The effective interaction will trigger the movement and entry of dendriplexes into the cell cytosol and up to the nucleus either through passive transportation due perturbations of membrane or due to endocytosis (Dufes et al., 2005). Figure 9.3 illustrates the mechanistic aspects of dendrimeric gene delivery vectors.

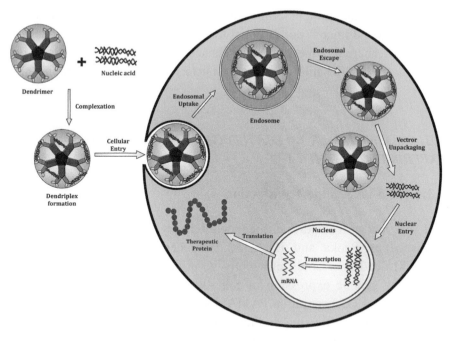

FIGURE 9.3   Mechanistic aspects of dendrimer-mediated gene delivery.

The report suggested that dendrimers interact with the cellular membranes by forming the transient pores and this effect mainly dependent upon charged cationic surface groups present on the dendrimers. This charged density of dendrimer increases with the number of generation, therefore G7 dendrimer generation was found to be more active than G5 (Behr, 1994; Sadekar et al., 2012).

Dendriplex further being released into the cytosol after endosomal entrapment. The release of dendriplex into cytosol occurs before degradation by enzymatic DNAse or acidification inside the lysosomal cavity. The drugs that inhibit the endosomal acidification can also enhance the dendrimer-mediated gene transfection (Zhang et al., 2016). It has also been signifying that presence of amine groups impart essential characteristics to dendrimers, which is beneficial in retarding degradation caused by acidification.

After the release from the endosome, DNA penetrates and translocate to the nucleus within 30 min post transfection. Further gene expression occurs in the nuclear membrane as DNA converts into mRNA by transcription. Lastly, the formed mRNA will be entered into the nucleus and translated into the therapeutic protein. Some studies also suggested that the release of DNA from dendrimer is not essential for its transportation into the nuclear membrane. Therefore, there might be chances that dendrimer might also be linked to the DNA while crossing the nuclear membrane (Godbey et al., 1999).

9.3.4   DENDRIMER-MEDIATED GENE DELIVERY IN EXPERIMENTAL THERAPY

9.3.4.1   *In Vitro* Gene Transfection

The first report of dendrimer-mediated delivery of genetic material was published by Szoka and Haensler. They demonstrated that Starburst PAMAM dendrimer could efficiently be utilized as delivery systems to induce the genetic expression in cell suspension cultures. Their results suggested that G5 generation PAMAM dendriplex of DNA enhances the gene transfection efficiency in the various cell culture medium of human and monkeys (Haensler et al., 1993). The reports indicated that PAMAM dendrimer has the ability for endosomal escape due to alteration of pH in endosomal membrane based on the presence of amino functionalities (Behr, 1997).

The investigation further revealed that gene transfection through dendriplex was mediated through degraded dendrimers. Heat treatment increased the gene transfection capability of the dendrimer in various solvents. Degradation by heating was induced at amide linkage that causes hetero dispersion of the population of degraded compounds ranging in minuscule to large molecular weights. Fractured dendrimer that is constituent of the degraded product with high molecular weight exhibits an increased apparent change and protonation of terminal amine groups to reduce its pH that will improve the transfection efficiency. The degree of increase in transfection efficiency due to heating was mainly due to more flexibility of the degraded fractured dendrimer. It remains compact when bound and combined with DNA and swelling occurs when released (Tang et al., 1996).

The general mechanism to facilitate endosomal escape is by the accumulation of Cl⁻ ions has been proposed, which could lead to osmotic swelling of the endosomal membrane (Sonawane et al., 2003). The study by Bielinska and coworkers exhibited the antisense capability of PAMAM dendrimers for oligonucleotides and to create the cells lines, which constitutively expresses the genes. The same report also investigated the transfection efficiency of different generation of dendrimers in Rat2 cells and observed the exponential increase in the efficiency of gene transfection with the application of higher generation of dendrimers (Bielinska et al., 1996).

PAMAM dendrimers are highly useful in gene transfection in primary cell lines including fibroblasts and lung epithelial cell lines of human in comparison to other nonviral gene delivery agents (Kurmi et al., 2010). Each cell lines have different abilities to react with the dendriplexes for transfection because of their different physiological properties. Therefore, always there is a need to determine the best suitable generation experimentally for the particular type of cell line. Sometimes, the addition of other agents to the dendriplexes modifies the gene effectiveness of gene delivery like the incorporation of chloroquine inhibits the endosomal acidification and markedly improves the gene delivery ability. Also, the addition of (diethylamino) ethyl dextran change the nature of dendriplexes by dispersing complex aggregates thereby increases the gene expression (Erbacher et al., 1996). The significant advantage associated with the application of dendrimer-mediated nonviral gene delivery vector is that it can be complexed with even very large-sized DNA. The example of such gene transfection was successfully performed using PAMAM dendrimers on 60 Mb artificial mammalian chromosome (Gebhart, 2001).

Recently, the study of amphiphilic dendrimeric molecules for improved gene transfection of siRNA and *in vitro* gene knockdown was performed. The nonviral delivery vector bears the lipotropic chain of alkyl groups and a hydrophilic dendrons of PAMAM with overall thirteen amine groups at the interior and terminal parts. This amphiphilic PAMAM dendrimer has delivered the Hsp27 siRNA in prostate cancer models efficiently. The gene silencing and resultant antiproliferation effect on human prostate carcinoma PC-3 cells were achieved (Yu et al., 2012).

9.3.4.2 *Ex Vivo* Gene Transfection

The efficiency of dendrimers to transfect profoundly *in vitro* has made dendrimers a vital and routine tool as the non-viral delivery vector for gene delivery. Although there is small impact produced by dendrimer-mediated gene transfection due to its lower *in vitro* conversion into the clinic, and it remains the major challenge for researchers. Consequently, dendrimer-mediated gene delivery systems have been used for their *ex vivo* applications.

The group of scientists further demonstrated that PAMAM dendrimeric architecture efficiently transfect the animal and human eye part in *ex vivo* studies. The study proved that activated PAMAM dendrimers could efficiently transfect corneal endothelial cells in *ex vivo* and resulted into 6-10% corneal endothelial cells expressing the marker gene. This approach also induces the formation of TNF fusion protein, which inhibits TNF-mediated cytotoxicity in a bioassay (Hudde et al., 1999). The pDNA should be delivered at the particular retinal site for effective gene transfection. The administration of carboplatin encapsulated with PAMAM dendrimer into subconjunctival space can be able to reduce the retinoblastoma significantly (Kang et al., 2009). The key target in ocular gene therapy is retinal pigment epithelium (RPE). Mastorakos and coworkers have demonstrated the use of hydroxyl-terminated PAMAM dendrimers with functionalized amine groups for efficient gene transfection in human RPE cells (Mastorakos et al., 2015).

PAMAM dendrimers were used effectively and efficiently for transfection of genetic material in murine cardiac grafts. The use of starburst PAMAM dendrimeric architectures and its combination with viral IL-10 gene significantly prolong the graft survival and enhancing immunosuppressive effect in comparison to naked pDNA (Gothwal et al., 2015). In another study, G5 ethylenediamine core dendrimeric architectures was studied for efficient transfection of genetic material in the same model. The DNA/dendrimer complex was perfused through the coronary arteries, and further gene expression was studied through quantitative X-Gal staining. Various variables were studied for their effect on gene delivery such as charge ratio of DNA:dendrimer, the size of DNA, preservation solutions, time of ischemia and enhancing vascular permeability. The results showed that charge ratio of 1:20 DNA:dendrimer produces highest gene expression in the grafts. The observed results also confirmed that prolonged ischemic time and increasing vascular permeability enhances the gene expression (Wang et al., 2000). Further, the effect of electroporation combination with dendriplex was studied for improving gene transfer into murine cardiac transplants. Gene expression was enhanced 10–45 fold in murine cardiac grafts after the intracoronary transfer of dendriplexes (Tevaearai et al., 2014).

9.3.4.3 *In Vivo* **Gene Transfection**

9.3.4.3.1 *Dendriplexes Biocompatibility*

The formation of DNA:dendrimer complex significantly reduces the cytotoxicity. The charge ratio of DNA:dendrimer also significantly affects the complement activation. The results of the investigation have shown that by utilizing suitable dendriplex formation, the complement activation can be avoided or could be significantly reduced (Plank et al., 1996).

Malik and coworkers have investigated the effect of several dendrimers on hemolytic and cytotoxic reactions and observed that PAMAM dendrimers can cause generation-dependent hemolysis. Dendrimers with carboxylate functionality were neither found hemolytic nor cytotoxic. In the same study anionic dendrimers have shown longer circulation time as compared to ^{125}I-labeled PAMAM dendrimers. Also, the study reports that lower generation of dendrimers circulates for the longer time in Wistar rats after i.v. administration. Surface functional groups are not the only factor which affects the biocompatibility of the dendrimers, and the aromatic dendrimeric core has also significantly produced hemolysis in rat blood cells. The reaction might proceed through the hydrophobic membrane contact of the aromatic core group (Malik et al., 2000).

Recently, 2,2-bis(hydroxymethyl)propionic acid (MPA) dendrimeric architecture along with four different PAMAM dendrimers have been synthesized. These formulated dendrimers have been decorated with negatively charged hydrophilic functionalities, which improves the biocompatibility of dendrimers. The MPA dendrimer with ionic functionality was found to produce very low cytotoxicity against U251MG and mesenchymal stem cells compared to PAMAM dendrimers, which produces higher cytotoxic reactions at the same concentration (Movellan et al., 2015). The PEGylated hyperbranched PAMAM dendrimers have been utilized to increase gene delivery efficiency as well as to reduce the toxicity of the cells (Gajbhiye et al., 2007; Sun et al., 2014).

9.3.4.3.2 *Brain-Targeted Gene Delivery*

The major limiting factor for brain targeted delivery of genetic material is the blood–brain barrier (BBB). Dendrimers have the ability to transport the genetic material and transfect it through the biological membrane by endocytosis. PAMAM dendrimers show minimal brain uptake after systemic administration in healthy animals and rapidly cleared from systemic circulation. PEGylated PAMAM dendrimers had longer circulation time and distributed mainly to the peripheral organs. BBB permeability increases in case of neuroinflammation and that may lead to increase in dendrimer transfection (Dwivedi et al., 2007; Posadas et al., 2016).

Conjugating different ligands on the surface of dendrimer causes significant BBB-targeted gene delivery. Surface modification of dendrimer is necessary to improve BBB crossing and gene transfection for brain targeted delivery. Transferrin-conjugated PEG-modified PAMAM dendrimer was evaluated for efficient gene delivery for brain targeting. In the study, transferrin (Trf) was utilized as targeting ligand for conjugation to PEG-modified PAMAM dendrimer and characterized further by NMR to confirm its synthesis. The biodistribution studies were evaluated using fluorimetry, flow cytometry, and radiolabeling techniques. The transfection efficiency was assessed by measuring the tissue luciferase action in Balb/c mice upon systemic intravenous injection. Comparative studies of PAMAM-PEG-Trf/DNA, PAMAM/DNA, and PAMAM-PEG/DNA complex performed for evaluating potency for transfection efficiency. The results have shown 2.25-fold larger uptake in brain and very high transfection efficiency for PAMAM-PEG-Trf/DNA complex. Tissue expression experiment also showed widespread exogenous expression in mouse brain with PAMAM-PEG-Trf/DNA complex (Huang et al., 2007). Dual targeting fourth generation G4 PAMAM dendrimers conjugated with Trf and wheat germ agglutinin (WA) on the exterior surface and loaded with doxorubicin (Dx) in the inner part was formulated and evaluated for its tumor-targeting properties in the brain. The PAMAM-PEG-WA-Trf/Dx showed enhanced transport and accumulation of Dx at the site of brain tumor (He et al., 2011).

Lactoferrin (Ltf), is a glycoprotein bound to iron molecules. It can be utilized as a brain targeting ligand with PEG-modified PAMAM dendrimer. The synthesized complex of PAMAM-PEG-Ltf/DNA displayed 2.2-fold larger brain delivery than the complex of PAMAM-PEG-Trf *in vivo*. The transfection ability was also found to be higher along with more widespread gene expression in PAMAM-PEG-Ltf/DNA complex (Huang et al., 2008).

Nanoglobular PLL dendrimer of third generation G3 was employed to conjugate RGD peptide with PEG spacer coupled to Dx. The synthesized G3-[PEG-RGD]-[Dx] was again combined with siRNA and utilized for efficient genetic transfection of Dx and siRNA (Kaneshiro et al., 2009). Angiopep as brain-targeting ligand has been further utilized. In the investigation, the synthesized PAMAM-PEG-angiopep/DNA complex showed greater penetration and accumulation in BBB. Further, the complex also showed higher gene expression efficiency in the brain (Ke et al., 2009). Further, the use of epidermal growth factor (EGF) conjugated boron-treated dendrimer (BRD) for the molecular targeting of boron neutron capture therapy (BNCT) have been demonstrated. The results of biodistribution data showed that the efficacy of BNCT increases significantly following intracerebral convection-enhanced delivery of BRD-EGF compared to intratumoral injection (Yang et al., 2009).

Recently, numerous ligands grafted dendrimers have been developed and compared for their brain targeting capability. Different ligands utilized for conjugation are sialic acid (SLA), glucosamine (GLA), and concanavalin A (CCA) with PPI dendrimers and evaluated anticancer drug Paclitaxel for its targeted delivery to the brain. The order of targeting potential found was SLA > GLA > CCA. The report concluded that these ligands are potential anchoring candidates for improved targeted delivery of drugs in brain tumor for better therapeutic outcome (Patel et al., 2016).

9.3.4.3.3   Gene Delivery in Murine Cardiac Grafts

Effective gene transfection into murine cardiac grafts using Starburst PAMAM dendrimers was first reported in 1998. They have demonstrated augmented pDNA-mediated gene delivery into the graft model. In the investigation after sacrificing neonatal donor mice, their hearts were removed and placed in the ear prima of recipients. The synthesized G5 and G9 ethylenediamine (EDA) dendrimers were injected during the transplantation into the graft. The X-Gal staining was used to measure the genetic expression. The results showed the significant increase in transfection efficiency and prolonged transgene expression with G5 EDA dendrimer in comparison with naked pDNA (Qin et al., 1998).

Further, the increase in transfection efficiency by combining electroporation with dendrimeric vector in cardiac transplants have been demonstrated. The study was performed in nonvascularized as well as in vascularized transplantation model. In the nonvascularized model, the heart was removed from mice, immersed in dendrimer/DNA complex solution.

After immersion electroporation, wave pulses were applied. In the case of vascularized transplantation model dendrimer/DNA complex was infused into the aortic root of the inferior vena cava and the heart was removed to be placed in a cuvette. The electroporation wave pulses were further applied. B-galactoside genetic expression was further studied using X-Gal staining. The results showed significantly improved gene transfer and expression using the combination of electroporation with dendrimers (Wang et al., 2001).

9.3.4.3.4 Gene Delivery to Lungs

Gene transfection to the lungs can be beneficial to treat related disorders like pulmonary fibrosis, cystic fibrosis or asthma. Transfection efficiency of the delivery vectors can be investigated estimating the gene expression level of chloramphenicol acetyltransferase (CPAT) and luciferase (Luc) in the lungs (Raczka et al., 1998). Kukowska-Latallo and coworkers have investigated the gene delivery of G9 PAMAM dendrimers through intravascular and endobronchial routes and compared it with naked pDNA. The sample was inserted through intravascular delivery in female BALB/c mice. For endobronchial delivery, intratracheal and intranasal instillation was performed, and differences in transfection between trachea and lung tissues were measured. Further, the gene expression study was carried out by measuring the levels of CPAT by ELISA method. The results of intravascular delivery showed higher CPAT gene expression in lungs, and in comparison, very low levels of CPAT transgene expression was observed after intravascular administration of naked pDNA. Endobronchial delivery by intranasal and intratracheal routes achieves quite lower level of CPAT in than naked pDNA. The results also demonstrated that intravascular delivery achieves gene expression in parenchymal lung cells, while endobronchial delivery produces gene delivery specifically to the bronchial epithelium (Kukowska et al., 2000).

The investigation on branched polyethyleneimine (PEI) and PAMAM dendrimers for their gene transfection abilities to lungs was done in BALB/c mice. The dendriplexes was administered via cannula intubation, and Luc activity was examined for gene expression. The results showed that branched PEI efficiently mediate transfer of genes to the murine lungs after intratracheal administration and having superior gene transfer ability than fractured

PAMAM dendrimer (Rudolph et al., 2000). In the recent study, chemically modified, dendrimeric structures were synthesized for RNA transfection to the specific cells. The chemical modification of free amines with alkyl chains on PPI and PAMAM dendrimers have the significant contribution for targeting Tie2-expressing lung endothelial cells (Khan et al., 2015).

9.3.5 ADVANTAGES OFFERED BY DENDRIMERS OVER OTHER GENE CARRIERS

Dendrimers offers several advantages over other delivery vectors. The nano-dimensional size makes them less susceptible for reticuloendothelial system uptake. Lower polydispersity index due to stringent control during synthesis is also an important advantage of dendrimer over other carrier system. In addition, the formation of hollow core due to multiple branches with density variation creates a region that can be utilized for drug encapsulation (Kesharwani et al., 2014a). The presence of multiple functional groups at the exterior area can be utilized to conjugate vector devices for targeting to particular site in the body. Also, dendrimers can be modified as stimuli responsive to release the drug from its core. The important advantage of dendrimers is that they can be fabricated and tailored for targeted and specialized applications (Tekade et al., 2008; Kesharwani et al., 2011; Tekade et al., 2015; Tekade et al., 2016). They are ideal drug-delivery systems due to their feasible topology, functionality, and dimensions; and also, their size is very close to various important biological polymers and assemblies such as DNA and proteins which are physiologically ideal (Kesharwani et al., 2015a).

9.3.6 SURFACE ENGINEERED DENDRIMERS IN GENE DELIVERY

Dendrimer-mediated gene delivery had significant limitations of reduced transfection efficiency and severe cytotoxicity. The particular functional group containing ligands can be incorporated onto the surface of dendrimers. Various strategies can be employed for surface engineering such as modification of lipid chain, dendrimer fluorination, conjugation with amino acids, cyclodextrin, peptides, and polymeric ligands.

9.3.6.1   Lipid Chain Modification

Several attempts have been made by modifying the dendrimer with the addition of fusogenic lipid materials. The G5 PAMAM dendrimers have been amended functionally with hydrophobic chains of lauric, myristic, and palmitic fatty acids. The complex of functionalized dendrimer/DNA has demonstrated the remarkable capacity of transfection and low levels of cytotoxicity in mesenchymal stem cells (Santos et al., 2010a). In the separate study, lower generation G2 PAMAM dendrimer is functionalized with alkyl sulfonyl hydrophobic tails by aza-Michael type addition reaction. The synthesized functional dendrimer has been evaluated for its transfection ability in eukaryotic cells with the comparison to the unmodified PAMAM-G2 dendrimer. The results showed high gene transfection and very low cytotoxicity using modified alkyl sulfonyl PAMAM-G2 dendrimers (Morales et al., 2011).

Further lipid-modified triblock copolymeric system have been developed with G4-PAMAM, PEG, and DOPE (G4-PEG-DOPE). DOPE offered the hydrophobic environment and improved cell penetrability. PEG makes the system flexible for easy accessibility of siRNA. The synthesized triblock copolymeric system had demonstrated improved transfection efficiency for delivering anticancer drug doxorubicin (Biswas et al., 2013). In the recent study, cyclododeacetylated dendrimer with higher transfection efficiency and lower cytotoxicity than its dodeacetylated analog have been reported. The mechanism for increasing transfection efficiency is attributed to improved siRNA packaging and cellular uptake (Shen et al., 2016).

9.3.6.2   Dendrimer Fluorination

Fluorination of the compounds improves stability, therapeutic efficacy, pharmacokinetic properties, and affinity for intracellular vesicles and ensures improved cellular uptake. The improved transfection efficiency has been demonstrated in several cell lines with fluorinated dendrimers at very lower nitrogen/phosphorus ratio. Fluorination enhances the cellular uptake and improves serum stability, thereby exhibiting high gene transfection efficiency (Wang et al., 2014).

Fluorination strategy can also be employed for PPI dendrimeric structures for efficient gene delivery and reducing the cytotoxicity of these cationic polymers (Liu, H. et al., 2014). The fluorination of G4-PAMAM dendrimer yields excellent transfection efficiency and low cytotoxicity for TRAIL (TNF-related apoptosis-inducing ligands) gene therapy in cancer cells (Wang, et al., 2016). A recent experiment has been conducted in which bio-reducible fluorinated peptide dendrimers have been developed. The novel designed vector has various unique properties such as inactive surfaces that resist the protein interactions, improved cellular uptake with virus mimicking surface topography, cellular reuptake, endosomal escape, cytoplasmic trafficking, and entry to the nucleus. These all properties significantly contribute to enhance gene transfection efficiency, improved serum stability, and superior biocompatibility as compared to PEI-based system (Cai et al., 2016).

9.3.6.3 Amino Acid Conjugation

Conjugation of arginine or lysine bearing cationic functionality with PAMAM dendrimers has the significant effect of improving transfection efficiency and reducing cytotoxicity. Arginine- and lysine-based amino acids carry positive charges in their structures and beneficial to improve charge density of dendrimeric surface. The improved charge density is useful in stabilizing the dendrimer/DNA complex. Apart from that, the presence of guanidinium group with three nitrogen improves the affinity of dendrimer for biological membrane via enhanced ion pairing and hydrogen bonding (Pantos et al., 2008). Arginine-terminated G4-PAMAM dendrimers have been developed and proven to be successful delivery vector for efficient gene delivery with higher transfection efficiency than non-arginine-based delivery systems (Liu, et al., 2014). Arginine-grafted dendrimeric vectors also demonstrated lower cytotoxicity profile in human mesenchymal stem cells (Kim et al., 2015).

The latest experiment has demonstrated and explored the structural activity relation of 20 common amino acids conjugated dendrimers with respect to their gene transfection capability. Dendrimers conjugation with lipophilic and cationic amino acids improve gene delivery. Cationic charges on amino acids also increases the stability of complex. The imidazole ring of histidine improves buffering characteristics of the dendrimer, thereby facilitates the endosomal escape. The lipophilic amino acids (tyrosine and phenylalanine) have better cellular internalization. Dendrimers with hydrophilic and anionic amino acids have poor transfection efficiency. Dendrimers conjugated and modified using arginine, lysine, tyrosine, phenylalanine, and tryptophan showed very high cytotoxic reactions than other amino acids (Wang et al., 2016).

9.3.6.4 Cyclodextrin-Modified Dendrimers

Cyclodextrins are cyclic oligosaccharides with the α-D-glucopyranose central cavity as a hydrophobic part and having hydrophilic outer surface (Saenger et al., 1980). Cyclodextrins are most commonly known for their inclusion complex forming ability with various molecules in different states. In the Pioneer research, Generation 2, PAMAM dendrimers were modified by complexing with α, β, and γ-cyclodextrins. The results of luciferase gene expression have revealed that α-cyclodextrin conjugated dendrimers provided the maximum transfection efficiency in RAW264.7 cells (Arima et al., 2001). The relation of different generations of dendrimeric architectures (G2, G3, and G4) on conjugation with α-cyclodextrin for their gene transfection abilities have been investigated. The results revealed superior gene transfer ability of G3-α-cyclodextrin conjugated dendrimer and high interaction with a fluorescent probe confirms the greater inclusion ability of phospholipid after gene transfection (Kihara et al., 2002). Folate PEG-appended dendrimers of fourth generation with α-cyclodextrin conjugation (FP-G4-CYDα) have been prepared and investigated for their efficiency as tumor targeting siRNA carriers. The results confirmed noticeable RNAi effect with adequate physicochemical changes like folate-mediated endocytosis, improving endosomal escape and efficient siRNA delivery into the cytoplasm with reduced cytotoxicity. Apart from these, FP-G4-CYDα dendrimer complex showed better serum stability and longer circulation ability when compared with G3 generation of same complex (Ohyama et al., 2016). In the recent experiment, G5 PAMAM dendrimers conjugated β-cyclodextrin were utilized as a template to capture gold nanoparticles. The results showed that prepared gene delivery vector possess lower cytotoxicity and higher efficiency in gene delivery with β-cyclodextrin conjugation (Qiu et al., 2016).

9.3.6.5 Peptides Conjugation

Peptides have been effectively used as an anchored molecule that can be conjugated with dendrimers mostly because of its cancer targeted and site-specific gene delivery applications. The major role of peptide and protein conjugation with dendrimers is to improve transfection efficiency and cell penetration, brain, and cancer cell-specific targeting capabilities. The investigated proteins and peptides for successful dendrimer conjugation investigated includes epidermal growth factor (EGF) (Yoon et al., 2016), antibodies (Theoharis et al., 2009), transferrin (Alrobaian et al., 2014; Gao et al., 2015), lactoferrin (Somani et al., 2015), angiopeptide (Xu et al., 2016), mesenchymal stem cells (MSC) binding peptides (Santos et al., 2010b), transactivator of transcription (TAT) (Kang et al., 2005), rabies virus glycoprotein (RVG29) (Liu et al., 2009), Arginylglycylaspartic acid (RGD) peptide (Kong et al., 2015), leptin 30 (Liu et al., 2010), luteinizing hormone-releasing hormone (LHRH) peptide (Ghanghoria et al., 2016a, 2016b), and nuclear localizing signal peptide (Lee et al., 2014).

9.3.6.6 Polymer-Modified Dendrimers

PEG-modified dendrimeric structures have been studied extensively in gene delivery applications. PEGylation has the significant effect of reducing the cytotoxicity of dendrimers. The complexation of two PEG units of different molecular weights with PAMAM G5 dendrimers have been developed. The significant reduction in the cytotoxicity with PEGylation is due to decreased production of reactive oxygen species (ROS). The results also showed that with increasing the chain length and with higher molecular weight PEG, lower cytotoxicity can be achieved (Wang et al., 2010). PEGylation strategy has been employed in many studies to decrease the cytotoxicity of dendrimers with various bioactive molecules such as etoposide (Kojima et al., 2011), doxorubicin (Mecke et al., 2004), adriamycin (Kwon et al., 1996) and paclitaxel (Agashe et al., 2006). The PEGylation of PAMAM dendrimers is beneficial in reducing the hemolytic toxicity by shielding the positive surface charges. PEGylation reduced the hemolytic reactions by lowering the interactions of RBCs with positively charged amines groups (Zhu et al., 2010). PEGylation strategy also beneficial in delivering the PAMAM dendrimers loaded biomolecules to the site-specific target cells and have significant targeting potential for tumor cells of interests with little cytotoxicity and enhanced solubility (Prajapati et al., 2009; Jain et al., 2010; Luong et al., 2016). Further, various other polymeric systems were also utilized with dendrimers such as Pluronic P123 (Gu et al., 2015), poly-L-lysine (PLL) (Pan, S. et al., 2011), chitosan (Sarkar and Kundu, 2013; Tekade et al., 2014; Tekade et al., 2015c), and polyethyleneimine (PEI) (Pan et al., 2013).

9.3.6.7 Other Strategies

However, the research on potential use of dendrimers has a topic of interest from few decades, because of the variety of properties offered by this novel nanocarrier (Jain et al., 2015; Mansuri et al., 2016; Saleh et al., 2016). In context, employing the use of nanotechnology, the technique of delivery drugs or genetic materials through dendrimers further explore many areas in the field of nanomedicine (Dhakad et al., 2013; Tekade et al., 2013). In addition to the strategies mentioned above to improve gene transfection efficiency of dendrimers, various other methods can also be employed. The use of nanoparticles (gold nanostructures, graphene, quantum dots, carbon nanotubes and magnetic nanomaterial) has been demonstrated to improve gene transfection with reduced cytotoxicity for PAMAM and PPI dendrimers (Cheng et al., 2011). But the uses of nanoparticles are limited due to its reported pulmonary toxicity (Kayat et al., 2011; Chougule et al., 2014). Dendrimers can also be conjugated with positively charged groups, namely, oligoamine to increase the charge density (Uchida et al., 2011). Dendrimers can also be modified with imidazolium (Luo et al., 2011) guanidium (Tziveleka et al., 2007), phosphonium (Ornelas et al., 2012), hormone (dexamethasone and triamcinolone acetamide) (Jeon et al., 2015), folic acid (Ma et al., 2009), nucleobase (Wang et al., 2015), saponins (Glycyrrhizin) (Chopdey et al., 2015) and aminoglycoside antibiotics (neomycin and paromomycin) (Ghilardi et al., 2013; Yadav et al., 2016).

9.4 FUTURE PERSPECTIVES

The designing of advanced gene delivery vectors largely dependent upon the chemical structures of cationic complexation molecule and the physicochemical characteristics of the genetic material to further connect it with the clinical outcome. Although, this technology is still under development and required sequential research on the matter. Developmental strategies should include the complete analysis of the interaction between the nucleic acid complex and dendrimer, and their capability to facilitate transfection. Therefore, the researchers should focus more on upliftment of transfection ability.

Despite the known limitations, dendrimers still seem to be the most practical delivery tool. However, the developmental path required the complete understanding of dendrimer generation, functionalization, and chemical properties to improve the gene transfection and genetic expression through this novel nanocarriers. Further research will be necessary to obtain conclusive evidence on the suitability of dendrimers in gene delivery for biomedical applications. Instead of the independent studies done so far, a comprehensive evaluation of dendrimer-mediated gene delivery and different vectors should be performed to identify an optimized design for genetic materials intended for medical applications.

9.5 CONCLUSION

The transformation of gene delivery systems to clinical development was the primary focus and objective of the several published literatures. Targeted high-level transgen expression are very important to predict the efficacy of a formulation and additional complexity in *in vivo* determination. Dendritic architect provides a newer way to transfer the genetic material inside the host by offering advantages (i) avoid chemical degradation; (ii) higher transfection efficiency; (iii) safe and reliable method; (iv) targeted delivery; and (v) increased uptake. In addition, different generations of dendrimers with special structural composition provides opportunity to transfect variety of genetic materials. Recently, several reports have been published describing the delivery of siRNA through dendrimers for variety of clinical indications, including various types of carcinomas. Dendrimers-based delivery of genetic material is very important in particular for treating cancer, as targeted and very specific delivery is prerequisite in cancer treatment. However, understanding of dendrimer generation, physicochemical characteristics and surface modification aspects needed to be improved for better determination of transfection efficiency and clinical outcomes.

KEYWORDS

- cytoplasmic signaling
- dendriplexes
- gene therapy
- gene transfection
- nanosized architecture
- nucleic acids

REFERENCES

Abbasi, E., Aval, S. F., Akbarzadeh, A., Milani, M., Nasrabadi, H. T., Joo, S. W., Hanifehpour, Y., Nejati-Koshki, K., & Pashaei-Asl, R., (2014). Dendrimers: Synthesis, applications, and properties. *Nanoscale Res. Lett.*, *9*(1), 247.

Abdellatif, A. A., Pelt, J. L., Benton, R. L., Howard, R. M., Tsoulfas, P., Ping, P., Xu, X. M., & Whittemore, S. R., (2006). Gene delivery to the spinal cord: comparison between lentiviral, adenoviral, and retroviral vector delivery systems. *J. Neurosci. Res.*, *84*(3), 553–567.

Agashe, H. B., Dutta, T., Garg, M., & Jain, N. K., (2006). Investigations on the toxicological profile of functionalized fifth-generation poly (propylene imine) dendrimer. *J. Pharm. Pharmacol.*, *58*(11), 1491–1498.

Al Robaian, M., Chiam, K. Y., Blatchford, D. R., & Dufès, C., (2014). Therapeutic efficacy of intravenously administered transferrin-conjugated dendriplexes on prostate carcinomas. *Nanomedicine (Lond)*, *9*(4), 421–434.

Al-Dosari M. S., & Gao, X., (2009). Nonviral gene delivery: Principle, limitations, and recent progress. *AAPS J.*, *11*(4), 671.

Anson, D. S., (2004). The use of retroviral vectors for gene therapy-what are the risks? A review of retroviral pathogenesis and its relevance to retroviral vector-mediated gene delivery. *Genet. Vaccines Ther.*, *2*(1), 9.

Arima, H., Kihara, F., Hirayama, F., & Uekama, K., (2001). Enhancement of gene expression by polyamidoamine dendrimer conjugates with alpha-, beta-, and gamma-cyclodextrins. *Bioconjug. Chem.*, *12*(4), 476–484.

Baban, C. K., Cronin, M., O'Hanlon, D., O'Sullivan, G. C., & Tangney, M., (2010). Bacteria as vectors for gene therapy of cancer. *Bioeng. Bugs.*, *1*(6), 385–394.

Behr, J. P., (1994). Gene transfer with synthetic cationic amphiphiles: prospects for gene therapy. *Bioconjug. Chem.*, *5*(5), 382–389.

Behr, J. P., (1997). The proton sponge: a trick to enter cells the viruses did not exploit. *Chimia.*, *51*, 34–36.

Bensadoun, J. C., Déglon, N., Tseng, J. L., Ridet, J. L., Zurn, A. D., & Aebischer, P., (2000). Lentiviral vectors as a gene delivery system in the mouse midbrain: cellular and behavioral improvements in a 6-OHDA model of Parkinson's disease using GDNF. *Exp. Neurol.*, *164*(1), 15–24.

Bielinska, A. U., Kukowska-Latallo, J. F., & Baker, J. R. Jr., (1997). The interaction of plasmid DNA with polyamidoamine dendrimers: mechanism of complex formation and analysis of alterations induced in nuclease sensitivity and transcriptional activity of the complexed DNA. *Biochim. Biophys. Acta.*, *1353*(2), 180–190.

Bielinska, A. U., Kukowska-Latallo, J. F., Johnson, J., Tomalia, D. A., & Baker, J. R. Jr., (1996). Regulation of *in vitro* gene expression using antisense oligonucleotides or antisense expression plasmids transfected using starburst PAMAM dendrimers. *Nucleic Acids Res.*, *24*(11), 2176–2182.

Biswas, S., Deshpande, P. P., Navarro, G., Dodwadkar, N. S., & Torchilin, V. P., (2013). Lipid modified triblock PAMAM-based nanocarriers for siRNA drug co-delivery. *Biomaterials*, *34*(4), 1289–1301.

Bolhassani, A., (2011). Potential efficacy of cell-penetrating peptides for nucleic acid and drug delivery in cancer. *Biochim. Biophys. Acta.*, *1816*(2), 232–246.

Breau, W. C., Atwood, W. J., & Norkin, L. C., (1992). Class I major histocompatibility proteins are an essential component of the simian virus 40 receptor. *J. Virol.*, *66*(4), 2037–2045.

Buhleier, E., Wehner, W., & Vogtle, F., (1978). "Cascade"- and "nonskid-chain-like" synthesis of molecular cavity topologies. *Synthesis*, *1978*(2), 155–158.

Bushman, F. D., (2007). Retroviral integration and human gene therapy. *J. Clin. Invest.*, *117*(8), 2083–2086.

Butel, J. S., & Jarvis, D. L., (1986). The plasma-membrane-associated form of SV40 large tumor antigen: biochemical and biological properties. *Biochim. Biophys. Acta.*, *865*(2), 171–195.

Cai, X., Jin, R., Wang, J., Yue, D., Jiang, Q., Wu, Y., & Gu, Z., (2016). Bioreducible fluorinated peptide dendrimers capable of circumventing various physiological barriers for highly efficient and safe gene delivery. *ACS Appl. Mater. Interfaces*, *8*(9), 5821–5832.

Campos, S. K., & Barry, M. A., (2007). Current advances and future challenges in adenoviral vector biology and targeting. *Curr. Gene Ther.*, *7*(3), 189–204.

Chang, H., Wang, H., Shao, N., Wang, M., Wang, X., & Cheng, Y., (2014). Surface-engineered dendrimers with a diaminododecane core achieve efficient gene transfection and low cytotoxicity. *Bioconjug. Chem.*, *25*(2), 342–350.

Chen, H. H., Ho, Y. P., Jiang, X., Mao, H. Q., Wang, T. H., & Leong, K. W., (2008). Quantitative comparison of intracellular unpacking kinetics of polyplexes by a model constructed from quantum dot-FRET. *Mol. Ther.*, *16*(2), 324–332.

Cheng, Y., Zhao, L., Li, Y., & Xu, T., (2011). Design of biocompatible dendrimers for cancer diagnosis and therapy: current status and future perspectives. *Chem. Soc. Rev.*, *40*(5), 2673–2703.

Chopdey, P. K., Tekade, R. K., Mehra, N. K., Mody, N., & Jain, N. K., (2015). Glycyrrhizin conjugated dendrimer and multi-walled carbon nanotubes for liver specific delivery of doxorubicin. *J. Nanosci. Nanotechnol.*, *15*(2), 1088–1100.

Chougule, M. B., Tekade, R. K., Hoffmann, P. R., Bhatia, D., Sutariya, V. B., & Pathak, Y., (2014). Nanomaterial-based gene and drug delivery: Pulmonary toxicity considerations. In *Biointeractions of Nanomaterials*, pp. 225–248.

Coffin, J. M., (1979). Structure, replication, and recombination of retrovirus genomes: Some unifying hypotheses. *J. Gen. Virol.*, *42*(1), 1–26.

Dass, C. R., & Choong, P. F., (2006). Selective gene delivery for cancer therapy using cationic liposomes: *in vivo* proof of applicability. *J. Control. Release, 113*(2), 155–163.

Davtyan, H., Ghochikyan, A., Movsesyan, N., Ellefsen, B., Petrushina, I., Cribbs, D. H., Hannaman, D., Evans, C. F., & Agadjanyan, M. G., (2012). Delivery of a DNA vaccine for Alzheimer's disease by electroporation versus gene gun generates potent and similar immune responses. *Neurodegener. Dis., 10*(1–4), 261–264.

Daya, S., & Berns, K. I., (2008). Gene therapy using adeno-associated virus vectors. *Clin. Microbiol. Rev., 21*(4), 583–593.

De Almeida, L. P., Zala, D., Aebischer, P., & Déglon, N., (2001). Neuroprotective effect of a CNTF-expressing lentiviral vector in the quinolinic acid rat model of Huntington's disease. *Neurobiol. Dis., 8*(3), 433–446.

Dean, D. A., Machado, A. D., Blair, P. K., Yeldandi, A. V., & Young, J. L., (2003). Electroporation as a method for high level nonviral gene transfer to the lung. *Gene Ther., 10*, 1608–1615.

Dennig, J., & Duncan, E., (2002). Gene transfer into eukaryotic cells using activated polyamidoamine dendrimers. *J. Biotechnol., 90*(3–4), 339–347.

Dhakad, R. S., Tekade, R. K., & Jain, N. K., (2013). Cancer targeting potential of folate targeted nanocarrier under comparative influence of tretinoin and dexamethasone. *Curr. Drug Del., 10*, 477–491.

Dufès, C., Uchegbu, I. F., & Schätzlein, A. G., (2005). Dendrimers in gene delivery. *Adv. Drug Deliv. Rev., 57*(15), 2177–21202.

During, M. J., (1997). Adeno-associated virus as a gene delivery system. *Adv. Drug Deliv. Rev., 27*(1), 83–94.

Dwivedi, P., Tekade, R. K., & Jain, N. K., (2013). Nanoparticulate carrier mediated intranasal delivery of insulin for the restoration of memory signaling in Alzheimer's disease. *Curr. Nanosci., 9*(1), 46–55.

Eastham, J. A., Hall, S. J., Sehgal, I., Wang, J., Timme, T. L., Yang, G., Connell-Crowley, L., Elledge, S. J., Zhang, W. W., & Harper, J. W., (1995). *In vivo* gene therapy with p53 or p21 adenovirus for prostate cancer. *Cancer Res., 55*(22), 5151–5155.

El-Aneed, A., (2004). An overview of current delivery systems in cancer gene therapy. *J. Control Release, 94*(1), 1–14.

Eming, S. A., Krieg, T., & Davidson, J. M., (2007). Gene therapy and wound healing. *Clin. Dermatol., 25*(1), 79–92.

Erbacher, P., Roche, A. C., Monsigny, M., & Midoux, P., (1996). Putative role of chloroquine in gene transfer into a human hepatoma cell line by DNA/lactosylated polylysine complexes. *Exp. Cell Res., 225*(1), 186–194.

Erbacher, P., Roche, A. C., Monsigny, M., & Midoux, P., (1997). The reduction of the positive charges of polylysine by partial gluconoylation increases the transfection efficiency of polylysine/DNA complexes. *Biochim. Biophys. Acta., 1324*(1), 27–36.

Federici, T., Kutner, R., Zhang, X. Y., Kuroda, H., Tordo, N., Boulis, N. M., & Reiser, J., (2009). Comparative analysis of HIV-1-based lentiviral vectors bearing lyssavirus glycoproteins for neuronal gene transfer. *Genet. Vaccines Ther., 7*, 1.

Felgner, P. L., Gadek, T. R., Holm, M., Roman, R., Chan, H. W., & Wenz, M., (1987). Lipofection: A highly efficient, lipid-mediated DNA-transfection procedure. *Proc. Natl. Acad. Sci. USA, 84*, 7413–7417.

Freytag, S. O., Khil, M., Stricker, H., Peabody, J., Menon, M., DePeralta-Venturina, M., Nafziger, D., Pegg, J., Paielli, D., Brown, S., Barton, K., Lu, M., Aguilar-Cordova, E., & Kim, J. H., (2002). Phase I study of replication-competent adenovirus-mediated double suicide gene therapy for the treatment of locally recurrent prostate cancer. *Cancer Res.*, *62*(17), 4968–4976.

Friedmann, T., (1994). Gene therapy for neurological disorders. *Trends Genet.*, *10*(6), 210–214.

Fuchs, S., Kapp, T., Otto, H., Schöneberg, T., Franke, P., Gust, R., & Schlüter, A. D., (2004). A surface-modified dendrimer set for potential application as drug delivery vehicles: synthesis, *in vitro* toxicity, and intracellular localization. *Chemistry*, *10*(5), 1167–1192.

Gajbhiye, V., Palanirajan, V. K., Tekade, R. K., & Jain, N. K., (2009a). Dendrimers as therapeutic agents: a systematic review. *J. Pharm. Pharmacol.*, *61*(8), 989–1003.

Gajbhiye, V., Vijayaraj, K. P., Tekade, R. K., & Jain, N. K., (2007). Pharmaceutical and biomedical potential of PEGylated dendrimers. *Curr. Pharm. Des.*, *13*(4), 415–429.

Gajbhiye, V., Vijayaraj, K. P., Tekade, R. K., & Jain, N. K., (2009b). PEGylated PPI dendritic architectures for sustained delivery of H2 receptor antagonist. *Eur. J. Med. Chem.*, *44*(3), 1155–1166.

Gandhi, N. S., Tekade, R. K., & Chougule, M. B., (2014). Nanocarrier mediated delivery of siRNA/miRNA in combination with chemotherapeutic agents for cancer therapy: Current progress and advances. *J. Control Release*, *194*, 238–256.

Gao, S., Li, J., Jiang, C., Hong, B., & Hao, B., (2015). Plasmid pORF-hTRAIL targeting to glioma using transferrin-modified polyamidoamine dendrimer. *Drug Des. Devel. Ther.*, *10*, 1–11.

Gao, X., & Huang, L., (1995). Cationic liposome-mediated gene transfer. *Gene Ther.*, *2*(10), 710–722.

Gebhart, C. L., & Kabanov, A. V., (2001). Evaluation of polyplexes as gene transfer agents. *J. Control. Release.*, *73*(2–3), 401–416.

Gerrard, A. J., Hudson, D. L., Brownlee, G. G., & Watt, F. M., (1993). Towards gene therapy for haemophilia B using primary human keratinocytes. *Nat. Genet.*, *3*(2), 180–183.

Ghanghoria, R., Kesharwani, P., Tekade, R. K., & Jain, N. K., (2016b). Targeting luteinizing hormone-releasing hormone: A potential therapeutics to treat gynecological and other cancers. *J. Control. Release.*, pii: S0168–3659(16)31169–5.

Ghanghoria, R., Tekade, R. K., Mishra, A. K., Chuttani, K., & Jain, N. K., (2016a). Luteinizing hormone-releasing hormone peptide tethered nanoparticulate system for enhanced antitumoral efficacy of paclitaxel. *Nanomedicine*, *11*(7) 797–816.

Ghilardi, A., Pezzoli, D., Bellucci, M. C., Malloggi, C., Negri, A., Sganappa, A., Tedeschi, G., Candiani, G., & Volonterio, A., (2013). Synthesis of multifunctional PAMAM-aminoglycoside conjugates with enhanced transfection efficiency. *Bioconjug. Chem.*, *24*(11), 1928–1936.

Gilboa, E., & Smith, C., (1994). Gene therapy for infectious diseases: the AIDS model. *Trends Genet.*, *10*(4), 139–144.

Gillies, E. R., Dy, E., Fréchet, J. M., & Szoka, F. C., (2005). Biological evaluation of polyester dendrimer: poly(ethylene oxide) "bow-tie" hybrids with tunable molecular weight and architecture. *Mol. Pharm.*, *2*(2), 129–138.

Godbey, W. T., Wu, K. K., & Mikos, A. G., (1999). Tracking the intracellular path of poly(ethylenimine)/DNA complexes for gene delivery. *Proc. Natl. Acad. Sci. USA*, *96*, 5177–5181.

Gómez-Navarro, J., Curiel, D. T., & Douglas, J. T., (1999). Gene therapy for cancer. *Eur. J. Cancer*, *35*(6), 867–885.

Gómez, C. E., Nájera, J. L., Krupa, M., & Esteban, M., (2008). The poxvirus vectors MVA and NYVAC as gene delivery systems for vaccination against infectious diseases and cancer. *Curr. Gene Ther.*, *8*(2), 97–120.

Goss, J. R., Harley, C. F., Mata, M., O'Malley, M. E., Goins, W. F., Hu, X., Glorioso, J. C., & Fink, D. J., (2002). Herpes vector-mediated expression of proenkephalin reduces bone cancer pain. *Ann. Neurol.*, *52*(5), 662–665.

Gothwal, A., Kesharwani, P., Gupta, U., Khan, I., Iqbal, M. A. M. C., Banerjee, S., & Iyer, A. K., (2015). Dendrimers as an effective nanocarrier in cardiovascular disease. *Curr. Pharm. Des.*, *21*(30), 4519–4526.

Goula, D., Benoist, C., Mantero, S., Merlo, G., Levi, G., & Demeneix, B. A., (1998a). Polyethylenimine-based intravenous delivery of transgenes to mouse lung. *Gene Ther.*, *5*(9), 1291–1295.

Goula, D., Remy, J. S., Erbacher, P., Wasowicz, M., Levi, G., Abdallah, B., & Demeneix, B. A., (1998b). Size, diffusibility and transfection performance of linear PEI/DNA complexes in the mouse central nervous system. *Gene Ther.*, *5*(5), 712–717.

Grossman, M., Raper, S. E., Kozarsky, K., Stein, E. A., Engelhardt, J. F., Muller, D., Lupien, P. J., & Wilson, J. M., (1994). Successful ex vivo gene therapy directed to liver in a patient with familial hypercholesterolaemia. *Nat. Genet.*, *6*(4), 335–341.

Gu, Z., Wang, M., Fang, Q., Zheng, H., Wu, F., Lin, D., Xu, Y., & Jin, Y., (2015). Preparation and *in vitro* characterization of pluronic-attached polyamidoamine dendrimers for drug delivery. *Drug Dev. Ind. Pharm.*, *41*(5), 812–818.

Haensler, J., & Szoka, F. C. Jr., (1993). Polyamidoamine cascade polymers mediate efficient transfection of cells in culture. *Bioconjug. Chem.*, *4*(5), 372–379.

Harper, S. Q., Hauser, M. A., DelloRusso, C., Duan, D., Crawford, R. W., Phelps, S. F., Harper, H. A., Robinson, A. S., Engelhardt, J. F., Brooks, S. V., & Chamberlain, J. S., (2002). Modular flexibility of dystrophin: implications for gene therapy of Duchenne muscular dystrophy. *Nat. Med.*, *8*(3), 253–261.

Harvie, P., Wong, F. M., & Bally, M. B., (2000). Use of poly (ethylene glycol)-lipid conjugates to regulate the surface attributes and transfection activity of lipid-DNA particles. *J. Pharm. Sci.*, *89*(5), 652–663.

Haskins, M., (2009). Gene therapy for lysosomal storage diseases (LSDs) in large animal models. *ILAR J.*, *50*(2), 112–121.

Hatakeyama, H., Ito, E., Akita, H., Oishi, M., Nagasaki, Y., Futaki, S., & Harashima, H., (2009). A pH-sensitive fusogenic peptide facilitates endosomal escape and greatly enhances the gene silencing of siRNA-containing nanoparticles *in vitro* and *in vivo*. *J. Control. Release*, *139*(2), 127–132.

He, H., Li, Y., Jia, X. R., Du, J., Ying, X., Lu, W. L., Lou, J. N., & Wei, Y., (2011). PEGylated Poly (amidoamine) dendrimer-based dual-targeting carrier for treating brain tumors. *Biomaterials*, *32*(2), 478–487.

Heller, L. C., Ugen, K., & Heller, R., (2005). Electroporation for targeted gene transfer. *Expert Opin. Drug Deliv.*, *2*, 255–268.

Hemmati, M., Najafi, F., Shirkoohi, R., Moghimi, H. R., Zarebkohan, A., & Kazemi, B., (2016). Synthesis of a novel PEGDGA-coated hPAMAM complex as an efficient and biocompatible gene delivery vector: An *in vitro* and *in vivo* study. *Drug Deliv.*, 1–14.

Herweijer, H., & Wolff, J. A., (2003). Progress and prospects: Naked DNA gene transfer and therapy. *Gene Ther.*, *10*(6), 453–458.

Horn, N. A., Meek, J. A., Budahazi, G., & Marquet, M., (1995). Cancer gene therapy using plasmid DNA: purification of DNA for human clinical trials. *Hum. Gene Ther., 6*(5), 565–573.

Hu, J., Hu, K., & Cheng, Y., (2016). Tailoring the dendrimer core for efficient gene delivery. *Acta. Biomater., 35*, 1–11.

Huang, R. Q., Qu, Y. H., Ke, W. L., Zhu, J. H., Pei, Y. Y., & Jiang, C., (2007). Efficient gene delivery targeted to the brain using a transferrin-conjugated polyethyleneglycol-modified polyamidoamine dendrimer. *FASEB J., 21*(4), 1117–1125.

Huang, R., Ke, W., Liu, Y., Jiang, C., & Pei, Y., (2008). The use of lactoferrin as a ligand for targeting the polyamidoamine-based gene delivery system to the brain. *Biomaterials, 29*(2), 238–246.

Hudde, T., Rayner, S. A., Comer, R. M., Weber, M., Isaacs, J. D., Waldmann, H., Larkin, D. F., & George, A. J., (1999). Activated polyamidoamine dendrimers, a non-viral vector for gene transfer to the corneal endothelium. *Gene Ther., 6*(5), 939–943.

Hyde, S. C., Gill, D. R., Higgins, C. F., Trezise, A. E., MacVinish, L. J., Cuthbert, A. W., Ratcliff, R., Evans, M. J., & Colledge, W. H., (1993). Correction of the ion transport defect in cystic fibrosis transgenic mice by gene therapy. *Nature, 362*(6417), 250–255.

Immonen, A., Vapalahti, M., Tyynelä, K., Hurskainen, H., Sandmair, A., 1 Vanninen, R., Langford, G., Murray, N., & Ylä-Herttuala, S., (2004). AdvHSV-tk gene therapy with intravenous ganciclovir improves survival in human malignant glioma: a randomised, controlled study. *Mol. Ther., 10*(5), 967–972.

Jain, K., Kesharwani, P., Gupta, U., & Jain, N. K., (2010). Dendrimer toxicity: Let's meet the challenge. *Int. J. Pharm., 394*(1–2), 122–142.

Jain, N. K., & Tekade, R. K., (2013). Dendrimers for enhanced drug solubilization, In: *Drug Delivery Strategies for Poorly Water-Soluble Drugs,* Douroumis, D., Fahr, A., (ed.), John Wiley & Sons Ltd, Oxford, UK, Ch. 13, vol. *12*, pp. 373–409.

Jain, S., Kesharwani, P., Tekade, R. K., & Jain, N. K., (2015). One platform comparison of solubilization potential of dendrimer with some solubilizing agents. *Drug Dev. Ind. Pharm., 41*, 722–727.

Jeon, P., Choi, M., Oh, J., & Lee, M., (2015). Dexamethasone-conjugated polyamidoamine dendrimer for delivery of the heme oxygenase-1 gene into the ischemic brain. *Macromol. Biosci., 15*(7), 1021–1028.

Judge, A. D., Bola, G., Lee, A. C., & MacLachlan, I., (2006). Design of noninflammatory synthetic siRNA mediating potent gene silencing *in vivo. Mol. Ther., 13*(3), 494–505.

Kabanov, V. A., Sergeyev, V. G., Pyshkina, O. A., Zinchenko, A. A., Zezin, A. B., Joosten, J. G. H., Brackman, J., & Yoshikawa, K., (2000). Interpolyelectrolyte complexes formed by DNA and astramol poly (propylene imine) dendrimers. *Macromolecules, 33*, 9587–9593.

Kaneshiro, T. L., & Lu, Z. R., (2009). Targeted intracellular codelivery of chemotherapeutics and nucleic acid with a well-defined dendrimer-based nanoglobular carrier. *Biomaterials, 30*(29), 5660–5666.

Kang, H., DeLong, R., Fisher, M. H., & Juliano, R. L., (2005). Tat-conjugated PAMAM dendrimers as delivery agents for antisense and siRNA oligonucleotides. *Pharm. Res., 22*(12), 2099–2106.

Kang, S. J., Durairaj, C., Kompella, U. B., O'Brien, J. M., & Grossniklaus, H. E., (2009). Subconjunctival nanoparticle carboplatin in the treatment of murine retinoblastoma. *Arch. Ophthalmol., 127*(8), 1043–1047.

Kayat, J., Gajbhiye, V., Tekade, R. K., & Jain, N. K., (2011). Pulmonary toxicity of carbon nanotubes: A systematic report. *Nanomedicine*, *7*(1), 40–49.

Ke, W., Shao, K., Huang, R., Han, L., Liu, Y., Li, J., Kuang, Y., Ye, L., Lou, J., & Jiang, C., (2009). Gene delivery targeted to the brain using an Angiopep-conjugated polyethyleneglycol-modified polyamidoamine dendrimer. *Biomaterials*, *30*(36), 6976–6985.

Kennedy, P. G., (1997). Potential use of herpes simplex virus (HSV) vectors for gene therapy of neurological disorders. *Brain*, *120*(Pt 7), 1245–1259.

Kesharwani, P., Jain, K., & Jain, N. K., (2014a). Dendrimer as nanocarrier for drug delivery. *Prog. Polym. Sci.*, *39*, 268–307.

Kesharwani, P., Tekade, R. K., & Jain, N. K., (2014b). Generation dependent cancer targeting potential of poly (propyleneimine) dendrimer. *Biomaterials*, *35*, 5539–5548.

Kesharwani, P., Tekade, R. K., & Jain, N. K., (2015a). Dendrimer generational nomenclature: The need to harmonize. *Drug Discov. Today*, *20*, 497–499.

Kesharwani, P., Tekade, R. K., & Jain, N. K., (2015b). Generation dependent safety and efficacy of folic acid conjugated dendrimer based anticancer drug formulations. *Pharm. Res.*, *32*, 1438–1450.

Kesharwani, P., Tekade, R. K., Gajbhiye, V., Jain, K., & Jain, N. K., (2011). Cancer targeting potential of some ligand-anchored poly (propylene imine) dendrimers: a comparison. *Nanomedicine*, *7*, 295–304.

Kesmodel, S. B., & Spitz, F. R., (2003). Gene therapy for cancer and metastatic disease. *Expert Rev. Mol. Med.*, *5*(17), 1–18.

Khan, M. K., Nigavekar, S. S., Minc, L. D., Kariapper, M. S. T., Nair B. M., Lesniak, W. G., & Balogh, L. P., (2005). Biodistribution of dendrimers and dendrimer nanocomposites implications for cancer imaging and therapy. *Technol. Cancer Res. Treat.*, *4*(6), 603–613.

Khan, O. F., Zaia, E. W., Jhunjhunwala, S., Xue, W., Cai, W., Yun, D. S., Barnes, C. M., Dahlman, J. E., Dong, Y., Pelet, J. M., Webber, M. J., Tsosie, J. K., Jacks, T. E., Langer, R., & Anderson, D. G., (2015). Dendrimer-inspired nanomaterials for the *in vivo* delivery of siRNA to lung vasculature. *Nano Lett.*, *15*(5), 3008–3016.

Kihara, F., Arima, H., Tsutsumi, T., Hirayama, F., & Uekama, K., (2002). Effects of structure of polyamidoamine dendrimer on gene transfer efficiency of the dendrimer conjugate with alpha-cyclodextrin. *Bioconjug. Chem.*, *13*(6), 1211–1219.

Kim, H., Nam, K., Nam, J. P., Kim, H. S., Kim, Y. M., Joo, W. S., & Kim, S. W., (2015). VEGF therapeutic gene delivery using dendrimer type bio-reducible polymer into human mesenchymal stem cells (hMSCs). *J. Control. Release*, *220*(Pt A), 222–228.

Kofler, P., Wiesenhofer, B., Rehrl, C., Baier, G., Stockhammer, G., & Humpel, C., (1998). Liposome-mediated gene transfer into established CNS cell lines, primary glial cells, and *in vivo*. *Cell Transplant*, *7*(2), 175–185.

Kohn, D. B., Kantoff, P. W., Eglitis, M. A., McLachlin, J. R., Moen, R. C., Karson, E., Zwiebel, J. A., Nienhuis, A., Karlsson, S., & O'Reilly, R., (1987). Retroviral-mediated gene transfer into mammalian cells. *Blood Cells*, *13*(1–2), 285–298.

Kojima, C., Turkbey, B., Ogawa, M., Bernardo, M., Regino, C. A., Bryant, L. H. Jr., Choyke, P. L., Kono, K., & Kobayashi, H., (2011). Dendrimer-based MRI contrast agents: the effects of PEGylation on relaxivity and pharmacokinetics. *Nanomedicine*, *7*(6), 1001–1008.

Kolhe, P., Khandare, J., Pillai, O., Kannan, S., Lieh-Lai, M., & Kannan, R. M., (2006). Preparation, cellular transport, and activity of polyamidoamine-based dendritic nanodevices with a high drug payload. *Biomaterials*, *27*(4), 660–669.

Kong, L., Alves, C. S., Hou, W., Qiu, J., Möhwald, H., Tomás, H., & Shi, X., (2015). RGD peptide-modified dendrimer-entrapped gold nanoparticles enable highly efficient and specific gene delivery to stem cells. *ACS Appl. Mater. Interfaces.*, *7*(8), 4833–4843.

Kotterman, M. A., & Schaffer, D. V., (2014). Engineering adeno-associated viruses for clinical gene therapy. *Nat. Rev. Genet.*, *15*(7), 445–451.

Kotterman, M. A., Vazin, T., & Schaffer, D. V., (2015). Enhanced selective gene delivery to neural stem cells *in vivo* by an adeno-associated viral variant. *Development*, *142*(10), 1885–1892.

Kukowska-Latallo, J. F., Raczka, E., Quintana, A., Chen, C., Rymaszewski, M., & Baker, J. R. Jr., (2000). Intravascular and endobronchial DNA delivery to murine lung tissue using a novel, nonviral vector. *Hum. Gene Ther.*, *11*(10), 1385–1395.

Kurmi, B. D., Kayat, J., Gajbhiye, V., Tekade, R. K., & Jain, N. K., (2010). Micro-and nano-carrier-mediated lung targeting. *Expert Opin. Drug Deliv.*, *7*(7), 781–794.

Kwon, G. S., & Okano, T., (1996). Polymeric micelles as new drug carriers. *Adv. Drug Del. Rev.*, *21*, 107–116.

Laufs, S., Gentner, B., Nagy, K. Z., Jauch, A., Benner, A., Naundorf, S., Kuehlcke, K., Schiedlmeier, B., Ho, A. D., Zeller, W. J., & Fruehauf, S., (2003). Retroviral vector integration occurs in preferred genomic targets of human bone marrow-repopulating cells. *Blood*, *101*(6), 2191–2198.

Lechardeur, D., Sohn, K. J., Haardt, M., Joshi, P. B., Monck, M., Graham, R. W., Beatty, B., Squire, J., O'Brodovich, H., & Lukacs, G. L., (1999). Metabolic instability of plasmid DNA in the cytosol: a potential barrier to gene transfer. *Gene Ther.*, *6*(4), 482–497.

Lee, J., Jung, J., Kim, Y. J., Lee, E., & Choi, J. S., (2014). Gene delivery of PAMAM dendrimer conjugated with the nuclear localization signal peptide originated from fibroblast growth factor 3. *Int. J. Pharm.*, *459*(1–2), 10–18.

Li, Q., Bolli, R., Qiu, Y., Tang, X. L., Guo, Y., & French, B. A., (2001). Gene therapy with extracellular superoxide dismutase protects conscious rabbits against myocardial infarction. *Circulation*, *103*(14), 1893–1898.

Li, S. D., & Huang, L., (2010). Stealth nanoparticles: high density but sheddable PEG is a key for tumor targeting. *J. Control. Release.*, *145*(3), 178–81.

Liang, H. D., Lu, Q. L., Xue, S. A., Halliwell, M., Kodama, T., Cosgrove, D. O., Stauss, H. J., Partridge, T. A., & Blomley, M. J., (2004). Optimisation of ultrasound-mediated gene transfer (sonoporation) in skeletal muscle cells. *Ultrasound Med. Biol.*, *30*(11), 1523–1529.

Lin, M. T., Pulkkinen, L., Uitto, J., & Yoon, K., (2000). The gene gun: current applications in cutaneous gene therapy. *Int. J. Dermatol.*, *39*(3), 161–170.

Liu, C., Liu, X., Rocchi, P., Qu, F., Iovanna, J. L., & Peng, L., (2014). Arginine-terminated generation 4 PAMAM dendrimer as an effective nanovector for functional siRNA delivery *in vitro* and *in vivo*. *Bioconjug. Chem.*, *25*(3), 521–532.

Liu, F., Song, Y., & Liu, D., (1999). Hydrodynamics-based transfection in animals by systemic administration of plasmid DNA. *Gene Ther.*, *6*(7), 1258–1266.

Liu, H., Wang, Y., Wang, M., Xiao, J., & Cheng, Y., (2014). Fluorinated poly (propylenimine) dendrimers as gene vectors. *Biomaterials*, *35*(20), 5407–5413.

Liu, W., Liu, Z., Liu, L., Xiao, Z., Cao, X., Cao, Z., Xue, L., Miao, L., He, X., & Li, W., (2008). A novel human foamy virus mediated gene transfer of GAD67 reduces neuropathic pain following spinal cord injury. *Neurosci. Lett.*, *432*(1), 13–18.

Liu, X., Liu, C., Catapano, C. V., Peng, L., Zhou, J., & Rocchi, P., (2014). Structurally flexible triethanolamine-core poly(amidoamine) dendrimers as effective nanovectors to deliver RNAi-based therapeutics. *Biotechnol. Adv.*, *32*(4), 844–852.

Liu, Y., Huang, R., Han, L., Ke, W., Shao, K., Ye, L., Lou, J., & Jiang, C., (2009). Brain-targeting gene delivery and cellular internalization mechanisms for modified rabies virus glycoprotein RVG29 nanoparticles. *Biomaterials*, *30*(25), 4195–4202.

Liu, Y., Li, J., Shao, K., Huang, R., Ye, L., Lou, J., & Jiang, C., (2010). A leptin derived 30-amino-acid peptide modified pegylated poly-L-lysine dendrigraft for brain targeted gene delivery. *Biomaterials. 31*(19), 5246–5257.

Losordo, D. W., Vale, P. R., Symes, J. F., Dunnington, C. H., Esakof, D. D., Maysky, M., Ashare, A. B., Lathi, K., & Isner, J. M., (1998). Gene therapy for myocardial angiogenesis: initial clinical results with direct myocardial injection of phVEGF165 as sole therapy for myocardial ischemia. *Circulation*, *98*(25), 2800–2804.

Lundberg, C., Björklund, T., Carlsson, T., Jakobsson, J., Hantraye, P., Déglon, N., & Kirik, D., (2008). Applications of lentiviral vectors for biology and gene therapy of neurological disorders. *Curr. Gene Ther., 8*(6), 461–473.

Luo, K., Li, C., Wang, G., Nie, Y., He, B., Wu, Y., & Gu, Z., (2011). Peptide dendrimers as efficient and biocompatible gene delivery vectors: Synthesis and *in vitro* characterization. *J. Control. Release*, *155*(1), 77–87.

Luong, D., Kesharwani, P., Deshmukh, R., Mohd Amin, M. C., Gupta, U., Greish, K., & Iyer, A. K., (2016). PEGylated PAMAM dendrimers: Enhancing efficacy and mitigating toxicity for effective anticancer drug and gene delivery. *Acta Biomater.*, *43*, 14–29.

Lurquin, P. F., (1997). Gene transfer by electroporation. *Mol. Biotechnol.*, *7*, 5–35.

Ma, K., Hu, M. X., Qi, Y., Zou, J. H., Qiu, L. Y., Jin, Y., Ying, X. Y., & Sun, H. Y., (2009). PAMAM-triamcinolone acetonide conjugate as a nucleus-targeting gene carrier for enhanced transfer activity. *Biomaterials., 30*(30), 6109–6118.

Maheshwari, R., Tekade, M., Sharma, P. A., & Tekade, R. K., (2015a). Nanocarriers Assisted siRNA Gene Therapy for the Management of Cardiovascular Disorders. *Curr. Pharm. Des., 21*, 4427–4440.

Maheshwari, R., Tekade, R. K., Sharma, P. A., Darwhekar, G., Tyagi, A., Patel, R. P., & Jain, D. K., (2012). Ethosomes and ultradeformable liposomes for transdermal delivery of clotrimazole: A comparative assessment. *Saudi Pharm. J.*, *20*, 161–170.

Maheshwari, R., Thakur, S., Singhal, S., Patel, R. P., Tekade, M., & Tekade, R. K., (2015b). Chitosan encrusted nonionic surfactant based vesicular formulation for topical administration of ofloxacin. *Sci. Adv. Mater.*, *7*, 1163–1176.

Mali, S., (2013). Delivery systems for gene therapy. *Indian J. Hum. Genet.*, *19*(1), 3–8.

Malik, N., Wiwattanapatapee, R., Klopsch, R., Lorenz, K., Frey, H., Weener, J. W., Meijer, E. W., Paulus, W., & Duncan, R., (2000). Dendrimers: Relationship between structure and biocompatibility *in vitro*, and preliminary studies on the biodistribution of 125-I-labelled polyamidoamine dendrimers *in vivo*. *J. Control. Release*, *65*(1–2), 133–148.

Mansuri, S., Kesharwani, P., Tekade, R. K., & Jain, N. K., (2016). Lyophilized mucoadhesive-dendrimer enclosed matrix tablet for extended oral delivery of albendazole. *Eur. J. Pharm. Biopharm.*, *102*, 202–213.

Martello, F., Piest, M., Engbersen, J. F., & Ferruti, P., (2012). Effects of branched or linear architecture of bioreducible poly(amido amine)s on their *in vitro* gene delivery properties. *J. Control. Release*, *164*(3), 372–379.

Mastorakos, P., Kambhampati, S. P., Mishra, M. K., Wu, T., Song, E., Hanes, J., & Kannan, R. M., (2015). Hydroxyl PAMAM dendrimer-based gene vectors for transgene delivery to human retinal pigment epithelial cells. *Nanoscale.*, *7*(9), 3845–3856.

Mc Mahon, J. M., & Wells, D. J., (2004). Electroporation for gene transfer to skeletal muscles: Current status. *Biol. Drugs.*, *18*, 155–165.

McMurtry, M. S., Archer, S. L., Altieri, D. C., Bonnet, S., Haromy, A., Harry, G., Bonnet, S., Puttagunta, L., & Michelakis, E. D., (2005). Gene therapy targeting survivin selectively induces pulmonary vascular apoptosis and reverses pulmonary arterial hypertension. *J. Clin. Invest.*, *115*(6), 1479–1491.

Mecke, A., Uppuluri, S., Sassanella, T. M., Lee, D. K., Ramamoorthy, A., Baker, J. R. Jr., Orr, B. G., & Banaszak, H. M. M., (2004). Direct observation of lipid bilayer disruption by poly(amidoamine) dendrimers. *Chem. Phys. Lipids.*, *132*(1), 3–14.

Midoux, P., Breuzard, G., Gomez, J. P., & Pichon, C., (2008). Polymer-based gene delivery: a current review on the uptake and intracellular trafficking of polyplexes. *Curr. Gene Ther.*, *8*(5), 335–352.

Mody, N., Tekade, R. K., Mehra, N. K., Chopdey, P., & Jain, N. K., (2014). Dendrimer, liposomes, carbon nanotubes and PLGA nanoparticles: one platform assessment of drug delivery potential. *AAPS Pharm. Sci. Tech.*, *15*, 388–399.

Morales-Sanfrutos, J., Megia-Fernandez, A., Hernandez-Mateo, F., Giron-Gonzalez, M. D., Salto-Gonzalez, R., & Santoyo-Gonzalez, F., (2011). Alkyl sulfonyl derivatized PAMAM-G2 dendrimers as nonviral gene delivery vectors with improved transfection efficiencies. *Org. Biomol. Chem.*, *9*(3), 851–864.

Morgan, M. T., Carnahan, M. A., Immoos, C. E., Ribeiro, A. A., Finkelstein, S., Lee, S. J., & Grinstaff, M. W., (2003). Dendritic molecular capsules for hydrophobic compounds. *J. Am. Chem. Soc.*, *125*(50), 15485–15489.

Moss, B., (1996). Genetically engineered poxviruses for recombinant gene expression, vaccination, and safety. *Proc. Natl. Acad. Sci. USA*, *93*(21), 11341–11348.

Movellan, J., González-Pastor, R., Martín-Duque, P., Sierra, T., De la Fuente, J. M., & Serrano, J. L., (2015). New ionic bis-MPA and PAMAM dendrimers: A study of their biocompatibility and DNA-complexation. *Macromol. Biosci.*, *15*(5), 657–667.

Muzzonigro, T. S., Ghivizzani, S. C., & Robbins, P. D., (1999). The role of gene therapy. Fact or fiction? *Clin. Sports Med.*, *18*(1), 223–239, vii–viii.

Nagano, M., Shinohara, T, Avarbock, M. R., & Brinster, R. L., (2000). Retrovirus-mediated gene delivery into male germ line stem cells. *FEBS Lett.*, *475*(1), 7–10.

Navarro, J., Risco, R., Toschi, M., & Schattman, G., (2008). Gene therapy and intracytoplasmatic sperm injection (ICSI)-A review. *Placenta.*, *29* Suppl B, 193–199.

Neumann, E., Schaefer-Ridder, M., Wang, Y., & Hofschneider, P. H., (1982). Gene transferin to mouse lyoma cells by electroporation in high electric fields. *EMBO. J.*, *1*, 841–845.

Nishikawa, M., Takakura, Y., & Hashida, M., (1996). Pharmacokinetic evaluation of polymeric carriers. *Adv. Drug Del. Rev.*, *21*, 135–155.

Nomani, A., Haririan, I., Rahimnia, R., Fouladdel, S., Gazori, T., Dinarvand, R., Omidi, Y., & Azizi, E., (2010). Physicochemical and biological properties of self-assembled antisense/poly (amidoamine) dendrimer nanoparticles: the effect of dendrimer generation and charge ratio. *Int. J. Nanomedicine.*, *5*, 359–369.

Ohyama, A., Higashi, T., Motoyama, K., & Arima, H., (2016). *In vitro* and *in vivo* tumor-targeting siRNA delivery using folate-PEG-appended dendrimer (G4)/α-Cyclodextrin conjugates. *Bioconjug. Chem.*, *27*(3), 521–532.

Ornelas-Megiatto, C., Wich, P. R., & Fréchet, J. M., (2012). Polyphosphonium polymers for siRNA delivery: an efficient and nontoxic alternative to polyammonium carriers. *J. Am. Chem. Soc.*, *134*(4), 1902–1905.

Ottaviani, M. F., Bossmann, S., Turro, N. J., & Tomalia, D. A., (1994). Characterization of starburst dendrimers by the EPR technique. 1. Copper complexes in water solution. *J. Am. Chem. Soc.*, *116*, 661–671.

Ottaviani, M. F., Furini, F., Casini, A., Turro, N. J., Jockusch, S., Tomalia, D. A., & Messori, L., (2000). Formation of supramolecular structures between DNA and starburst dendrimers studied by EPR, CD, UV, and melting profiles. *Macromolecules.*, *33*(21), 7842–7851.

Pack, D. W., Hoffman, A. S., Pun, S., & Stayton, P. S., (2005). Design and development of polymers for gene delivery. *Nat. Rev. Drug Discov.*, *4*(7), 581–593.

Pages, J. C., & Danos, O., (2003). Retrovectors go forward, In *Pharmaceutical Gene Delivery Systems*, Rolland, A., Sullivan, S. M., (ed.), Eastern Hemisphere Distribution, USA, Chapter 9.

Pan, S., Cao, D., Huang, H., Yi, W., Qin, L., & Feng, M., (2013). A Serum-resistant low-generation polyamidoamine with PEI 423 outer layer for gene delivery vector. *Macromol. Biosci.*, *13*(4), 422–436.

Pan, S., Wang, C., Zeng, X., Wen, Y., Wu, H., & Feng, M., (2011). Short multi-armed polylysine-graft-polyamidoamine copolymer as efficient gene vectors. *Int. J. Pharm.*, *420*(2), 206–215.

Pan, X., Thompson, R., Meng, X., Wu, D., & Xu, L., (2011). Tumor-targeted RNA-interference: functional non-viral nanovectors. *Am. J. Cancer Res.*, *1*(1), 25–42.

Pantos, A., Tsogas, I., & Paleos, C. M., (2008). Guanidinium group: A versatile moiety inducing transport and multicompartmentalization in complementary membranes. *Biochim. Biophys. Acta.*, *1778*(4), 811–823.

Patel, H. K., Gajbhiye, V., Kesharwani, P., & Jain, N. K., (2016). Ligand anchored poly (propyleneimine) dendrimers for brain targeting: Comparative *in vitro* and *in vivo* assessment. *J. Colloid Interface Sci.*, *482*, 142–150.

Pawliuk, R., Westerman, K. A., Fabry, M. E., Payen, E., Tighe, R., Bouhassira, E. E., Acharya, S. A., Ellis, J., London, I. M., Eaves, C. J., Humphries, R. K., Beuzard, Y., Nagel, R. L., & Leboulch, P., (2001). Correction of sickle cell disease in transgenic mouse models by gene therapy. *Science*, *294*, 2368–2371.

Pillai, O., & Panchagnula, R., (2001). Polymers in drug delivery. *Curr. Opin. Chem. Biol.*, *5*(4), 447–451.

Plank, C., Mechtler, K., Szoka, F. C. Jr., & Wagner, E., (1996). Activation of the complement system by synthetic DNA complexes: A potential barrier for intravenous gene delivery. *Hum. Gene Ther.*, *7*(12), 1437–1446.

Plank, C., Schillinger, U., Scherer, F., Bergemann, C., Rémy, J. S., Krötz, F., Anton, M., Lausier, J., & Rosenecker, J., (2003). The magnetofection method: Using magnetic force to enhance gene delivery. *Biol. Chem.*, *384*(5), 737–747.

Posadas, I., Monteagudo, S., & Ceña, V., (2016). Nanoparticles for brain-specific drug and genetic material delivery, imaging and diagnosis. *Nanomedicine (Lond)*, *11*(7), 833–849.

Prajapati, R. N., Tekade, R. K., Gupta, U., Gajbhiye, V., & Jain, N. K., (2009). Dendrimer-mediated solubilization, formulation development and *in vitro-in vivo* assessment of piroxicam. *Mol. Pharm.*, *6*(3), 940–950.

Prokop, A., & Davidson, J. M. (2014). Gene delivery into cells and tissues, In: *Principles of Tissue Engineering,* Lanza, R., Langer, R., Vacanti, J., (ed.), Elsevier Academic Press, ABD, pp. 493–515.

Qin, L., Pahud, D. R., Ding, Y., Bielinska, A. U., Kukowska-Latallo, J. F., Baker, J. R. Jr., & Bromberg, J. S., (1998). Efficient transfer of genes into murine cardiac grafts by Starburst polyamidoamine dendrimers. *Hum. Gene Ther., 9*(4), 553–560.

Qiu, J., Kong, L., Cao, X., Li, A., Tan, H., & Shi, X., (2016). Dendrimer-entrapped gold nanoparticles modified with β-cyclodextrin for enhanced gene delivery applications. *RSC Adv., 6,* 25633–25640.

Quinonez, R., & Sutton, R. E., (2002). Lentiviral vectors for gene delivery into cells. *DNA Cell Biol., 21*(12), 937–951.

Raczka, E., Kukowska-Latallo, J. F., Rymaszewski, M., Chen, C., & Baker, J. R. Jr., (1998). The effect of synthetic surfactant exosurf on gene transfer in mouse lung *in vivo. Gene Ther., 5*(10), 1333–1339.

Raja, J. V., Rachchh, M. A., & Gokani, R. H., (2012). Recent advances in gene therapy for thalassemia. *J. Pharm. Bioallied. Sci., 4*(3), 194–201.

Reynolds, P. N., Feng, M., & Curiel, D. T., (1999). Chimeric viral vectors--the best of both worlds? *Mol. Med. Today, 5*(1), 25–31.

Roth, J. A., Swisher, S. G., & Meyn, R. E., (1999). p53 tumor suppressor gene therapy for cancer. *Oncology (Williston Park), 13*(10 Suppl 5), 148–154.

Rudolph, C., Lausier, J., Naundorf, S., Müller, R. H., & Rosenecker, J., (2000). *In vivo* gene delivery to the lung using polyethylenimine and fractured polyamidoamine dendrimers. *J. Gene Med., 2*(4), 269–278.

Sadekar, S., & Ghandehari, H., (2012). Transepithelial transport and toxicity of PAMAM dendrimers: implications for oral drug delivery. *Adv. Drug Deliv. Rev., 64*(6), 571–588.

Saenger, I. W., (1980). Cyclodextrin inclusion compounds in research and industry. *Angrew. Chem. Int. Ed. Engl., 19*, 344–362.

Saleh, T. A., Al-Shalalfeh, M. M., & Al-Saadi, A. A., (2016). Graphene Dendrimer-stabilized silver nanoparticles for detection of methimazole using Surface-enhanced Raman scattering with computational assignment. *Sci. Rep., 6*, 32185.

Santos, J. L., Oliveira, H., Pandita, D., Rodrigues, J., Pego, A. P., Granja, P. L., & Tomás, H., (2010a). Functionalization of poly(amidoamine) dendrimers with hydrophobic chains for improved gene delivery in mesenchymal stem cells. *J. Control. Release, 144*(1), 55–64.

Santos, J. L., Pandita, D., Rodrigues, J., Pêgo, A. P., Granja, P. L., Balian, G., & Tomás, H., (2010b). Receptor-mediated gene delivery using PAMAM dendrimers conjugated with peptides recognized by mesenchymal stem cells. *Mol. Pharm., 7*(3), 763–774.

Sarkar, K., & Kundu, P. P., (2013). PAMAM conjugated chitosan through naphthalimide moiety for enhanced gene transfection efficiency. *Carbohydr. Polym., 98*(1), 495–504.

Schaffer, D. V., Fidelman, N. A., Dan, N., & Lauffenburger, D. A., (2000). Vector unpacking as a potential barrier for receptor-mediated polyplex gene delivery. *Biotechnol. Bioeng., 67*(5), 598–606.

Sharma, P. A., Maheshwari, R., Tekade, M., & Tekade, R. K., (2015). Nanomaterial based approaches for the diagnosis and therapy of cardiovascular diseases. *Curr. Pharm. Des., 21*(30), 4465–4478.

Shen, W., Liu, H., Ling-Hu, Y., Wang, H., & Cheng, Y., (2016). Enhanced siRNA delivery of a cyclododecylated dendrimer compared to its linear derivative. *J. Mater. Chem. B., 4,* 5654–5658.

Singh, J., Jain, K., Mehra, N. K., & Jain, N. K., (2016). Dendrimers in anticancer drug delivery: Mechanism of interaction of drug and dendrimers. *Artif. Cells Nanomed. Biotechnol., 44*(7), 1626–34.

Somani, S., Robb, G., Pickard, B. S., & Dufès, C., (2015). Enhanced gene expression in the brain following intravenous administration of lactoferrin-bearing polypropylenimine dendriplex. *J. Control. Release, 217,* 235–242.

Sonawane, N. D., Szoka, F. C. Jr., & Verkman, A. S., (2003). Chloride accumulation and swelling in endosomes enhances DNA transfer by polyamine-DNA polyplexes. *J. Biol. Chem., 278*(45), 44826–44831.

Stang, E., Kartenbeck, J., & Parton, R. G., (1997). Major histocompatibility complex class I molecules mediate association of SV40 with caveolae. *Mol. Biol. Cell, 8*(1), 47–57.

Strayer, D. S., (1999). Viral gene delivery. *Expert. Opin. Investig. Drugs, 8*(12), 2159–2172.

Suh, J., Wirtz, D., & Hanes, J., (2003). Efficient active transport of gene nanocarriers to the cell nucleus. *Proc. Natl. Acad. Sci. USA, 100*(7), 3878–3882.

Sullivan, S. M., (2003). Introduction to gene therapy and guidelines to pharmaceutical development. In: *Pharmaceutical Gene Delivery Systems.* Eastern Hemisphere Distribution, Sullivan, S. M., Rolland, A., (ed.), USA, pp. 17–31.

Sun, Y., Jiao, Y., Wang, Y., Lu, D., & Yang, W., (2014). The strategy to improve gene transfection efficiency and biocompatibility of hyperbranched PAMAM with the cooperation of PEGylated hyperbranched PAMAM. *Int. J. Pharm., 465*(1–2), 112–119.

Svenson, S., & Tomalia, D. A., (2005). Dendrimers in biomedical applications reflections on the field. *Adv. Drug Del. Rev., 57,* 2106–2129.

Szymanski, P., Markowicz, M., & Mikiciuk-Olasik, E., (2011). Nanotechnology in pharmaceutical and biomedical applications: *Dendrimers. Nano. Brief Rep. Rev., 6,* 509–539.

Tang, M. X., & Szoka, F. C., (1997). The influence of polymer structure on the interactions of cationic polymers with DNA and morphology of the resulting complexes. *Gene Ther., 4*(8), 823–832.

Tang, M. X., Redemann, C. T., & Szoka, F. C. Jr., (1996). *In vitro* gene delivery by degraded polyamidoamine dendrimers. *Bioconjug. Chem., 7*(6), 703–714.

Tekade, R. K., (2015). Editorial: Contemporary siRNA therapeutics and the current state-of-art. *Curr. Pharm. Des., 21*(31), 4527–4528.

Tekade, R. K., & Chougule, M. B., (2013). Formulation development and evaluation of hybrid nanocarrier for cancer therapy: Taguchi orthogonal array based design. *Bio. Med. Res. Int., 2013,* 712678.

Tekade, R. K., Dutta, T., Gajbhiye, V., & Jain, N. K., (2009a). Exploring dendrimer towards dual drug delivery: pH responsive simultaneous drug-release kinetics. *J. Microencapsul., 26*(4), 287–296.

Tekade, R. K., Dutta, T., Tyagi, A., Bharti, A. C., Das, B. C., & Jain, N. K., (2008). Surface-engineered dendrimers for dual drug delivery: a receptor up-regulation and enhanced cancer targeting strategy. *J. Drug Target, 16*(10), 758–72.

Tekade, R. K., Kumar, P. V., & Jain, N. K., (2009b). Dendrimers in oncology: an expanding horizon. *Chem. Rev., 109*(1), 49–87.

Tekade, R. K., Maheshwari, R., Sharma, P. A., Tekade, M., & Chauhan, A. S., (2015a). siRNA therapy, challenges and underlying perspectives of dendrimer as delivery vector. *Curr. Pharm. Des., 21,* 4614–4636.

Tekade, R. K., Tekade, M., Kesharwani, P., & D'Emanuele A., (2016). RNAi-combined nano-chemotherapeutics to tackle resistant tumors. *Drug Discov. Today.*, *21*(11), 1761–1774.

Tekade, R. K., Tekade, M., Kumar, M., & Chauhan, A. S., (2015b). Dendrimer-stabilized smart-nanoparticle (DSSN) platform for targeted delivery of hydrophobic antitumor therapeutics. *Pharm. Res.*, *32*, 910–928.

Tekade, R. K., Youngren-Ortiz, S. R., Yang, H., Haware, R., & Chougule, M. B., (2014). Designing hybrid onconase nanocarriers for mesothelioma therapy: a Taguchi orthogonal array and multivariate component driven analysis. *Mol. Pharm.*, *11*(10), 3671–3683.

Tekade, R. K., Youngren-Ortiz, S. R., Yang, H., Haware, R., & Chougule, M. B., (2015c). Albumin-chitosan hybrid onconase nanocarriers for mesothelioma therapy. *Cancer Res.*, *75*(15 Supplement), 3680–3680.

Tevaearai, H. T., Gazdhar, A., Giraud, M. N., & Flück, M., (2014). *In vivo* electroporation-mediated gene delivery to the beating heart. *Methods Mol. Biol.*, *1121*, 223–229.

Theoharis, S., Krueger, U., Tan, P. H., Haskard, D. O., Weber, M., & George, A. J., (2009). Targeting gene delivery to activated vascular endothelium using anti E/P-Selectin antibody linked to PAMAM dendrimers. *J. Immunol. Methods*, *343*(2), 79–90.

Tomalia, D. A., Berry, V., Hall, M., & Hedstrand, D. M., (1987). Dendrimers with hydrophobic cores and formationof supramolecular dendrimer surfactant assemblies. *Macromolecules*, *20*, 1164–1169.

Trobridge, G. D., (2009). Foamy virus vectors for gene transfer. *Expert. Opin. Biol. Ther.*, *9*(11), 1427–1436.

Tsurumi, Y., Takeshita, S., Chen, D., Kearney, M., Rossow, S. T., Passeri, J., Horowitz, J. R., Symes, J. F., & Isner, J. M., (1996). Direct intramuscular gene transfer of naked DNA encoding vascular endothelial growth factor augments collateral development and tissue perfusion. *Circulation*, *94*(12), 3281–3290.

Tziveleka, L. A., Psarra, A. M., Tsiourvas, D., & Paleos, C. M., (2007). Synthesis and characterization of guanidinylated poly(propylene imine) dendrimers as gene transfection agents. *J. Control. Release*, *117*(1), 137–146.

Uchida, H., Miyata, K., Oba, M., Ishii, T., Suma, T., Itaka, K., Nishiyama, N., & Kataoka, K., (2011). Odd-even effect of repeating aminoethylene units in the side chain of N-substituted polyaspartamides on gene transfection profiles. *J. Am. Chem. Soc.*, *133*(39), 15524–15532.

Verma, I. M., & Somia, N., (1997). Gene therapy-promises, problems and prospects. *Nature*, *389*(6648), 239–242.

Vile, R. G., Tuszynski, A., & Castleden, S., (1996). Retroviral vectors. From laboratory tools to molecular medicine. *Mol. Biotechnol.*, *5*(2), 139–158.

Wakimoto, H., Yoshida, Y., Aoyagi, M., Hirakawa, K., & Hamada, H., (1997). Efficient retrovirus-mediated cytokine-gene transduction of primary-cultured human glioma cells for tumor vaccination therapy. *Jpn. J. Cancer Res.*, *88*(3), 296–305.

Wang, F., Hu, K., & Cheng, Y., (2016). Structure-activity relationship of dendrimers engineered with twenty common amino acids in gene delivery. *Acta. Biomater.*, *29*, 94–102.

Wang, H., Wei, H., Huang, Q., Liu, H., Hu, J., Cheng, Y., & Xiao, J., (2015). Nucleobase-modified dendrimers as nonviral vectors for efficient and low cytotoxic gene delivery. *Colloids Surf. B. Biointerfaces*, *136*, 1148–1155.

Wang, M., Liu, H., Li, L., & Cheng, Y., (2014). A fluorinated dendrimer achieves excellent gene transfection efficacy at extremely low nitrogen to phosphorus ratios. *Nat. Commun.*, *5*, 3053.

Wang, W., Xiong, W., Zhu, Y., Xu, H., & Yang, X., (2010). Protective effect of PEGylation against poly(amidoamine) dendrimer-induced hemolysis of human red blood cells. *J. Biomed. Mater. Res. B Appl. Biomater.*, *93*(1), 59–64.

Wang, Y., Bai, Y., Price, C., Boros, P., Qin, L., Bielinska, A. U., Kukowska-Latallo, J. F., Baker, J. R. Jr., & Bromberg, J. S., (2001). Combination of electroporation and DNA/dendrimer complexes enhances gene transfer into murine cardiac transplants. *Am. J. Transplant*, *1*(4), 334–338.

Wang, Y., Boros, P., Liu, J., Qin, L., Bai, Y., Bielinska, A. U., Kukowska-Latallo, J. F., Baker, J. R. Jr., & Bromberg, J. S., (2000). DNA/dendrimer complexes mediate gene transfer into murine cardiac transplants ex vivo. *Mol. Ther.*, *2*(6), 602–608.

Wang, Y., Wang, M., Chen, H., Liu, H., Zhang, Q., & Cheng, Y., (2016). Fluorinated dendrimer for TRAIL gene therapy in cancer treatment. *J. Mater. Chem. B.*, *4*, 1354–1360.

Wickham, T. J., Carrion, M. E., & Kovesdi, I., (1995). Targeting of adenovirus penton base to new receptors through replacement of its RGD motif with other receptor-specific peptide motifs. *Gene Ther.*, *2*(10), 750–756.

Wolff, J. A., & Budker, V., (2005). The mechanism of naked DNA uptake and expression. *Adv. Genet.*, *54*, 3–20.

Xiao, F., Wei, Y., Yang, L., Zhao, X., Tian, L., Ding, Z., Yuan, S., Lou, Y., Liu, F., Wen, Y., Li, J., Deng, H., Kang, B., Mao, Y., Lei, S., He, Q., Su, J., Lu, Y., Niu, T., Hou, J., & Huang, M. J., (2002). A gene therapy for cancer based on the angiogenesis inhibitor, vasostatin. *Gene Ther.*, *9*(18), 1207–1213.

Xu, Z., Wang, Y., Ma, Z., Wang, Z., Wei, Y., & Jia, X., (2016). A poly (amidoamine) dendrimer-based nanocarrier conjugated with Angiopep-2 for dual-targeting function in treating glioma cells. *Polym. Chem.*, *7*, 715–721.

Yadav, S., Deka, S. R., Jha, D., Gautam, H. K., & Sharma, A. K., (2016). Amphiphilic azo-benzene-neomycin conjugate self-assembles into nanostructures and transports plasmid DNA efficiently into the mammalian cells. *Colloids Surf. B. Biointerfaces*, *148*, 481–486.

Yang, N. S., Burkholder, J., Roberts, B., Martinell, B., & McCabe, D., (1990). *In vivo* and *in vitro* gene transfer to mammalian somatic cells by particle bombardment. *Proc. Natl. Acad. Sci. USA*, *87*(24), 9568–9572.

Yang, W., Barth, R. F., Wu, G., Huo, T., Tjarks, W., Ciesielski, M., Fenstermaker, R. A., Ross, B. D., Wikstrand, C. J., Riley, K. J., & Binns, P. J., (2009). Convection enhanced delivery of boronated EGF as a molecular targeting agent for neutron capture therapy of brain tumors. *J. Neurooncol.*, *95*(3), 355–365.

Yaron, Y., Kramer, R. L., Johnson, M. P., & Evans, M. I., (1997). Gene therapy. Is the future here yet? *Obstet. Gynecol. Clin. North Am.*, *24*(1), 179–199.

Yenari, M. A., Fink, S. L., Sun, G. H., Chang, L. K., Patel, M. K., Kunis, D. M., Onley, D., Ho, D. Y., Sapolsky, R. M., & Steinberg, G. K., (1998). Gene therapy with HSP72 is neuroprotective in rat models of stroke and epilepsy. *Ann. Neurol.*, *44*(4), 584–591.

Yenari, M. A., Zhao, H., Giffard, R. G., Sobel, R. A., Sapolsky, R. M., & Steinberg, G. K., (2003). Gene therapy and hypothermia for stroke treatment. *Ann. New York. Acad. Sci.*, *993*, 54–68, discussion 79–81.

Ylä-Herttuala, S., & Martin, J. F., (2000). Cardiovascular gene therapy. *Lancet.*, *355*(9199), 213–222.

Yoon, A. R., Kasala, D., Li, Y., Hong, J., Lee, W., Jung, S. J., & Yun, C. O., (2016). Antitumor effect and safety profile of systemically delivered oncolytic adenovirus complexed with

EGFR-targeted PAMAM-based dendrimer in orthotopic lung tumor model. *J. Control. Release*, *231*, 2–16.

Youngren, S. R., Tekade, R. K., Gustilo, B., Hoffmann, P. R., & Chougule, M. B., (2013b). STAT6 siRNA matrix-loaded gelatin nanocarriers: formulation, characterization, and ex vivo proof of concept using adenocarcinoma cells. *Bio. Med. Res. Int.*, *2013*, 858946.

Youngren, S. R., Tekade, R. K., Hoffmann, P. R., & Chougule, M. B., (2013a). Biocompatible nanocarrier mediated delivery of STAT-6 siRNA to cancer cells. *Cancer Res.*, *73*(8 Supplement), 3313–3313.

Yu, T., Liu, X., Bolcato-Bellemin, A. L., Wang, Y., Liu, C., Erbacher, P., Qu, F., Rocchi, P., Behr, J. P., & Peng, L., (2012). An amphiphilic dendrimer for effective delivery of small interfering RNA and gene silencing *in vitro* and *in vivo*. *Angew. Chem. Int. Ed. Engl.*, *51*(34), 8478–8484.

Zabner, J., Couture, L. A., Gregory, R. J., Graham, S. M., Smith, A. E., & Welsh, M. J., (1993). Adenovirus-mediated gene transfer transiently corrects the chloride transport defect in nasal epithelia of patients with cystic fibrosis. *Cell*, *75*(2), 207–216.

Zelphati, O., & Szoka, F. C. Jr., (1996). Mechanism of oligonucleotide release from cationic liposomes. *Proc. Natl. Acad. Sci. USA*, *93*(21), 11493–11498.

Zhang, C., & Zhou, D., (2016). Adenoviral vector-based strategies against infectious disease and cancer. *Hum. Vaccin. Immunother.*, 1–11.

Zhang, J., Liu, D., Zhang, M., Sun, Y., Zhang, X., Guan, G., Zhao, X., Qiao, M., Chen, D., & Hu, H., (2016). The cellular uptake mechanism, intracellular transportation, and exocytosis of polyamidoamine dendrimers in multidrug-resistant breast cancer cells. *Int. J. Nanomedicine*, *11*, 3677–3690.

Zhou, W., & Freed, C. R., (2009). Adenoviral gene delivery can reprogram human fibroblasts to induced pluripotent stem cells. *Stem Cells*, *27*(11), 2667–2674.

Zhu, S., Hong, M., Tang, G., Qian, L., Lin, J., Jiang, Y., & Pei, Y., (2010). Partly PEGylated polyamidoamine dendrimer for tumor-selective targeting of doxorubicin: the effects of PEGylation degree and drug conjugation style. *Biomaterials*, *31*(6), 1360–1371.

Zinselmeyer, B. H., Mackay, S. P., Schatzlein, A. G., & Uchegbu, I. F., (2002). The lower-generation polypropylenimine dendrimers are effective gene-transfer agents. *Pharm. Res.*, *19*(7), 960–967.

CHAPTER 10

DENDRIMERS AS NANOCARRIERS FOR ANTICANCER DRUGS

MICHAŁ GORZKIEWICZ[1] and BARBARA KLAJNERT-MACULEWICZ[1,2]

[1]Department of General Biophysics, Faculty of Biology and Environmental Protection, University of Lodz, 141/143 Pomorska Street, 90-236, Lodz, Poland

[2]Leibniz Institute of Polymer Research Dresden, Hohe Str. 6, 01069 Dresden, Germany

CONTENTS

ABSTRACT

Modern methods of chemotherapy bring promising results in the treatment of cancer; however, they are still far from perfect. Patients subjected to this type of therapy usually suffer from various side effects, caused by systemic toxicity, unfavorable biodistribution or low specificity of action of anticancer drugs. As the development of new therapeutics is extremely expensive and time-consuming, it is worth to take a closer look at the opportunities to improve the existing therapies. Nanotechnology offers the possibility to develop an efficient drug-delivery system, based on the innovative macromolecules. Such nanostructures enable the targeted transport of traditional therapeutic agents, improve their solubility, and shield them from environmental conditions. They can also contribute to overcoming multidrug resistance and reduce the detrimental side effects. Among all the nanoparticles studied for their delivery capabilities, dendrimers attract particular attention, thanks to the well-defined three-dimensional (3D) architecture, biocompatibility, high reactivity, and drug-loading capacity. These polymers show a potential of becoming clinically relevant nanocarriers for several anticancer drugs in the nearest future.

10.1   INTRODUCTION

"Cancer" is a term used to describe a group of related diseases of genetic origin, involving atypical cell growth with the potential to invade or spread to other surrounding tissues. The development of cancer is random and disorganized. Characteristics such as uncontrolled cell growth, invasiveness, as well as the ability to avoid programmed cell death and to evade the activity of immune system, which normally removes invalid or damaged cells of the organism, are most frequently associated with mutations or epigenetic alteration of proto-oncogenes, tumor suppressor genes, and those responsible for DNA repair processes (NCI, 2016). Apart from the mutations and abnormal proliferation, a hallmark of cancer cells is the expression of a variety of surface markers and proteins, which facilitate their growth and supplement the cells with necessary nutrients and oxygen at the expense of normal cells. The atypical signal proteins may enhance the proliferation even further.

The majority of cancer types are caused by environmental factors, which include tobacco smoking (25–30%), diet and obesity (30–35%), infections (15–20%), ionizing or non-ionizing radiation (up to 10%), stress, lack of physical activity and pollutants. Only 5–10% is associated with inherited genetics (Anand et al., 2008). The risk of cancer increases significantly with age, and many forms of cancer occur more frequently in the developed countries.

Cancer is considered as one of the leading causes of death worldwide. According to National Cancer Institute (NCI), there were 14 million new cases and 8.2 million cancer-related deaths in 2012. The number of new cancer cases is predicted to rise up to 22 million within the next 20 years (NCI, 2016).

Several different methods of cancer treatment, which are currently in use or under development, can be generally divided into five groups: radiation therapy, surgery, chemotherapy, targeted therapy, and immunotherapy (Feng and Chien, 2003). The choice of the therapy depends on the location, grade, and stage of the illness, as well as the patient's general condition. Surgery and radiotherapy dominated the field of cancer treatment until the 1960s, when it became clear that these methods do not provide sufficient survival rate (DeVita Jr. and Chu, 2008).

Among the above-mentioned strategies of cancer treatment, chemotherapy is the most common, but still far from satisfactory. Its efficiency is usually limited by low specificity of action of the drugs, which is responsible for various side effects, as well as unfavorable biodistribution and low solubility. Variable microenvironment and drug resistance of cancer cells constitute another obstacle toward drug effectiveness. Therefore, the intensive research is carried out to develop a capable delivery system, which could improve the efficacy of the traditional chemotherapy, at the same time reducing side effects and overcoming the multidrug resistance. To date, a series of nanoparticles, including nanotubes, liposomes, and dendrimers have been tested for their potential use as drug-delivery devices. Numerous studies indicate that the latter seem to be the best nanocarriers for many therapeutic compounds, mainly due to their well-defined structure, high reactivity, biocompatibility, and solubility.

In this chapter, we summarize available data on the possibility of application of dendrimers as carriers for anticancer drugs and highlight the recent achievements in this field.

10.2 ANTICANCER DRUGS: HISTORY, PRINCIPLE OF ACTION AND LIMITATIONS

The term "chemotherapy" was proposed by Paul Ehrlich, German physician and scientist, winner of the Nobel Prize in Physiology or Medicine in 1908. Chemotherapy is a method of cancer treatment that utilizes one or a combination of anticancer drugs as part of a standardized treatment, which is meant to cure the patient or prolong his or her life and reduce symptoms of the illness. Because other ways of treatment often do not bring promising results, cancer chemotherapy continues to play an increasingly important role in the management of malignancies, either directly or as an adjuvant to surgery and/or radiotherapy.

Numerous currently used chemotherapeutics have been discovered as a result of both *in vitro* (in murine and/or human cancer cell lines) and *in vivo* (in rodent tumor models) screening for compounds with proper cytotoxic activity. Typically, anticancer drugs are natural products or synthetic agents designed to target specific functions of tumor cells.

The beginnings of chemotherapy date back to 1861, when the antitumor effect of an extract from the roots of *Podophyllum peltatum* has been described for the first time by Robert Bentley. Twenty years later, in 1881, the active element of this extract was discovered and named "picropodophyllin." The mechanism of action of this compound, which involves inhibition of mitotic spindle formation, and thus, cell cycle arrest, was characterized in 1946. Further research on picropodophyllin led to the synthesis of two structural analogs – etoposide and teniposide, being even more effective than the original compound against many kinds of tumors (Jones, 2014).

The era of cancer chemotherapy began with the report on the impact of mustard gas on human organism (Krumbhaar and Krumbhaar, 1919). The autopsies of soldiers killed upon exposure to this agent during the war, revealed the substantial lymphoid and myeloid suppression. This discovery ultimately led to elaboration of the group of anticancer compounds known as "nitrogen mustards" and their implementation to chemotherapy. This includes mechlorethamine (used in the treatment of non-Hodgkin lymphoma), cyclophosphamide (leukemia, lymphoma, breast cancer, lung cancer, prostatic cancer, and ovarian cancer), chlorambucil (chronic lymphocytic leukemia), and melphalan (myeloma and ovarian cancer) (Jones, 2014). Another discovery that accelerated the development of anticancer compounds concerned folic acid, which was found to stimulate the proliferation of acute lymphoblastic leukemia (ALL) cells (Farber, 1949). On the other hand, it was also shown that deficiency of folic acid may be responsible for the bone marrow abnormalities similar to those caused by nitrogen mustard. These findings have contributed to the synthesis of folate antagonists, aminopterin and amethopterin (methotrexate), which proved to be effective against both leukemias (Farber et al., 1948) and solid tumors (Wright et al., 1951).

As a result of those discoveries, by 1955 the US National Cancer Institute (NCI) had established a screening program, designed to analyze chemical compounds for their anticancer properties on a selected panel of cancer cell lines. To date, numerous substances have been identified and evaluated for their antitumor activity. This includes both natural (e.g., mitomicyn C from *Streptomyces caespitosus*, daunorubicin and doxorubicin from *Streptomyces peucetius*, or vinblastine and vincristine from *Catharanthus roseus*) and synthetic compounds. The latter were largely developed based on naturally occurring substances (e.g., topotecan and irinotecan—the derivatives of camptothecin from *Camptotheca acuminata*, or docetaxel—the derivative of paclitaxel from *Taxus brevifolia*) (Gordaliza, 2007; Jones, 2014). Anticancer drugs can be generally classified according to the specificity and mechanism of action (Table 10.1).

Traditional chemotherapeutic agents have been designed to influence the processes involved in DNA replication and regulation of cell division, thus eliminating the rapidly dividing cells more efficiently. Unfortunately, such an approach turned out to be ineffective, due to the numerous limitations, possible side effects and systemic toxicity.

Chemotherapy was meant to selectively eliminate tumor cells, leaving healthy tissues intact. However, conventional anticancer drugs target all highly proliferative cells and have little influence on other aspects of tumor progression, such as tissue invasion, metastasis, or progressive loss of differentiation (Payne and Miles, 2008). Thus, normal cells such as blood cells in the bone marrow or epithelial cells of the gastrointestinal tract are also affected by those drugs (Liauw, 2013). Side effects associated with systemic toxicity can occur during the treatment, immediately after, or even months or years after the end of chemotherapy. These include, for example, bone marrow suppression, nausea and vomiting, diarrhea, and weight changes (Payne and Miles, 2008). In many cases, chemotherapy may cause various organ damage, leading to severe pain, and sometimes, death.

Anticancer drugs are typically administered intravenously and distributed throughout the whole organism via the bloodstream. This can result in an unfavorable drug biodistribution and enhance the deleterious effects on healthy tissues. Anticancer drugs, especially hydrophobic particles, may also be quickly removed from the bloodstream by monocytes and macrophages, which are the part of the reticuloendothelial system (RES)

TABLE 10.1 Classification of Anticancer Drugs

Type of drug	Principle of action	Example
Phase-specific toxicity:		
Phase-specific/ cell cycle-specific	Destruction of proliferating cells only during a specific stage or stages of the cell cycle/destruction of all rapidly proliferating cells regardless of the growth phase	Methotrexate
Cell cycle-nonspecific	Equal effect on tumor and normal cells regardless of the growth phase or the rate of cell division	Platinum derivatives (e.g., carboplatin)
Mechanism of action:		
Alkylating agents	Attachment of the alkyl group to the 7 nitrogen atom of the guanine's purine ring in DNA, leading to inhibition of DNA replication and induction of apoptosis	Nitrogen mustards (e.g., melphalan)
Heavy metals	Formation of cross-links between the DNA strands, leading to inhibition of DNA replication and induction of apoptosis	Cisplatin
Antimetabolites	Competition with the natural substrate for the active site on an essential enzyme or receptor, leading to the induction of apoptosis	Cytarabine
Cytotoxic antibiotics	Impact on the function and synthesis of nucleic acids	Doxorubicin
Spindle poisons	Impact on assembly and disassembly of microtubules	Vinca alcaloids (e.g., vincristine)
Topoisomerase inhibitors	Inhibition of topoisomerases activity, involved mainly in DNA replication	Camptothecin

(Kakde et al., 2011). Hydrophilic carriers can prevent drug recognition by macrophages.

Furthermore, intravenous administration may contribute to the acquisition of multidrug resistance by cancer cells (Brigger et al., 2002), including fast metabolism of the drugs, decreased expression of target proteins, inefficient uptake and accumulation of therapeutic agents inside the cancer cells. Classical anticancer drugs are usually characterized by poor solubility, due to their hydrophobic nature. Traditionally used solvents contain polyoxyethylated castor oil and dehydrated alcohol (Feng and Chien, 2003), which may cause severe side effects, including neurotoxicity, nephrotoxicity and cardiotoxicity, hyperlipidaemia, abnormal lipoprotein patterns, erythrocyte aggregation, and hypersensitivity reactions (Weiss et al., 1990).

The development of targeted therapy was meant to provide a solution for the problems of traditional chemotherapy. Such an approach aimed to block the growth of cancer cells by interfering with specific molecules essential for carcinogenesis and tumor growth, rather than by affecting all rapidly dividing cells. This may result in reduced toxicity, increased therapeutic index and improved biodistribution of the drug, which are the major factors of cancer chemotherapy's success.

Two of the most common methods of targeted therapy involve the use of monoclonal antibodies and tyrosine kinase inhibitors. Other popular molecular targets for chemotherapy include the cell cycle regulators (e.g., cyclin-dependent kinases or p21 gene) (Shapiro and Harper, 1999) and apoptosis modulators (e.g., Ras protein or tumor suppressor p53) (Isoldi et al., 2005).

The development of monoclonal antibodies was possible due to elaboration of the hybridoma technique in the 1970s. These high molecular weight glycoproteins are designed to target and selectively bind to the specific antigens on cancer cells. Their anticancer activity is based on various actions such as the influence on the receptor-mediated signaling pathways, inhibition of target molecule function, induction of apoptosis or antibody-dependent cytotoxicity. Presently, monoclonal antibodies constitute about a quarter of all drugs under development, and about 30 compounds are in use or during clinical trials (Breedveld, 2000). Examples include, among others, rituximab, a monoclonal antibody against the protein CD20, used in various types of leukemias and lymphomas, and trastuzumab, which interacts with the HER2/neu receptor, applied mainly in breast cancer treatments.

Monoclonal antibodies are also used to develop more complex therapeutic agents for targeted anticancer therapy, such as immunotoxins—chimeric proteins consisting of a tumor-specific antibody fragment linked to a toxic moiety (e.g., bacterial toxins, of which the most popular are exotoxin A from *Pseudomonas aeruginosa* and diphtheria toxin from *Corynebacterium diphtheriae*). Such compounds specifically bind to a surface antigen on a cancer cells and enter the cells by endocytosis, resulting in cell death (Choudhary et al., 2011).

Tyrosine kinases are important regulators of signal transduction pathways, cell proliferation, differentiation, migration, metabolism, and apoptosis. Tyrosine kinases constitute a family of enzymes which primary action is ATP-dependent phosphorylation of selected tyrosine residues in the sequence of target proteins. This reaction is an essential element of cellular signaling and maintenance of homeostasis (Schlessinger, 2000). Tyrosine kinases are usually classified into two subclasses: receptor tyrosine kinases (RTKs), for example, EGFR, PDGFR, FGFR-related kinases, and non-receptor tyrosine kinases (NRTKs), for example, SRC, ABL, FAK, and Janus kinase (Paul and Mukhopadhyay, 2004).

Tyrosine kinases and their signaling pathways are often altered during the tumor development and progression (Blume-Jensen and Hunter, 2001), which makes them a fine target for chemotherapy. Of all the tyrosine kinase inhibitors, the most successful are Gleevec/Imatinib (inhibitor of BCR-ABL protein, which is responsible for the constitutive proliferative signaling), Iressa/Gefitinib and Tarceva/Erlotinib (selective inhibitors of EGF receptor tyrosine kinase) (Paul and Mukhopadhyay, 2004). What is more, the extra-cellular domains of the tyrosine kinase receptors provide an excellent target for monoclonal antibodies (Bennasroune et al., 2004).

Several receptor tyrosine kinases are involved in angiogenesis, a process of development, growth and expansion of new vessels. Tumor growth greatly depends on angiogenesis in order to ensure a continuous supply of oxygen and nutrients, as well as removal of metabolic waste products. Cancer cells typically follow a sigmoid-shaped growth curve, according to which tumor doubling size varies with tumor size (Payne and Miles, 2008). Solid tumors can grow up to 1–2 mm without requirement of blood supply, since at this stage the diffusion processes are sufficient for the transport of compounds between the cells and the environment. As solid tumor become larger than 2 mm, it enters a stage of cellular hypoxia, thus initiating the process of angiogenesis, which is critical for cancer progression and metastasis (Folkman et al., 1971). Therefore, the inhibition of angiogenesis is one of the main issues in the modern anticancer therapy. Numerous studies have led to the identification of several therapeutic molecules, which are currently under clinical trials (Kerbel, 2000; Brannon-Peppas and Blanchette, 2004).

A key target for inhibition of angiogenesis is a class of growth factors, including fibroblast growth factor (FGF), platelet-derived growth factor (PDGF), and vascular endothelial growth factor (VEGF). These growth factors and corresponding receptor tyrosine kinases play essential role in proliferation of endothelial cells, the main component of blood vessels. The latter is the key factor involved in the development of cancer and metastasis, therefore the most thoroughly tested. Three anti-angiogenic tyrosine kinase inhibitors correlated with VEGF – sunitinib, sorafenib, and pazopanib – were recently approved for treatment of patients with advanced cancers. Other important targets include a family of matrix metalloproteinases and cell adhesion molecules (Gotink and Verheul, 2010).

Unfortunately, the targeted therapy based on the inhibition of angiogenesis has its drawbacks. Angiogenesis inhibitors can hamper the delivery of therapeutics to cancer cells, and hence decrease the drugs' therapeutic benefits (Van der Veldt et al., 2012). What is more, tumors can adapt to antiangiogenic therapy by accumulating particularly aggressive cells (Conley et al., 2012). Drug resistance is another limitation of the use of angiogenesis inhibitors. It was initially postulated (Boehm et al., 1997) that this issue will not relate to antiangiogenic drugs as they target blood vessels rather than tumor cells. However, subsequent studies revealed that tumors can indeed adapt and become resistant to antiangiogenic treatment (Bergers and Hanahan, 2008).

Targeted therapy is not without side effects, which are usually drug-specific (Widakowich et al., 2007). What is more, the response to the traditional chemotherapy is usually determined by the change in the size of tumor (Payne and Miles, 2008), and many of targeted therapy's drugs do not reduce it. Instead, they inhibit its proliferation, thus being referred to as cytostatics.

Therefore, the development of an efficient system to eliminate cancer cells without deleterious side effects remains a challenge to modern anticancer therapies.

10.3 NANOTECHNOLOGY IN OVERCOMING THE LIMITATIONS OF ANTICANCER DRUGS

Despite the large number of research and the great interest in the development of effective anticancer therapy, the invention of a new drug is not an easy task. It is estimated that every new therapeutic takes 12–15 years to elaborate, at a cost of over 800 million dollars (DiMasi et al., 2003). As the development and implementation of a new anticancer drug is an extremely time-consuming, laborious and expensive process, the improvement of existing therapies seems to be a promising alternative. One of the strategies involves the use of carrier systems which will not only provide an efficient transport of the drug directly to the cancer cells but will also eliminate the adverse side effects of classical chemotherapy. In order to achieve this goal, drug-delivery systems must hold adequate amount of drug and remove the problems like drug resistance based on cellular or non-cellular mechanisms, altered biodistribution or clearance of anticancer drugs from the organism. The delivery systems should meet the requirements like prolonged blood circulation, sufficient tumor accumulation, efficient uptake by cancer cells and controlled drug release (Kakde et al., 2011).

Recent years have brought an exceptional growth of research and applications in the area of nanoscience and nanotechnology. It is believed that introduction of nanotechnology in medicine will improve the diagnosis and treatment of disease. Presently, many compounds are under investigation for drug delivery, particularly for cancer therapy (De Jong and Borm, 2008). Nanostructures have been reported to protect the drugs from degradation, improve their solubility, prolong blood circulation time, and modulate pharmacokinetic properties, as well as to provide the targeted delivery and controlled release of therapeutics (Emeje et al., 2012), allowing to overcome several limitations of conventional anticancer therapy. In this strategy, a pro-drug model or an active targeting approach may be used, with the drug being converted into an inactive form or attached to a linker that is removed only upon encountering cancer cells. This tactic exploits the pathophysiological differences between normal and malignant cells, such as acidic pH and expression of tumor specific antigens (Patri et al., 2002).

What is more, it has been found that nanoparticles can accumulate in tumors by passive retention mechanism, known as Enhanced Permeability and Retention Effect (EPR), even in the absence of targeting ligands (Duncan and Sat, 1998). This is because tumor vasculature is characterized by irregular epithelium, weakened lymphatic drainage and decreased uptake of the interstitial fluid, which favor the passive accumulation of nanoparticles in tumors (Bertrand et al., 2014).

All these features make the drug delivery nanosystems a promising alternative to conventional chemotherapy. Numerous molecules such as liposomes, solid lipid nanomolecules, dendrimers, silicon nanostructures, micelles, or carbon nanomaterials have been tested for drug delivery. However, only a limited number of liposomes and polymers were clinically approved as nanocarriers for anticancer chemotherapeutics (Piktel et al., 2016). The first are easy to prepare, but are mostly unstable, incapable of entering tumorous vasculature and rapidly removed by RES. Therefore, the attention of scientists turned toward the polymeric systems, of which dendrimers are of particular interest.

10.3.1 DENDRIMERS AS NANODEVICES FOR DRUG DELIVERY

Dendrimers, which constitute a class of highly branched polymers, appeared to be superior to the conventional, linear macromolecules in terms of drug delivery due to several features. Thanks to the highly optimized methods of synthesis, dendrimers are characterized by well-defined structure, monodispersity and high purity (Tomalia et al., 1985) as well as multivalency and biocompatibility (Fischer et al., 1999; Reddy et al., 1999). These polymers provide improved solubility of the drugs, high loading capacity (Kojima et al., 2000), and controllable biodistribution patterns. The nanometric size of dendrimers is beneficial for their entry into hyper-permeable tumor blood vessels, while the globular shape significantly decreases their renal filtration rate (Kakde et al., 2011), resulting in prolonged circulation time. What is more, high molecular weight favors the tumor localization of dendrimers, preventing their escape due to the EPR effect (Klajnert and Bryszewska, 2000).

The blood half-life may be further extended by surface modification of the dendrimer particles, for example, by PEGylation (Kobayashi et al., 2001), which has been also shown to prevent RES uptake by increasing the hydrophilic character of the macromolecules (Peng and Hsu, 2001). In addition, such modifications may enhance the targeting potential, for example, by ligand/receptor-mediated endocytosis, and further improve biocompatibility of dendrimers (Tomalia et al., 2012). Dendrimers maintain blood concentrations of the drugs above the minimal therapeutic dose, and ensure their protection from environmental degradation. The use of dendrimers as drug carriers may also facilitate the circumvention of resistance mechanisms (Mintzer and Grinstaff, 2011).

The dendrimers' three-dimensional structure with reactive surface moieties and several internal cavities enable the transport of therapeutics either physically entrapped inside the dendritic scaffold or bound to the terminal groups (attached covalently or complexed by electrostatic and van der Waals forces) (Figure 10.1). Among these possibilities, the use of drug-dendrimer covalent conjugates appears to be the most promising approach. Numerous research on non-covalent complexes between dendrimers and drugs, as well as *in vitro* studies show that dendrimers can serve as an effective nanocarriers without the necessity of formation of covalent chemical bonds with the therapeutic particles. The preparation of such complexes is much easier (either by equilibrium dialysis method or direct mixing), and what

FIGURE 10.1 Possible ways of drug-dendrimer binding: A. Encapsulation; B. Covalent bonding; C. Non-covalent interactions with charged functional groups.

is more, they provide pH-triggered drug release (Boas et al., 2004), which may be widely exploited for tumor specific delivery. However, as the anti-cancer drugs interact with dendrimers mainly due to the electrostatic forces, the complexes may be unstable in different environmental conditions and pH, which may significantly decrease the transport efficacy *in vivo*. More-over, the nature of surface modifications and their density may influence the number of electric charges and hamper the formation of complexes. Cova-lent conjugates provide higher stability due to the permanent chemical bond between the drug and dendrimer. However, this method requires the selec-tion of proper linker, which could undergo specific digestion inside the target cell, thus allowing release of the drug. In some cases, a covalent bond may reduce the therapeutic activity of the attached compound.

10.4   DENDRIMERS AS NANOCARRIERS FOR ANTICANCER DRUGS: *IN VITRO* AND *IN VIVO* STUDIES

10.4.1   DOXORUBICIN

Probably the most comprehensive study on the use of dendrimers as nano-carriers for anticancer drug delivery concerns doxorubicin (DOX) (Figure 10.2), commonly used in the treatment hematological malignancies, carci-nomas and soft tissue sarcomas (Tacar et al., 2013). DOX has been a result of the research on *Streptomyces* spp., which in 1950s was found to produce an antibiotic (daunorubicin) with anticancer properties. Daunorubicin was successfully applied in the therapy of leukemias and lymphomas; however, it was found to cause fatal cardiac toxicity. In order to reduce the detrimental

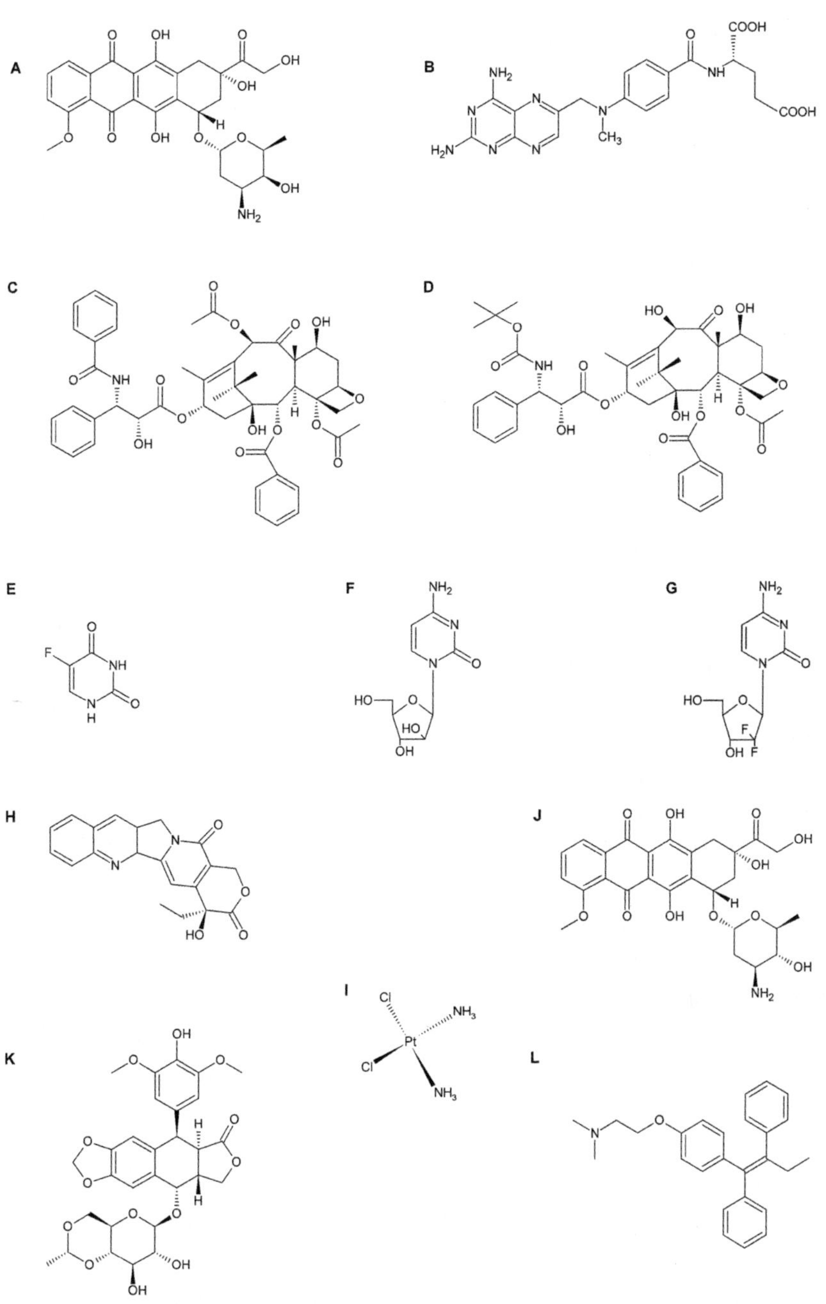

FIGURE 10.2 Chemical structures of the drugs tested for the possibility of transport by dendrimers. A. Doxorubicin; B. Methotrexate; C. Paclitaxel; D. Docetaxel; E. 5-fluorouracil; F. Cytarabine; G. Gemcitabine; H. Camptothecin; I. Cisplatin; J. Epirubicin; K. Etoposide; L. Tamoxifen.

effect, *Streptomyces* spp. has been subjected to the series of genetic modifications, leading to the production of doxorubicin by these bacteria. Unfortunately, though doxorubicin had a higher therapeutic index in comparison to daunorubicin, cardiotoxic activity remained to be a major problem, along with several others common side effects. Nevertheless, these compounds became a prototype for subsequent research, and nowadays there are more than 2,000 analogs of doxorubicin (Rivankar, 2014).

The anticancer activity of doxorubicin is primarily based on two mechanisms: intercalation into DNA and disruption of topoisomerase II-mediated DNA repair, or generation of reactive oxygen species (ROS) and their damage to cellular membranes, DNA and proteins (Thorn et al., 2011).

Because DOX is characterized by good solubility in water, a combination with dendrimers may bring several other advantages. What is more, DOX also has fluorescent properties, which facilitates its trafficking both *in vivo* and *in vitro* and makes it a good choice for research on the dendrimer-based anticancer delivery systems.

Unmodified poly (amidoamine) (PAMAM) dendrimers demonstrate exceptional abilities to encapsulate doxorubicin and transport it efficiently both *in vitro* and *in vivo*, as PAMAM G3 were found to enhance oral bioavailability of DOX in rats (Ke et al., 2008). But since the cytotoxic effect of cationic dendrimers increases with the generation, it is often necessary to modify their surface, which can also facilitate the formation of complexes and provide the targeting properties. Zhang et al. tested PAMAM G5 dendrimers with acetyl, glycidol hydroxyl or succinamic acid terminal groups for encapsulation of doxorubicin. All three types of dendrimers were able to effectively encapsulate DOX and display anticancer effect *in vitro*. The relatively stronger interactions of DOX with acetyl- and glycidol hydroxyl-terminated dendrimers were responsible for the sustained release of DOX from the complexes, indicating that these types of modification may be a proper option for further *in vivo* studies (Zhang et al., 2014). The ability to encapsulate the drug was also studied for PEGylated PAMAM G3 and G4 dendrimers, and it was shown to increase with the generation and length of poly (ethylene glycol) (PEG) chains. The highest ability was detected for PAMAM G4 with the PEG chain having a molecular weight of 2000, which could retain an average of 6,5 DOX molecules (Kojima et al., 2000). Interestingly, PEGylated PAMAM G5 dendrimers had a similar loading capacity, regardless of the PEGylation degree. Nevertheless, such complexes were found to be stable and water soluble. What is more, they were able to release the drug in a sustained manner *in vitro* (Liao et al., 2014). PEGylated PAMAM G4 dendrimers further modified with transferrin and wheat germ agglutinin (He et al., 2011) or tamoxifen (Li et al., 2012) as the targeting moieties were synthesized and loaded with DOX in order to enable blood–brain barrier (BBB) penetration and tumor targeting. These complexes significantly reduced the cytotoxicity of DOX to the normal cells, while efficiently delivering the drug across BBB and inhibiting the growth of the glioma cells. Also, acetylated PAMAM G5 dendrimers modified with folic acid were able to transport DOX specifically to the cancer cells overexpressing folic acid receptor (Wang et al., 2011).

Other types of dendrimers were also tested for possible transport of DOX in the complexed form. DOX-loaded poly(propylenimine) (PPI) G5 dendrimers conjugated with dextran exhibited enhanced uptake by A549 cancer cell line as well as sustained drug release profile and decreased hemolytic activity (Agrawal et al., 2009). Cationic poly-lysine G6 dendrimers improved the penetration of DOX into prostate 3D multicellular tumor spheroids, resulting in the increased cytotoxic activity of the drug and delaying the growth of cancer cells. The retention of the complexes and enhanced therapeutic efficacy was also achieved in tumor-bearing mice (Al-Jamal et al., 2013).

Also, the delivery system based on an amphiphilic dendrons, which could generate supramolecular micelles and encapsulate doxorubicin with high loading capacity was elaborated. Such nanocarriers were found to abolish the systemic toxicity, enhance anticancer potency of the drug and overcome resistance mechanisms in breast cancer models by improving cellular uptake, at the same time decreasing the efflux of DOX (Wei et al., 2015).

In order to devise the most efficient transport systems for doxorubicin, dendrimers were used in combinations with different nanoparticles. For instance, DOX-PAMAM G4 complexes were introduced to the interior of liposomes, and such constructs were able to modulate the release of the drug and enhance its cytotoxic activity against human cancer cell lines (Papagiannaros et al., 2005). PAMAM dendrimers were also used to stabilize various molecules, such as the magnetite Fe_3O_4 nanoparticles, which enabled high DOX-loading efficiency with controllable enzymatic release exploiting the activity of cathepsin B, a lysosomal cysteine protease overexpressed in many tumor cells and tumor (Chandra et al., 2015). The DOX-loaded mesoporous silica nanoparticles modified with PAMAM G2 dendrimers and complexed with siRNA exhibited improved transport into multidrug-resistant cancer cells with minimal premature release, significantly enhancing the efficacy of the treatment (Chen et al., 2009). Attachment of PAMAM-OH G4 dendrimers and folic acid on the surface of graphene oxide particles resulted in efficient internalization of these nanoparticles and delivery of DOX into HeLa cells (Siriviriyanun et al., 2015).

DOX-dendrimer complexes have been directly compared with the covalent conjugates for cytotoxic effects of the combined therapy with ionizing radiation. The anticancer activity and intracellular accumulation of PAMAM G2 dendrimers covalently conjugated with doxorubicin and modified with a vector protein (alpha-fetoprotein recombinant third domain) were studied in breast adenocarcinoma cells, together with the free drug, unmodified PAMAM G2 and DOX-loaded PAMAM G2 alone or in combination with ionizing radiation. For MCF-7/MDR1 cell line, synergistic effects were shown after the combined treatment for both complexes and conjugates (Zamulaeva et al., 2016), suggesting that both systems can transport doxorubicin to a comparable extent. High anticancer activity and selectivity against human ovarian adenocarcinoma SKOV3 cells, as well as cytotoxicity against human peripheral blood lymphocytes have been observed for similar DOX-PAMAM G2 conjugates modified with vector protein (Yabbarov et al., 2013).

PEGylated DOX-PAMAM G3 conjugates have been developed for the preparation of propellant-based metered-dose inhalers with enhanced aerosol characteristics. Such constructs exhibited improved cellular entry kinetics in comparison to the free drug, as well as intracellular, pH-dependent release of the drug and cytotoxicity against A549 cell line. The PEG chains ensured high dispersibility and lung deposition of the conjugates (Zhong and da Rocha, 2016). The PEGylation degree has been also reported to affect the pharmacokinetics of DOX-PAMAM conjugates, along with the way of the drug attachment (Zhu et al., 2010). PAMAM G4 dendrimers modified with PEG and Arg-Gly-Asp (RGD) with a high affinity to integrins were conjugated with DOX via acid-sensitive cis-aconityl linker. The conjugates showed specific targeting by binding with the integrin receptors overexpressed by tumor cells, and controlled release of DOX in weakly acidic lysosomes in C6 cell line. The second effect was not observed for conjugates prepared using acid-insensitive succinic linker. *In vivo* evaluation in an orthotopic murine model of C6 glioma indicated significantly prolonged half-life and higher accumulation in brain tumor in comparison to normal brain tissue (Zhang et al., 2011a). In another study, PEG chains were attached to every surface moiety of PAMAM G4, and then DOX was conjugated with the nanocarrier through amide or hydrazone bond formed with PEG. For the dendrimers bearing doxorubicin through amide linkage, the drug was released to a small degree at pH 7.4 and 5.5. A significant amount of DOX was released at pH 5.5 (which corresponds to the pH of late endosomes) from dendrimers with hydrazone linkage. The latter also exhibited higher cytotoxicity, suggesting the importance of pH-sensitive linkage. The conjugates exhibited similar anticancer activity for both DOX-resistant and DOX-sensitive cell lines (Kono et al., 2008). DOX-PAMAM G4.5 conjugates were combined with different photochemical internalization strategies to evaluate the cytotoxic effects. It was shown that the "light after" strategy efficiently released DOX from the conjugates, resulted in higher nuclear accumulation of the drug and increased rate of cell death through synergistic effects (Lai et al., 2007).

pH-sensitive DOX-dendrimer conjugates were also used in combination with different nanoparticles. PEG-PAMAM G4 were covalently linked on the surface of gold nanorod, with subsequent conjugation of DOX to dendrimer layer. The drug release was inconsiderable under physiological pH, but it was enhanced at a weakly acidic pH. Furthermore, the combined photothermal-chemo treatment of cancer cells using these nanoformulations was demonstrated both *in vitro* and *in vivo* to exhibit higher therapeutic efficacy than the single treatments (Li et al., 2014a). PEGylated PAMAM G2.5 dendrimers linked with DOX were shown to stabilize Fe_3O_4 nanoparticles and exhibit their cytotoxic activity via EPR effect and pH-dependent release in lysosomes (Chang et al., 2011).

Kojima et al. placed various DOX-PAMAM conjugates inside collagen gels for application as metastasis-associated drug delivery system. The highly invasive MDA-MB-231 cells were found to be more sensitive than the MCF-7 cells to the dendrimer-embedded gels, and the cytotoxic activity was dependent on their chemical compositions. What is more, the collagen gels demonstrated *in vivo* anticancer effects in metastatic murine model (Kojima et al., 2013a, 2013b).

Much attention is being paid to lysine dendrimers for chemical conjugation of doxorubicin. Such polymers were used for the synthesis of dendronized heparine DOX conjugates. These nanoparticles exhibited strong antitumor activity, inhibition of angiogenesis and induced apoptosis in the 4T1 breast tumor model. What is more, no significant systemic toxicity in both tumor-bearing and healthy mice was observed (She et al., 2013).

The same group of scientists linked doxorubicin with the lysine-based alkyne-boc-dendrimers through Gly-Phe-Leu-Gly (GFLG) peptide linker, a substrate for mentioned earlier cathepsin B. Such linkage ensures intralysosomal drug release, at the same time providing the stability of nanocarrier in plasma and serum during the transport. The DOX-GFLG-dendrimer demonstrated higher accumulation and retention within SKOV-3 ovarian tumor tissue, resulting in a higher cytotoxic activity (Zhang et al., 2014a). The conjugates were also able to induce apoptosis in the 4T1 breast tumor model and prevent the DOX-related systemic toxicity (Zhang et al., 2014b). Similar enzyme-sensitive amphiphilic PEGylated DOX-GFLD-dendrimers also inhibited proliferation and induced apoptosis of the 4T1 breast cancer–bearing mice (Li et al., 2014b). Moreover, lysine dendrimers were used to achieve liver-specific delivery of DOX. For this purpose, PEGylated, galactose-functionalized dendrimer-DOX conjugates with hydrazone drug-polymer bonds were elaborated. Thanks to the sugar moieties, nanocarriers were able to specifically target the drug and destroy HepG2 cells *in vitro* (She et al., 2015). PEGylated poly-lysine dendrimer also significantly increased the recovery of doxorubicin in the thoracic lymph after both intravenous and subcutaneous dosing and increased the lymphatic concentration of DOX (Ryan et al., 2013).

Poly-lysine G3 dendrimers with a silsesquioxane cubic core (so-called "nanoglobules") with compact globular structure and highly functionalized surface were used for delivery of DOX. They showed higher cytotoxicity than free drug in glioblastoma U87 cells (Kaneshiro and Lu, 2009).

Of particular note is the research performed by the team of LM Kaminskas. PEGylated poly-lysine dendrimers with the surface generation of L-lysine or succinimyldipropyldiamine (SPN), conjugated with DOX via pH-sensitive linker showed similar cytotoxic properties to the free drug during *in vitro* studies. However, the SPN dendrimers demonstrated reduced metabolic lability and increased uptake into RES organs in comparison to all-lysine dendrimers. *In vivo* assessment in Walker 256 tumor-bearing rats revealed enhanced accumulation in the cancer tissue (Kaminskas et al., 2011), as well as prolonged plasma exposure and lower systemic toxicity of DOX-dendrimer conjugates (Kaminskas et al., 2012). Most importantly, PEGylated DOX-poly-lysine dendrimers exhibited the controlled and prolonged exposure of lung cancers to the drug. After intratracheal instillation to rats, approximately 15% of the dose remained in the lungs after 7 days. Twice weekly intratracheal instillation of the dendrimer caused over 95%

reduction of tumor size in the syngeneic rodent model of lung metastasized breast cancer in comparison to administration of free doxorubicin, which reduced lung tumor only in 30-50%. DOX solution also caused extensive lung-related toxicity and death within several days after a single dose. The data suggest that PEGylated dendrimers have potential as inhalable drug delivery systems to promote the prolonged exposure of lung cancers to chemotherapeutic drugs and to improve anticancer activity (Kaminskas et al., 2014).

Among the non-classical dendrimers conjugated with doxorubicin, an asymmetric polyester dendrimer should be mentioned. This dendrimer consists of two hemispheres with one of them functionalized with poly(ethylene oxide) chains. The new type of dendritic polymer was found to optimize blood circulation time and drug release, as well as to improve the cell uptake in mice bearing C-26 colon carcinomas. Surprisingly, the DOX-dendrimer conjugate caused complete tumor regression and 100% survival of mice over the 60-day experiment after a single dose (Lee et al., 2006). DOX was also covalently bound to a 3-arm poly(ethylene oxide)-dendrimer hybrid. Regardless the increased half-life in serum and sustained release of the drug, the cytotoxicity of the conjugate was significantly reduced, indicating that a portion of doxorubicin may have been hydrolyzed under physiological conditions (Padilla De Jesús et al., 2002).

10.4.2 METHOTREXATE

As mentioned before, methotrexate (MTX, formerly known as amethopterin) (Figure 10.2) has been developed based on the studies indicating a significant role of folic acid in the development of leukemia. This folate analog is used in the treatment of cancer (both hematological malignancies and solid tumors) and autoimmune diseases (e.g., rheumatoid arthritis). Anticancer activity of MTX involves competitive inhibition of dihydrofolate reductase (DHFR), the rate-limiting enzyme in the conversion of dihydrofolate to the active tetrahydrofolate. The resulting decrease in *de novo* production of purines and pyrimidines leads to the obstructions of DNA synthesis. What is more, reduced production of tetrahydrofolate and methyltetrahydrofolate, the cellular donors of methyl moieties, may cause the inhibition of methionine and S-adenosylmethionine synthesis (Chan and Cronstein, 2013).

Problems of MTX-based chemotherapy are associated with low stability of the drug, as well as the resistance mechanisms (due to the decreased uptake by the membrane reduced folate carrier (RFC) and shorter retention time or increased dihydrofolate reductase activity) (Bertino et al., 1996) and various side effects (including development of hepatic fibrosis or cirrhosis, bone marrow suppression, alopecia or stomatitis) (Chan and Cronstein, 2013).

Methotrexate was one of the first anticancer therapeutics tested for the possibility of dendrimer-mediated transport, both in the complexed and conjugated forms. Patri et al. compared the transport capacity of MTX-PAMAM G5 complexes and conjugates. Surface of the dendrimers was additionally modified with folic acid as targeting agent. Because methotrexate is a hydrophobic molecule, non-covalent complexes were unstable and exerted the cytotoxic effect similar to this of a free drug. In contrast, drug-dendrimer covalent conjugates were able to efficiently and specifically kill the receptor-expressing cells by intracellular delivery of the drug through receptor-mediated endocytosis. This study demonstrates that MTX-PAMAM covalent conjugates are better suited for the targeted drug delivery (Patri et al., 2005). Interestingly, PAMAM dendrimers were found to facilitate the encapsulation of MTX inside liposomes. This phenomenon is due to the ability of positively charged dendrimers to interact with phospholipids, leading to the solubilization of the drug in the interior of liposomes. The entrapment was found to be generation-dependent, and resulted in increased concentration and sustained release of MTX (Khopade et al., 2002).

In order to improve the complexation ability of PAMAM dendrimers, their surface have been modified with PEG moieties of various molecular weight (550 or 2000). The encapsulation capacity increased with the dendrimer generation and chain length of PEG grafts. The best parameters were achieved for PAMAM G4 with the longest poly(ethylene glycol) chains (similarly to the encapsulation of doxorubicin), as it was able to bind 26 methotrexate molecules per dendrimer molecule. The methotrexate-loaded PEG-modified dendrimers showed delayed release of the drug in an aqueous solution of low ionic strength (Kojima et al., 2000).

And interesting approach was developed by Tekade et al., who proposed a combined strategy of dual drug delivery, receptor up-regulation, and drug targeting to improve the efficiency of methotrexate transport to cancer cells. The team synthesized folate-modified PPI G5 dendrimers loaded with MTX and retinoic acid (ATRA), which were designed to transport the drug specifically to the tumor cells, characterized by over-expression of folic acid receptors. What is more, the addition of ATRA was meant to selectively up-regulate the folate receptors, additionally sensitizing the target cells for this type of carrier system. Such compound was also able to hold methotrexate inside the dendritic scaffold, due to the non-solvent effect. This method proved to be a step toward the improvement of anticancer therapy, as the prepared constructs showed the lowest cytotoxicity compared to the free drug, MTX-ATRA and PPI-MTX-ATRA combinations (Tekade et al., 2008).

Surface-functionalized and non-classical dendrimers proved to interact with metotrexate more efficiently. Melamine-based dendrimers loaded with methotrexate were found to exert reduced hepatotoxicity in comparison to the free drug in C3H mice (Neerman et al., 2004), and the polyether-copolyester dendrimers modified with D-glucosamine were able to facilitate the transport of MTX through the blood-brain barrier and enhance its cytotoxic activity against glioma cells (Dhanikula et al., 2007). Nevertheless, most of the research on the possibility of transport of methotrexate by dendrimers refers to covalent conjugates.

PAMAM dendrimers and their derivatives modified with folic acid and conjugated with MTX are being thoroughly tested for enhanced cytotoxic activity, giving promising results, both *in vitro* (Quintana et al., 2002; Soto-Castro et al., 2012) and *in vivo* (Kukowska Latallo et al., 2005). Another example of MTX-PAMAM targeting is the utilization of monoclonal antibodies. The specific binding and internalization of trastuzumab-coated MTX-PAMAM G5 conjugates were demonstrated in cell lines overexpressing HER2. In addition, binding and uptake of these constructs were completely blocked by the excess of free trastuzumab. The dendrimer conjugate also showed an unusually long residence time in the lysosomes and slow release of methotrexate (Shukla et al., 2008). Moreover, carboxylic acid-terminated (Gurdag et al., 2006) and D-glucohepton-O-1,4-lactone terminated PAMAM dendrimers were capable of efficient intracellular transport of the drug without the addition of targeting moieties (Zhang et al., 2011b).

Interestingly, van Dongen et al. synthesized PAMAM G5-PAMAM G5 dimer conjugated to a single particle of MTX, which exhibited enhanced binding with the cells. The observed mechanism was based on the interaction of MTX with the folate binding protein, inducing its structural rearrangement, followed by dendrimer-protein van der Waals interactions leading to tight binding (van Dongen et al., 2014). This solution is remarkable, as further research may lead to the development of carrier system not only actively transporting the drug, but greatly facilitating its interaction with the target cell.

10.4.3   TAXANES

One of the first taxanes, paclitaxel (PTX) (Figure 10.2), was discovered as a result of the National Cancer Institute screening program, being isolated from the bark of the Pacific yew, *Taxus brevifolia*. However, alternate sources, such as an approved semisynthetic process using a 10-deacetylbaccatin III precursor derived from the needles of more abundant yew species such as the European yew *Taxus baccata*, is currently meeting commercial demands. Docetaxel (DTX) (Figure 10.2) is a paclitaxel analog derived semisynthetically from 10-deacetylbaccatin III and is characterized by higher water solubility in comparison to PTX and exhibits higher cytotoxic activity *in vitro* and in tumor xenografts. Paclitaxel and docetaxel share a similar mechanism of action: they promote the microtubules assembly, at the same time inhibiting their disassembly. These taxanes decrease the lag time and shift the dynamic equilibrium between tubulin dimers and microtubules toward polymerization, thus stabilizing microtubules. This leads to the cell cycle arrest on the metaphase/anaphase boundary and the formation of an incomplete metaphase plate of chromosomes. Both drugs have also been shown to enhance the cytotoxic effects of ionizing radiation *in vitro*, to inhibit pro-angiogenic factors such as VEGF and to induce immunomodulatory and pro-inflammatory response. PTX and DTX are widely used in the treatment of various types of cancer, including breast, lung, ovarian, brain and prostate cancers. However, despite the great spectrum of activity, their clinical application is hampered by poor aqueous solubility, low bioavailability, non-specific toxicity leading to intolerable side effects, as well as the induction of drug resistance through various mechanisms (Pazdur et al., 1993; Rowinsky, 1997; Zhao and Astruc, 2012).

10.4.3.1   Paclitaxel

PAMAM dendrimers and their derivatives have been widely used for improvement of solubility and sustained release of paclitaxel. This includes either unmodified polymers (Devarakonda et al., 2007; Yang et al., 2016) or those with poly(butylene oxide) and poly(ethylene oxide) chains (Zhou et al.,

2013). The additional modification also increased the drug loading capacity of PAMAM dendrimers (Zhou et al., 2013). What is more, PTX-PAMAM complexes showed marked morphological differences in cell shape and size, along with enhanced apoptosis *in vitro* (Devarakonda et al., 2007), as well as more rapid lymphatic absorption, longer lymph nodes residence time and higher metastasis-inhibiting rate compared to free drug *in vivo* (Yang et al., 2016).

PAMAM dendrimers were also utilized for development of paclitaxel-transporting nanoparticles. For instance, dithiocarbamate-functionalized PAMAM dendrimers were used to cross-link the shell of arginine gold nanoparticles-stabilized nanocapsule. Such constructs were characterized by high stability and permeability *in vitro* and *in vivo*. They also exhibited controlled drug release and improved anticancer efficiency of PTX in comparison to the free drug (Jeong et al., 2016). Nanoparticles composed of PAMAM G4 dendrimers conjugated with 1,2-dipalmitoyl-sn-glycero-3-phosphocholine were shown to efficiently encapsulate paclitaxel and significantly prolong the survival of IGROV-1 ovarian tumor-bearing animals (Liu et al., 2015).

For the poly (propylenimine) dendrimers, PTX-loaded PPI G4 modified with folate, dextran and galactose were evaluated *in vivo* for their cytotoxic activity and targeting potential. Nanocarriers with folic acid showed the best targeting parameters and IC50 lower than in case of free drug (Kesharwani et al., 2011). The biodistribution studies of PPI G4.5 dendrimers modified with mesothelin-specific antibody (so-called "immunodendrimers") carrying paclitaxel confirmed the targeting efficiency and higher biodistribution of immunodendrimers in the mesothelin-expressing ovarian cancer cells. In addition, reduction of the tumor size has been observed (Jain et al., 2015).

Generation-dependent increase of PTX solubility was also observed in the case of non-classical, polyglycerol G4 and G5 dendrimers (Ooya et al., 2004).

A series of PTX-PAMAM conjugates have been synthesized and tested for anticancer effect as well. Increased cytotoxicity was observed for PAMAM G3 dendrimers with surface modified with lauryl chains (Teow et al., 2013) and hydroxyl-terminated PAMAM G4 dendrimers modified with bis(PEG) polymer (Khandare et al., 2006). He et al. evaluated the interaction between PTX-PAMAM G4 conjugates and biomembranes using coarse-grained molecular dynamics simulations, and discovered the relationship between the number of conjugated paclitaxel molecules and the ability of nanocarriers to cross the cell membrane (He et al., 2015). What is more, PTX-PAMAM G5 conjugates were found to affect the structure of microtubules not only via paclitaxel-mediated pathway, but also based on the dendrimer electrostatics. Both these actions arrested cell division, leading to the cell death (Cline et al., 2012; Cline et al., 2013).

Synthesis of water-soluble dendrimers based on melamine (Lim and Simanek, 2008) and triazine dendrimers (Lim et al., 2009) carrying 16 and 12 paclitaxel molecules, respectively, has also been described. The latter where further studied for biodistribution, biocompatibility, toxicity and therapeutic effect (Simanek and Lim, 2012).

10.4.3.2   Docetaxel

It has been shown that docetaxel can be transported in a complexed form by both the classical and non-classical dendritic polymers, giving hope for the improvement of the DTX-based chemotherapy. For example, trastuzumab-grafted PAMAM G4 dendrimers loaded with docetaxel were developed for targeted delivery of the drug to HER2-positive breast cancer cells. *In vitro* studies revealed that these constructs were more selective and displayed higher cellular internalization in the HER2-positive MDA-MB-453 human breast cancer cell line than HER2-negative MDA-MB-231 human breast cancer cells. Trastuzumab-conjugated dendrimers also exhibited higher antiproliferation activity and enhanced induction of apoptosis against MDA-MB-453 cells (Kulhari et al., 2016). For the specific treatment of liver cancer, DTX has been encapsulated in the lipoprotein-modified PPI G5 dendrimers. The purpose of the lipoprotein (LDL and HDL) modification was to improve the biocompatibility and biodegradability of nanocarriers, as well as to increase their loading capacity. Studied dendrimers, particularly DTX-LDL-PPI G5 showed excellent pharmacokinetic parameters, such as delayed drug release, reduced hemolytic activity, and the lowest IC50 value against HepG2 cell line. Longer blood circulation, liver-targeting potential and higher uptake of LDL- and HDL-conjugated dendrimers has been also observed *in vivo* (Jain et al., 2013).

Among non-classical dendrimers, H40-PLA nanoparticles deserve special attention. These dendritic copolymers, consisting of aliphatic polyester (H40) and poly(lactic acid) (PLA) were prepared and used as a model to evaluate whether the nanoparticles are sequestered by autophagy and fused with lysosomes. Besides being degraded through the endolysosomal pathway, the

DTX-loaded H40-PLA nanoparticles were also sequestered by autophago-somes and degraded through the autolysosomal pathway, which may lead to the decrease of their anticancer activity. Co-delivery of autophagy inhibitor and DTX by these dendrimers significantly enhanced the cytotoxicity *in vitro*, and decreased the volume of the tumors in immunodeficient murine models (Zhang et al., 2014c). Further studies involved nanoparticles additionaly modi-fied with D-α-tocopheryl poly(ethylene glycol) 1000 succinate (TPGS). DTX-H40-PLA-b-TPGS complexes exhibited faster drug release, as well as higher cellular uptake and cytotoxicity both *in vitro* and *in vivo*, as compared to the free drug, linear PLA and PLA-b-TPGS nanoparticles (Zeng et al., 2015).

For the possibility of use in the transport of docetaxel, β-cyclodextrin dendrimers were also examined, and contributed, e.g. to the improved solu-bility of the drug in aqueous solutions (Benito et al., 2004).

For the covalent DTX-dendrimer conjugates, PAMAM dendrimers grafted with semitelechelic N-(2-hydroxypropyl)methacrylamide (HPMA), containing docetaxel (DTX) attached by a pH-sensitive hydrazone bond were synthesized. *In vivo* studies showed that these dendrimers enhanced the EPR effect and exhibited pH-controlled liberation of DTX, thereby ensuring faster release of the drug in the mildly acidic tumor microenvironment. The conju-gate had a significantly higher maximum tolerated dose and antitumor activ-ity in comparison to the free DTX in C57BL/6 mice (Etrych et al., 2015). Furthermore, polysorbate 80 (P80)-modified PPI G5 dendrimers conjugated with docetaxel were found to efficiently cross BBB and destroy brain tumor cells. The *in vivo* anticancer activity assays revealed that DTX-P80-PPI G5 conjugates significantly reduced the tumor volume and extended the survival time of brain tumor bearing rats. Gamma scintigraphy and biodistribution studies confirmed the targeting efficiency and higher biodistribution of P80-anchored dendrimers in the brain (Gajbhiye and Jain, 2011).

Promising results of *in vivo* and *in vitro* studies resulted in phase I clini-cal trials using poly-lysine dendrimers modified with PEG and covalently attached to DTX. Animal studies showed that this carrier system developed by Starpharma, called DEP™ docetaxel, was significantly more efficient than free docetaxel in a breast cancer model, due to the longer circulation time, extended release of the drug and targeted delivery. In this study, 60% of animals treated with DEP™ docetaxel had no evidence of tumors after 94 days, whereas 100% of the docetaxel-treated mice showed significant tumor regrowth. For the clinical trials, DEP™ docetaxel administered intravenously caused no neutropenia or alopecia in comparison to severe neutropenia suffered by 75% of patients treated with DTX. What is more, DEP™ docetaxel exhibited anticancer activity at relatively low doses in prostate, lung and glioblastoma tumors (Starpharma, 2016).

10.4.4 NUCLEOSIDE ANALOGS

Nucleoside analogs (NAs) were among the first chemotherapeutics developed for the medical treatment of cancer. These compounds constitute a group of purine and pyrimidine nucleoside derivatives most commonly used for hematological malignancies treatment. This involves both adenosine analogs (e.g., fludarabine, cladribine) and cytidine analogs (e.g., cytarabine, decitabine). Moreover, gemcitabine (usually combined with other drugs, e.g., cisplatin), and uracil analogs (fluorouracil, capecitabine) have been successfully used in the treatment of solid tumors (Keating et al., 1993; Galmarini et al., 2002). Nucleoside analogs are antimetabolites, which primary cytotoxic activity involves disruption of metabolism of natural nucleotides or inhibition of synthesis of DNA and RNA (this applies both to an impact on the enzymes involved in the synthesis of nucleic acids and direct incorporation of the drug to the newly synthesized polynucleotide chain), leading to induction of apoptosis (Galmarini et al., 2002).

Most of the therapeutic nucleoside analogs are subjected to similar active transport through the cell membrane and activation by cellular kinases, resulting in the formation of active triphosphate forms (Figure 10.3). As hydrophilic molecules, NAs cannot penetrate the cell membrane passively, and require nucleoside transporters (NTs), which mediate the cell entry. These proteins are divided into two classes: concentrative (CNTs) and equilibrative (ENTs) nucleoside transporters (Baldwin et al., 1999).

The use of nucleoside analogs in anticancer therapy has its limitations, resulting from fast metabolism, as well as disadvantageous biodistribution, low solubility and specificity of interaction with the target cells. What is more, as NAs cannot be transported through the cell membrane in the form of active triphosphates, they have to be administered as prodrugs. The specific metabolism of nucleoside analogs creates a potential of several resistance mechanisms. These include decreased expression of nucleoside transporters or intracellular kinases, as well as alterations of proteins implicated in interactions with nucleoside analogs to exert cytotoxic effect, for example, DNA polymerases. Finally, nucleoside analogs are in danger of catabolic

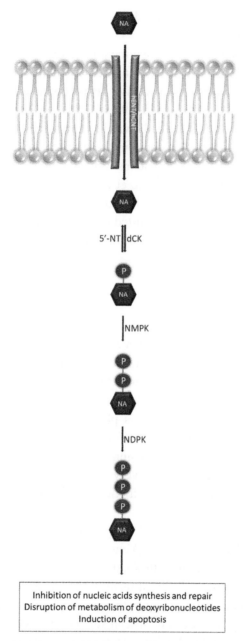

FIGURE 10.3 Metabolism and mechanisms of action of nucleoside analogs. NA: nucleoside analog; P: phosphate group; hENT/hCNT: human equilibrative/concentrative nucleoside transporter; dCK: deoxycytidine kinase; NMPK: monophosphate kinase; NDPK: diphosphate kinase; 5'-NT: 5'-nucleotidase.

degradation both in blood serum and inside the cells (Galmarini et al., 2001; Jordheim and Dumontet, 2007).

In order to overcome all of above-mentioned resistance mechanisms and to enhance cellular uptake of NAs, numerous studies on the use of dendrimers as drug delivery systems have been conducted. Several research studies using FPLC, NMR, spectrofluorimetric methods and isothermal titration calorimetry (ITC) have shown that dendrimers can actively interact with the nucleoside analogs both in the form of prodrugs (Palecz et al., 2016) or triphosphates. The compounds were complexed both on the surface and inside of the dendrimers (Szulc et al., 2013), and the process largely depended on the pH and concentration of ions. These studies indicate that nucleoside analogs could be transported to the target cells in an active phosphorylated form without the need of covalent bonding, thereby avoiding drug resistance (Szulc et al., 2012).

10.4.4.1 5-Fluorouracil

5-fluorouracil (5-FU) (Figure 10.2), a nucleoside analog commonly used in therapy of several malignancies for over 30 years, has been widely tested for use in transport systems based on dendrimers. 5-FU's cytotoxic activity is based on several mechanisms. Upon entering the cell, similarly to other nucleoside analogs, 5-fluorouracil is phosphorylated to its active forms, which inhibit thymidylate synthase (5-FdUMP) or disrupt standard DNA and RNA functions (5-FdUTP), causing replication, transcription and translation arrest. However, due to the very short half-life in blood, the use of delivery systems for proper biodistribution and circulation time extension is needed (Longley et al., 2003).

5-FU is a diprotic acid with pKa equal to 8 and 13. About 20% of its molecules is ionized at pH 7.4 (Venuganti and Perumal, 2008), which enables various interactions with dendrimers. Buczkowski et al. comprehensively studied the possibility of using PAMAM dendrimers as nanocarriers for 5-fluorouracil. According to their research, both the generation and the type of surface moieties are crucial for the interactions between 5-FU and PAMAM macromolecules. The spectroscopic and calorimetric measurements showed that the complexation of the drug with dendrimer is an exothermic process, followed by an advantageous change in entropy. PAMAM G3 dendrimer interacted with about 25±8 drug molecules, the majority of which (24±3) was bound to amide groups, while the rest (5±1) - to the surface amine groups. The same, but hydroxyl-terminated dendrimer bound significantly less molecules of 5-florouracil (6.0±1.6), principally through tertiary amine groups (Buczkowski et al., 2016). For the higher generations, about 30 molecules of 5-fluorouracil were complexed with PAMAM G4, but spectrophotometric measurements showed that the complete saturation of PAMAM G4 dendrimer may reach even 90 molecules of 5-FU (Buczkowski et al., 2011; Buczkowski et al., 2012a). PAMAM G5 interacted with approximately three times more molecules of 5-FU (n = 100), while its hydroxyl-terminated analog bound only 30 molecules (Buczkowski et al., 2015). The scientists also speculated that 5-FU can interact with the surface non-protonated amine and amide groups of PAMAM dendrimers via hydrogen bonds (Buczkowski et al., 2012b).

5-FU-PAMAM complexes in numerous variants were evaluated *in vitro* and *in vivo* for possibility of clinical application. PAMAM G4 facilitated the skin penetration by 5-FU, mainly by altering the skin barrier by dendrimer through transepidermal water loss and decrease of skin resistance, or by decreasing its solubility in the vehicle (Venuganti and Perumal, 2008). 5-fluorouracil encapsulated by PAMAM G5 dendrimer stabilized silver nanoparticles was found to have antiproliferative effect and to induce oxidative stress in A549 and MCF-7 cells (Matai et al., 2015a). 5-FU-PAMAM G5 complexes combined with antisense micro-RNA 21 also improved the cytotoxic activity of the drug in MCF-7 cell line (Mei et al., 2009). Furthermore, numerous variants of surface-modified PAMAM dendrimers showed a good complexation capacity in relation to 5-fluorouracil. Those modifications, involving addition of phospholipid (Tripathi et al., 2002), folic acid or PEG (Bhadra et al., 2003; Singh et al., 2008) significantly contributed to the improvement of lymphatic uptake, circulation time, tumor accumulation and drug release. Moreover, the surface modification of PAMAM G4 with poly(2-(N,N-diethylamino)ethyl methacrylate) (PDEA) resulted in highly pH-dependent release of the drug (Jin et al., 2011).

Recently, the interactions between 5-FU and polyglycerol dendrimers (PGD) have been reported (Lee and Ooya, 2012).

For the covalently-bound 5-fluorouracil nanocarriers, they are primarily based on the non-classical dendrimers, namely the dendritic polymers with a cyclic core of 1,4,7,10-tetraazacyclododecane and four poly(amido amine) branches (Zhuo et al., 1999) and a core structure of ascorbic acid and dicarboxylic acid (Lee et al., 2013). The first of them was found to undergo spontaneous hydrolysis in phosphate buffer solution (37 °C, pH 7.4), which resulted in the generation-dependent release of the drug over several days. The latter significantly reduced the toxicity of 5-FU against normal cell line, indicating that such constructs may restrain potential side effects. Generation-dependent anticancer activity and antiangiogenic properties has been observed as well both *in vitro* and *in vivo*.

5-fluorouracil was also covalently linked to the lysine dendrimers through amide bonds (Zhao et al., 2014). Such constructs showed stable drug release and preferential toxicity against the tumor cells, and are currently subjected to further *in vitro* and *in vivo* studies.

10.4.4.2   Cytarabine

Cytarabine (cytosine arabinoside, 1-D-arabinofuranosylcytosine, Ara-C) (Figure 10.2) is a structural analog of deoxycytidine used in the therapy of acute myeloid leukemia, acute lymphocytic leukemia (ALL) and in numerous lymphomas. Ara-C cytotoxic activity involve DNA polymerase inhibition and competition of Ara-CTP with deoxycytidine triphosphate (dCTP) for incorporation into the DNA molecule. The latter causes chain termination, leading to DNA replication arrest. Continued high cellular concentrations of Ara-CTP are thought to promote drug incorporation into newly synthesized DNA, thus initiating the process of apoptosis (Galmarini et al., 2001).

Cytarabine has been used to create both complexes and covalent conjugates with dendrimers. Szulc et al. demonstrated that maltose-modified PPI G4 dendrimers have the potential to form stable electrostatic complexes with active triphosphate form of Ara-C (Ara-CTP) and facilitate its delivery to the cancer cells. The cytotoxicity and apoptosis assays showed enhanced activity of Ara-CTP complexed with PPI-Mal G4 in comparison to free Ara-C and Ara-CTP against 1301 leukemic cell line. What is more, an enhanced uptake and cytotoxic activity of drug-dendrimer complexes in 1301 cells with blocked human equilibrative nucleoside transporter (hENT1) have been observed, indicating that such a system could overcome drug resistance associated with decreased expression of membrane transporters (Szulc et al., 2016).

A conjugate of Ara-C and PAMAM-OH G4, with the covalent bond between the primary hydroxyl group of cytarabine and dendrimer, was also created. Such conjugates were shown to continuously release the drug over

14 days in PBS, while the same process in human plasma was even faster. Conjugates showed four times higher cytotoxic activity against A549 cell line in comparison to the free Ara-C after 72 h of treatment. In addition, the covalent bonding between PAMAM-OH G4 and Ara-C protected the drug from deamination, which is one of the main processes of cell resistance against cytarabine (Sk et al., 2013).

10.4.4.3 Gemcitabine

Structural analog of Ara-C, gemcitabine (2',2'-difluoro-2'-deoxycytidine, dFdC) (Figure 10.2) has been elaborated to broaden the range of the drug's anticancer activity. Introduction of two fluorine atoms into the sugar moiety resulted in the increased lipophilicity of gemcitabine, facilitating its transport across the cell membrane, prolonging the retention time in tumor cells and enhancing the cytotoxicity of gemcitabine. dFdC is widely used in the therapy of both hematological malignancies and solid tumors (Plunkett et al., 1995; Galmarini et al., 2001).

Thermodynamic measurements showed that gemcitabine formed stable complexes with PAMAM-COONa G3.5 (44 ± 6 dFdC molecules per dendrimer molecule), PAMAM G4 (25 ± 8) and PAMAM-OH G4 (20 ± 9) dendrimers. Spontaneous and exothermal interactions between those compounds were found to be based, as expected, mainly on the electrostatic forces (Palecz et al., 2016). Furthermore, gemcitabine and retinoic acid loaded PAMAM dendrimer-coated magnetic iron oxide nanoparticles were prepared to exert the cytotoxic effect on pancreatic cancer cell lines. The loading efficiency of 10 μM gemcitabine reached 83% in PBS buffer, and the release of the drug in acetate buffer (imitating the endosomal conditions) after 10 hours equaled approximately 72%. It was also found that application of such constructs increased accumulation of gemcitabine inside the cells, thereby promoting its cytotoxic activity (Yalçin et al., 2014). dFdC was also efficiently transported to lung tumor cells by mannosylated PPI G4 dendrimers. Such complexes showed the lowest IC50 value in A549 cell line and longest residence time in lungs of albino rats compared to free drug and dFdC-PPI complexes (Soni et al., 2015).

Moreover, covalent conjugates of gemcitabine and PAMAM or oligo(ethylene glycol) (OEG)-based G4 dendrimers were constructed and evaluated during *in vitro* and *in vivo* studies. Both dFdC-OEG and dFdC-PAMAM conjugates had similar cytotoxicity; however, the dFdC-OEG with the longest surface PEG moieties showed the best parameters, including prolonged blood half-life and enhanced tumor accumulation, leading to considerably higher antitumor activity than dFdC-PAMAM (Wu et al., 2014).

10.4.5 OTHER EXAMPLES OF ANTICANCER DRUG-DENDRIMER NANOCARRIERS

The above-described medications are the most extensively studied in terms of the possible dendrimer-based transport. However, to fully exploit the opportunities posed by the dendritic polymers, scientific interest may not be limited to those most popular compounds. The examples of drugs investigated to a lesser extent are described below, and their structure are shown in Figure 10.2.

Camptothecin, a quinoline alkaloid isolated from *Camptotheca acuminata* and specific inhibitor of DNA topoisomerase I, was shown to form complexes with PAMAM (Sadekar et al., 2013; Kong et al., 2014) and poly(glycerol succinic acid) dendrimers (Morgan et al., 2003; Morgan et al., 2006), which enabled increased intracellular concentration and retention time (Morgan et al., 2003; Morgan et al., 2006), as well as enhanced oral absorption (Sadekar et al., 2013), high stability, increased water solubility, targeted delivery and high anticancer activity (Kong et al., 2014). Similar outcome has been proved for camptothecin-dendrimer conjugates, involving PEGylated PAMAM dendrimers (Zolotarskaya et al., 2015) and, interestingly, poly-lysine dendrimers with the drug attached inside the dendritic scaffold (Zhou et al., 2014).

PAMAM dendrimers and their derivatives are also comprehensively tested for the transport of cisplatin, an inorganic coordination compound, which forms crosslinks between DNA strands or between DNA and proteins, altering the processes of replication and inducing apoptosis. This includes studies on kinetics of drug-dendrimer complex formulation (Abderrezak et al., 2012; Kulhari et al., 2015) and biological properties of both complexes (Yellepeddi et al., 2011, 2012, 2013) and conjugates (Malik et al., 1999; Kesavan et al., 2015). Cisplatin-PAMAM formulations were shown to improve cellular accumulation and cytotoxicity *in vitro* (Yellepeddi et al., 2011, 2013; Kesavan et al., 2015). What is more, they also enhanced anticancer activity of the drug *in vivo* (Malik et al., 1999; Kesavan et al., 2015) increased its concentration in plasma, and prolonged lifespan of tumor-bearing mice (Yellepeddi et al., 2012).

Epirubicin, and analog of doxorubicin with a similar antitumor activity, but causing fewer side effects, has been complexed with PAMAM dendrimers in carbon dots/PAMAM hybrids (Matai et al., 2015b), and aptamer-based dendrimers (Taghdisi et al., 2016), which showed efficient drug delivery and pH-dependent release. The drug was also conjugated with amino adipic acid/PEG or β-glutamic acid/PEG dendritic structures, resulting in better stability in different pH buffers and in plasma, as well as prolonged residence time in blood (Pasut et al., 2005).

Already mentioned etoposide, a picropodophyllin analog and inhibitor of DNA topoisomerase II, has been studied in complexes with PAMAM-OH dendrimers modified with poly(caprolactone) and PEG (Wang et al., 2005), which exhibited moderate loading capacity, or PPI dendrimers modified with PEG and folic acid (Sideratou et al., 2010), which in turn enhanced the solubility and enabled specific targeting together with the slow release of the drug.

Tamoxifen, a selective estrogen receptor inhibitor used mainly in the breast cancer therapy, is also worth mentioning, because in nanotechnology it is used rather as a targeting agent than the drug. Recently, it was complexed with PAMAM dendrimers modified with myristic acid, which lead to sustained, pH-dependent release and high anticancer activity *in vitro* (Matai and Gopinath, 2016).

10.4.6 DENDRIMERS IN PHOTODYNAMIC THERAPY

Photodynamic therapy (PDT) in anticancer treatment is based on the use of the non-toxic, light-sensitive compounds, preferably with an affinity for tumor cells, which upon exposure to the light of specific wavelength acquire cytotoxic properties. These compounds contain chromophore molecules (e.g., cyclic tetrapyrrolic moiety), which as a result of irradiation transfer its energy to the cellular O_2 to form singlet oxygen and other ROS, which may cause significant damage to cell structures (Klajnert et al., 2012). Photodynamic therapy can also include the above-mentioned photochemical internalization (Lai et al., 2007).

Phototherapy has an advantage over chemotherapy resulting from significantly reduced systemic toxicity due to the administration of non-toxic compound, and subsequent stimulation of its anticancer properties after reaching the target cell. PDT as an alternative to traditional chemotherapy was proposed more than 30 years ago, but it still has its limitations, resulting from low tumor specificity leading to disadvantageous biodistribution and hampered cellular uptake (Klajnert et al., 2012). Dendrimers, thanks to their high loading capacity and utilization of EPR effect may serve as proper nanotools for optimizing anticancer PDT. Dendrimers can transport photosensitizers similarly to other drugs - in a form of complexes or covalent conjugates with the therapeutics attached to the surface moieties. However, because photosensitizers do not always have to be released from the structure of the polymer to exert its cytotoxic effect, the incorporation of the drug into the internal branches or the core on macromolecule is also possible.

Examples of photosensitizers used to create different formulations with dendrimers include 5-aminolevulinic acid (Battah et al., 2001), Zn-porphyrin (Nishiyama et al., 2003), pheophorbide a (Hackbarth et al., 2005), rose bengal and protoporphyrin IX (Kojima et al., 2007). Unfortunately, the use of dendrimers in photodynamic therapy also has its drawbacks, mainly due to the reduction of light-inducible cytotoxic properties of the drug caused by the interactions with the polymer (Hackbarth et al., 2005; Kojima et al., 2007).

10.4.7 BORON NEUTRON CAPTURE THERAPY

An interesting aspect of dendrimers in the transport of anticancer therapeutics concerns the improvement of boron neutron capture therapy (BNCT), a noninvasive method of therapy for the treatment of malignancies such as primary brain tumors or recurrent head and neck cancers. The strategy is based on nuclear reaction of stable isotope of boron (boron 10), which upon the absorption of low-energy thermal neutrons generates liable isotope (boron 11). Boron 11 immediately fissions, giving lithium 7 nuclei and α-particles, according to the following reaction:

$$^{10}B + n_{th} \rightarrow [^{11}B]^* \rightarrow \alpha + {}^7Li + 2.31 \text{ MeV}$$

Thanks to the limited propagation in tissues, α-particles may exert the cytotoxic activity only in the tumor area. The medical procedure involves injection of a tumor-localizing agent with boron 10 isotope, together with subsequent radiation and biologically destructive nuclear reaction (Barth et al., 2012).

Because it is necessary to specifically deliver the isotope to the tumor cells, dendrimers, particularly PAMAM, has been considered as boron carriers due to the high loading capacity and possibility of attachment of targeting moieties. It has been shown that one molecule of dendrimer may transport up to 1000 boron atoms (Wu et al., 2006). Boronated PAMAM dendrimers, modified with EGFR- (Wu et al., 2004) or VEGFR-specific molecules (Backer et al., 2005) showed targeting properties both *in vitro* and *in vivo* (Capala et al., 1996; Wu et al., 2007; Yang et al., 2009), and dendrimer-based neutron capture therapy in tumor-bearing rodent models resulted in extended life span of the animals (Barth et al., 2004; Wu et al., 2007).

10.5 CONCLUSION

The discovery of efficient anticancer drug remains a dream of scientists, medical doctors and patients. Development of new therapeutics is laborious and time-consuming, and current methods of chemotherapy, although often bringing promising results, are not without side effects, which often prevent full recovery. Thus, during the pursuit of new medicines, it is worth so slow down and lean on those already existing, trying to improve their performance. Modern nanotechnology enables the elaboration of efficient delivery devices, which can provide specific transport of pharmaceutics to cancer cells, while increasing their anticancer activity and preventing detrimental systemic toxicity. Cited scientific data on the use of dendrimers show that these compounds hold the potential of becoming the leading drug delivery systems in the nearest future.

KEYWORDS

- **nanostructure**
- **cancer chemotherapy**
- **biocompatibility**
- **drug delivery systems**
- **complexes**
- **conjugates**

REFERENCES

Abderrezak, A., Bourassa, P., Mandeville, J. S., Sedaghat-Herati, R., & Tajmir-Riahi, H. A., (2012). Dendrimers bind antioxidant polyphenols and cisplatin drug. *PLoS One.*, *7*(3), e33102.

Agrawal, A., Gupta, U., Asthana, A., & Jain, N. K., (2009). Dextran conjugated dendritic nanoconstructs as potential vectors for anti-cancer agent. *Biomaterials*, *30*, 3588–3596.

Al-Jamal, KT., Al-Jamal, WT., Wang, JT., Rubio, N., Buddle, J., Gathercole, D., Zloh, M., & Kostarelos, K., (2013). Cationic poly-L-lysine dendrimer complexes doxorubicin and delays tumor growth *in vitro* and *in vivo. ACS Nano.*, *7*(3), 1905–1917.

Anand, P., Kunnumakkara, A. B., Kunnumakara, A. B., Sundaram, C., Harikumar, K. B., Tharakan, S. T., Lai, O. S., Sung, B., & Aggarwal, B. B., (2008). Cancer is a preventable disease that requires major lifestyle changes. *Pharm. Res.*, *25*(9), 2097–2116.

Backer, M. V., Gaynutdinov, T. I., Patel, V., Bandyopadhyaya, A. K., Thirumamagal, B. T., Tjarks, W., Barth, R. F., Claffey, K., & Backer, J. M., (2005). Vascular endothelial growth factor selectively targets boronated dendrimers to tumor vasculature. *Mol Cancer Ther.*, *4*(9), 1423–1429.

Baldwin, S. A., Mackey, J. R., Cass, C. E., & Young, J. D., (1999). Nucleoside transporters: molecular biology and implications for therapeutic development. *Mol Med Today*, *5*, 216–224.

Barth, R. F., Vicente, M. G. H., Harling, O. K., Kiger, W. S., Riley, K. J., Binns, P. J., Wagner, F. M., Suzuki, M., Aihara, T., Kato, I., & Kawabata, S., (2012). Current status of boron neutron capture therapy of high grade gliomas and recurrent head and neck cancer. *Radiat Oncol.*, *7*, 146.

Barth, R. F., Wu, G., Yang, W., Binns, P. J., Riley, K. J., Patel, H., Coderre, J. A., Tjarks, W., Bandyopadhyaya, A. K., Thirumamagal, B. T., Ciesielski, M. J., & Fenstermaker, R. A., (2004). Neutron capture therapy of epidermal growth factor (+) gliomas using boronated cetuximab (IMC-C225) as a delivery agent. *Appl Radiat Isot.*, *61*, 899–903.

Battah, S. H., Chee, C. E., Nakanishi, H., Gerscher, S., MacRobert, A. J., & Edwards, C., (2001). Synthesis and biological studies of 5-aminolevulinic acidcointainig dendrimers for photodynamic therapy. *Bioconjugate Chem.*, *12*, 980–988.

Benito, J. M., Gómez-García, M., Ortiz Mellet, C., Baussanne, I., Defaye, J., García Fernández, J. M., (2004). Optimizing saccharide-directed molecular delivery to biological receptors: design, synthesis, and biological evaluation of glycodendrimer-cyclodextrin conjugates. *J. Am. Chem. Soc.*, *126*(33), 10355–10363.

Bennasroune, A., Gardin, A., Aunis, D., Crémel, G., & Hubert, P., (2004). Tyrosine kinase receptors as attractive targets of cancer therapy. *Crit. Rev. Oncol. Hematol.*, *50*(1), 23–38.

Bergers, G., & Hanahan, D., (2008). Modes of resistance to anti-angiogenic therapy. *Nat Rev Cancer*, *8*(8), 592–603.

Bertino, J. R., Göker, E., Gorlick, R., Li, W. W., & Banerjee, D., (1996). Resistance Mechanisms to Methotrexate in Tumors. *Oncologist*, *1*(4), 223–226.

Bertrand, N., Wu, J., Xu, X., Kamaly, N., & Farokhzad, O. C., (2014). Cancer nanotechnology: the impact of passive and active targeting in the era of modern cancer biology. *Adv. Drug Deliv. Rev.*, *66*, 2–25.

Bhadra, D., Bhadra, S., Jain, S., & Jain, N. K., (2003). A PEGylated dendritic nanoparticulate carrier of fluorouracil. *Int. J. Pharm.*, *257*(1–2), 111–124.

Blume-Jensen, P., & Hunter, T., (2001). Oncogenic kinase signalling. *Nature*, *411*(6835), 355–365.

Boas, U., & Heegaard, P. M., (2004). Dendrimers in drug research. *Chem. Soc. Rev.*, *33*(1), 43–63.

Boehm, T., Folkman, J., Browder, T., O'Reilly, M. S., (1997). Antiangiogenic therapy of experimental cancer does not induce acquired drug resistance. *Nature*, *390*(6658), 404–407.

Brannon-Peppas, L., & Blanchette, J. O., (2004). Nanoparticle and targeted systems for cancer therapy. *Adv Drug Deliv Rev.*, *56*, 1649–1659.

Breedveld, F. C., (2000). Therapeutic monoclonal antibodies. *Lancet.*, *355*(9205), 735–740.

Brigger, I., Dubernet, C., & Couvreur, P., (2002). Nanoparticles in cancer therapy and diagnosis. *Adv. Drug Deliv. Rev.*, *54*, 631–651.

Buczkowski, A., Olesinski, T., Zbicinska, E., Urbaniak, P., & Palecz, B., (2015). Spectroscopic and calorimetric studies of formation of the supramolecular complexes of PAMAM G5-NH2 and G5-OH dendrimers with 5-fluorouracil in aqueous solution, *Int. J. Pharm.*, *490*, 102–111.

Buczkowski, A., Piekarski, H., & Palecz, B., (2012b). Stoichiometry and equilibrium constant of the complex of PAMAM-NH2 G4 and 5-fluorouracil, *J. Mol. Liq.*, *173*, 8–12.

Buczkowski, A., Sekowski, S., Grala, A., Palecz, D., Milowska, K., Urbaniak, P., Gabryelak, T., Piekarski, H., & Palecz, B., (2011). Interaction between PAMAM-NH2 G4 dendrimer and 5-fluorouracil in aqueous solution. *Int. J. Pharm.*, 408(1–2), 266–270.

Buczkowski, A., Urbaniak, P., & Palecz, B., (2012a). Thermochemical and spectroscopic studies on the supramolecular complex of PAMAM-NH2 G4 dendrimer and 5- fluorouracil in aqueous solution. *Int. J. Pharm*, *428*, 178–182.

Buczkowski, A., Waliszewski, D., Urbaniak, P., & Palecz, B., (2016). Study of the interactions of PAMAM G3 NH2 and G3-OH dendrimers with 5-fluorouracil in aqueous solutions. *Int. J. Pharm.*, *505*(1–2), 1–13.

Capala, J., Barth, R. F., Bendayan, M., Lauzon, M., Adams, D. M., Soloway, A. H., Fernstermaker, R. A., & Carlssen, J., (1996). Boronated epidermal growth factor as a potential targeting agent for boron neutron capture therapy of brain tumors. *Bioconjugate. Chem.*, *7*(1), 7–15.

Chan, E. S., & Cronstein, B. N., (2013). Mechanisms of action of methotrexate. *Bull. Hosp. Jt. Dis.*, *71*(1), S5–8.

Chandra, S., Noronha, G., Dietrich, S., Lang, H., & Bahadur, D., (2015). Dendrimer-magnetic nanoparticles as multiple stimuli responsive and enzymatic drug delivery vehicle. *J. Magn. Magn. Mater.*, *380*, 7–12.

Chang, Y., Meng, X., Zhao, Y., Li, K., Zhao, B., Zhu, M., Li, Y., Chen, X., & Wang, J., (2011). Novel water-soluble and pH-responsive anticancer drug nanocarriers: doxorubicin-PAMAM dendrimer conjugates attached to superparamagnetic iron oxide nanoparticles (IONPs). *J. Colloid. Interface. Sci.*, *363*, 403–409.

Chen, A. M., Zhang, M., Wei, D., Stueber, D., Taratula, O., Minko, T., & He, H., (2009). Co-delivery ofdoxorubicin and Bcl-2 siRNA by mesoporous silica nanoparticles enhances the efficacy of chemotherapy in multidrug-resistant cancer cells. *Small*, *5*, 2673–2677.

Choudhary, S., Mathew, M., & Verma, R. S., (2011). Therapeutic potential of anticancer immunotoxins. *Drug Discov. Today*, *16*(11–12), 495–503.

Cline, E. N., Li, M. H., Choi, S. K., Herbstman, J. F., Kaul, N., Meyhöfer, E., Skiniotis, G., Baker, J. R., Larson, R. G., & Walter, N. G., (2013). Paclitaxel-conjugated PAMAM dendrimers adversely affect microtubule structure through two independent modes of action. *Biomacromolecules*, *14*(3), 654–664.

Cline, E., Li, M. H., Choi, S. K., Herbstman, J., Kaul, N., Meyhofer, E., Skiniotis, G., Baker, J. R., Larson, R. G., & Walter, N. G., (2012). Taxol-Conjugated Pamam Dendrimers Utilize Three Modes of Action on Microtubule Structure. *Biophys. J.*, *102*(3), 187a.

Conley, S. J., Gheordunescu, E., Kakarala, P., Newman, B., Korkaya, H., Heath, A. N., Clouthier, S. G., & Wicha, M. S., (2012). Antiangiogenic agents increase breast cancer stem cells via the generation of tumor hypoxia. *Proc. Natl. Acad. Sci. USA*, *109*(8), 2784–2789.

De Jong, W. H., & Borm, P. J., (2008). Drug delivery and nanoparticles: Applications and hazards. *International Journal of Nanomedicine*, *3*(2), 133–149.

Devarakonda, B., Judefeind, A., Chigurupati, S., Thomas, S., Shah, G. V., Otto, D. P., & De Villiers, M. M., (2007). The effect of polyamidoamine dendrimers on the *in vitro* cytotoxicity of paclitaxel in cultured prostate cancer (PC-3M) cells. *J. Biomed. Nanotechnol*, *3*(4), 384–393(10).

DeVita, V. T. Jr., & Chu, E., (2008). A history of cancer chemotherapy. *Cancer Res.*, *68*(21), 8643–8653.

Dhanikula, R. S., Argaw, A., Bouchard, J. F., & Hildgen, P., (2008). Methotrexate loaded polyether-copolyester dendrimers for the treatment of gliomas: enhanced efficacy and intratumoral transport capability. *Mol. Pharm.*, *5*(1), 105–116.

DiMasi, J. A., Hansen, R. W., & Grabowski, H. G., (2003). The price of innovation: new estimates of drug development costs. *J. Health Econ.*, *22*(2), 151–185.

Duncan, R., & Sat, Y. N., (1998). Tumour targeting by enhanced permeability and retention (EPR) effect. *Ann. Oncol.*, *9*(2), 39.

Emeje, M. O., Obidike, I. C, Akpabio, E. I., & Ofoefule, S. I., (2012). Nanotechnology in drug delivery. *Recent Advances in Novel Drug Carrier Systems*. Available from: https://www.intechopen.com/books/recent-advances-in-novel-drug-carrier-systems/nanotechnology-in-drug-delivery. Chapter 4.

Etrych, T., Strohalm, J., Sirova, M., Tomalová, B., Rossmann, P., Rihova, B., Ulbrich K., & Kovář, M., (2015). High-molecular weight star conjugates containing docetaxel with high anti-tumor activity and low systemic toxicity *in vivo*. *Polym. Chem.*, *6*, 160–170.

Farber, S., (1949). Some observations on the effect of folic acid antagonists on acute leukemia and other forms of incurable cancer. *Blood*, *4*, 160–167.

Farber, S., Diamond, L. K., Mercer, R. D., Sylvester, R. F. Jr., & Wolff, J. A., (1948). Temporary remissions in acute leukemia in children produced by folic acid antagonist, 4-aminopteroyl-glutamic acid (aminopterin). *N. Engl. J. Med*, *238*, 787–793.

Feng, S. S., & Chien, S., (2003). Chemotherapeutic engineering: Application and further development of chemical engineering principles for chemotherapy of cancer and other diseases. *Chem Engineering Science*, *58*, 4087–4114.

Fischer, D., Bieber T., Li, Y., Elsasser, H. P., & Kissel, T., (1999). A novel non-viral vector for DNA delivery based on low molecular weight, branched polyethylenimine: Effect of molecular weight on transfection and cytotoxicity. *Pharm Res.*, *16*, 1273–1279.

Folkman, J., Merler, E., Abernathy, C., & Williams, G., (1971). Isolation of a tumor factor responsible for angiogenesis. *J. Exp. Med.*, *133*, 275–288.

Gajbhiye, V., & Jain, N. K., (2011). The treatment of Glioblastoma Xenografts by surfactant conjugated dendritic nanoconjugates. *Biomaterials*, *32*(26), 6213–6225.

Galmarini, C. M., Mackey, J. R., & Dumontet, C., (2001). Nucleoside analogues: mechanisms of drug resistance and reversal strategies. *Leukemia, 15*(6), 875–890.

Galmarini, C. M., Mackey, J. R., & Dumontet, C., (2002). Nucleoside analogues and nucleobases in cancer treatment. *Lancet. Oncol., 3*(7), 415–424.

Gordaliza, M., (2007). Natural products as leads to anticancer drugs. *Clin. Transl. Oncol., 9*(12), 767–776.

Gotink, K. J., & Verheul, H. M. V., (2010). Anti-angiogenic tyrosine kinase inhibitors: what is their mechanism of action? *Angiogenesis, 13*(1), 1–14.

Gurdag, S., Khandare, J., Stapels, S., Matherly, L. H., & Kannan, R. M., (2006). Activity of dendrimer-methotrexate conjugates on methotrexate-sensitive and -resistant cell lines. *Bioconjug. Chem., 17*, 275–283.

Hackbarth, S., Ermilov, E. A., & Röder, B., (2005). Interaction of Pheophorbide a molecules covalently linked to DAB dendrimers. *Opt. Comm., 248*, 295–306.

He, H., Li, Y., Jia, X. R., Du, J., Ying, X., Lu, W. L., Lou, J. N., & Wei, Y., (2011). PEGylated Poly(amidoamine) dendrimer-based dual-targeting carrier for treating brain tumors. *Biomaterials, 32*(2), 478–487.

He, X., Lin, M., Lu, T., Qu, Z., & Xu, F., (2015). Molecular analysis of interactions between a PAMAM dendrimer–paclitaxel conjugate and a biomembrane. *Phys. Chem. Chem. Phys., 17*, 29507–29517.

Isoldi, M. C., Visconti, M. A., & Castrucci, A. M., (2005). Anti-cancer drugs: molecular mechanisms of action. *Mini. Rev. Med. Chem., 5*(7), 685–695.

Jain, A., Jain, K., Mehra, N. K., & Jain, N. K., (2013). Lipoproteins tethered dendrimeric nanoconstructs for effective targeting to cancer cells. *J. Nanopart. Res., 15*, 2003.

Jain, N. K., Tare, M. S., Mishra, V., & Tripathi, P. K., (2015). The development, characterization and *in vivo* anti-ovarian cancer activity of poly(propylene imine) (PPI)-antibody conjugates containing encapsulated paclitaxel. *Nanomedicine, 11*(1), 207–218.

Jeong, Y., Kim, S. T., Jiang, Y., Duncan, B., Kim, C. S., Saha, K., Yeh, Y. C., Yan, B., Tang, R., Hou, S., Kim, C., Park, M. H., & Rotello, V. M., (2016). Nanoparticle-dendrimer hybrid nanocapsules for therapeutic delivery. *Nanomedicine (Lond), 11*(12), 1571–1578.

Jin, Y., Ren, X., Wang, W., Ke, L., Ning, E., Du, L., Bradshaw J., (2011). A 5-fluorouracil-loaded pH-responsive dendrimer nanocarrier for tumor targeting. *Int. J. Pharm., 420*(2), 378–384.

Jones, G. B., (2014). *History of Anticancer Drugs*. eLS. John Wiley & Sons Ltd., Chichester. http://www.els.net [doi: 10.1002/9780470015902.a0003630.pub2].

Jordheim, L. P., & Dumontet, C., (2007). Review of recent studies on resistance to cytotoxic deoxynucleoside analogues. *Biochim. Biophys. Acta., 1776*(2), 138–159.

Kakde, D., Jain, D., Shrivastava, V., Kakde, R., & Patil, A. J., (2011). Cancer Therapeutics-Opportunities, Challenges and Advances in Drug Delivery. *JAPS, 01*(09), 01–10.

Kaminskas, L. M., Kelly, B. D., McLeod, V. M., Sberna, G., Owen, D. J., Boyd, B. J., & Porter, C. J., (2011). Characterisation and tumour targeting of PEGylated polylysine dendrimers bearing doxorubicin via a pH labile linker. *J. Control. Release, 152*(2), 241–248.

Kaminskas, L. M., McLeod, V. M., Kelly, B. D., Sberna, G., Boyd, B. J., Williamson, M., Owen, D. J., & Porter, C. J., (2012). A comparison of changes to doxorubicin pharmacokinetics, antitumor activity, and toxicity mediated by PEGylated dendrimer and PEGylated liposome drug delivery systems. *Nanomedicine, 8*(1), 103–111.

Kaminskas, L. M., McLeod, V. M., Ryan, G. M., Kelly, B. D., Haynes, J. M., Williamson, M., Thienthong, N., Owen, D. J., & Porter, C. J., (2014). Pulmonary administration of a doxorubicin-conjugated dendrimer enhances drug exposure to lung metastases and improves cancer therapy. *J. Control. Release*, *183*, 18–26.

Kaneshiro, T. L., & Lu, Z. R., (2009). Targeted intracellular codelivery of chemotherapeutics and nucleic acid with a well-defined dendrimer-based nanoglobular carrier. *Biomaterials*, *30*, 5660–5666.

Ke, W., Zhao, Y., Huang, R., Jiang, C., & Pei, Y., (2008). Enhanced oral bioavailability of doxorubicin in a dendrimer drug delivery system. *J. Pharm. Sci.*, *97*(6), 2208–2216.

Keating, M. J., O'Brien, S., Kantarjian, H., Robertson, L. B., Koller, C., Beran, M., & Estey, E., (1993). Nucleoside analogs in treatment of chronic lymphocytic leukemia. *Leuk. Lymphoma.*, *10*(l), 139–145.

Kerbel, R. S., (2000). Tumor angiogenesis: past, present and the near future. *Carcinogenesis*, *21*, 505–515.

Kesavan, A., Ilaiyaraja, P., Sofi Beaula, W., Veena Kumari, V., Sugin Lal, J., Arunkumar, C., Anjana, G., Srinivas, S., Ramesh, A., Rayala, S. K., Ponraju, D., & Venkatraman, G., (2015). Tumor targeting using polyamidoamine dendrimer-cisplatin nanoparticles functionalized with diglycolamic acid and herceptin. *Eur. J. Pharm. Biopharm.*, *96*, 255–263.

Kesharwani, P., Tekade, R. K., Gajbhiye, V., Jain, K., & Jain, N. K., (2011). Cancer targeting potential of some ligand-anchored poly (propylene imine) dendrimers: a comparison. *Nanomedicine*, *7*(3), 295–304.

Khandare, JJ., Jayant, S., Singh, A., Chandna, P., Wang, Y., Vorsa, N., & Minko, T., (2006). Dendrimer versus linear conjugate: Influence of polymeric architecture on the delivery and anticancer effect of paclitaxel. *Bioconjug. Chem.*, *17*(6), 1464–1472.

Khopade, A. J., Caruso, F., Tripathi, P., Nagaich, S., & Jain, N. K., (2002). Effect of dendrimer on entrapment and release of bioactive from liposomes. *Int. J. Pharm.*, *232*(1–2), 157–162.

Klajnert, B., & Bryszewska, M., (2000). Dendrimers, properties and applications. *Acta. Biochim. Polon.*, *48*, 199–208.

Klajnert, B., Rozanek, M., & Bryszewska, M., (2012). Dendrimers in photodynamic therapy. *Curr. Med. Chem.*, *19*(29), 4903–4912.

Kobayashi, H., Kawamoto, S., Saga, T., Sato, N., Hiraga, A., Ishimori, T., Konishi, J., Togashi, K., & Brechbiel, M. W., (2001). Positive effects of polyethylene glycol conjugation to generation-4 polyamidoamine dendrimers as macromolecular MR contrast agents. *Magn. Reson. Med.*, *46*(4), 781–788.

Kojima, C., Kono, K., Maruyama, K., & Takagishi, T., (2000). Synthesis of polyamidoamine dendrimers having poly(ethylene glycol) grafts and their ability to encapsulate anticancer drugs. *Bioconjug. Chem.*, *11*(6), 910–917.

Kojima, C., Nishisaka, E., Suehiro, T., Watanabe, K., Harada, A., Goto, T., Magata, Y., & Kono, K., (2013b). The synthesis and evaluation of polymer prodrug/collagen hybrid gels for delivery into metastatic cancer cells. *Nanomedicine*, *9*(6), 767–775.

Kojima, C., Suehiro, T., Watanabe, K., Ogawa, M., Fukuhara, A., Nishisaka, E., Harada, A., Kono, K., Inui, T., & Magata, Y., (2013a). Doxorubicin-conjugated dendrimer/collagen hybrid gels for metastasis-associated drug delivery systems. *Acta. Biomaterialia*, *9*, 5673–5680.

Kojima, C., Toi, Y., Harada, A., & Kono, K., (2007). Preparation of poly (ethylene glycol)-attached dendrimers encapsulating photosensitizers for application for photodynamic therapy. *Bioconjugate. Chem.*, *18*, 663–670.

Kong, X., Yu, K., Yu, M., Feng, Y., Wang, J., Li, M., Chen, Z., He, M., Guo, R., Tian, R., Li, Y., Wu, W., & Hong, Z., (2014). A novel multifunctional poly (amidoamine) dendrimeric delivery system with superior encapsulation capacity for targeted delivery of the chemotherapy drug 10-hydroxycamptothecin. *Int. J. Pharm.*, *465*(1–2), 378–387.

Kono, K., Kojima, C., Hayashi, N., Nishisaka, E., Kiura, K., Watarai, S., & Harada, A., (2008). Preparation and cytotoxic activity of poly(ethylene glycol)-modified poly(amidoamine) dendrimers bearing adriamycin. *Biomaterials*, *29*(11), 1664–1675.

Krumbhaar, E. B., & Krumbhaar, H. D., (1919). The blood and bone marrow in yellow cross gas (mustard gas) poisoning. Changes produced in the bone marrow of fatal cases. *J. Med. Res.*, *40*, 497–507.

Kukowska-Latallo, J. F., Candido, K. A., Cao, Z., Shraddha, S., Nigavekar, S., Majoros, I. J., Thomas, T. P., Balogh, L. P., Khan, M. K., & Baker, J. R. Jr., (2005). Nanoparticles of anti-cancer drug improves therapeutic response in animal model of human epithelial cancer. *Cancer Res.*, *65*, 5317–5324.

Kulhari, H., Pooja, D., Shrivastava, S., Kuncha, M., Naidu, V. G. M., Bansal, V., Sistla, R., & Adams, D. J., (2016). Trastuzumab-grafted PAMAM dendrimers for the selective delivery of anticancer drugs to HER2-positive breast cancer. *Sci. Rep.*, *6*, 23179.

Kulhari, H., Pooja, D., Singh, M. K., & Chauhan, A. S., (2015). Optimization of carboxylate-terminated poly(amidoamine) dendrimer-mediated cisplatin formulation. *Drug Dev. Ind. Pharm.*, *41*(2), 232–238.

Lai, P. S., Lou, P. J., Peng, C. L., Pai, C. L., Yen, W. N., Huang, M. Y., Young, T. H., & Shieh, M. J., (2007). Doxorubicin delivery by polyamidoamine dendrimer conjugation and photochemical internalization for cancer therapy. *J. Control. Release*, *122*(1), 39–46.

Lee, C. C., Gillies, E. R., Fox, M. E., Guillaudeu, S. J., Fréchet, J. M., Dy, E. E., & Szoka, F. C., (2006). A single dose of doxorubicin-functionalized bow-tie dendrimer cures mice bearing C-26 colon carcinomas. *Proc. Natl. Acad. Sci. USA*, *103*(45), 16649–16654.

Lee, H., & Ooya, T., (2012). 19 F-NMR, 1 H-NMR, and fluorescence studies of interaction between 5-fluorouracil and polyglycerol dendrimers. *J. Phys. Chem. B.*, *116*, 12263–12267.

Lee, S. M., Bala, Y. S., Lee, W. K., Jo, N. J., & Chung, I., (2013). Antitumor and antiangiogenic active dendrimer/5-fluorouracil conjugates. *J. Biomed. Mater. Res. A.*, *101*(8), 2306–2312.

Li, N., Li, N., Yi, Q., Luo, K., Guo, C., Pan, D., & Gu, Z., (2014b). Amphiphilic peptide dendritic copolymer-doxorubicin nanoscale conjugate self-assembled to enzyme-responsive anti-cancer agent. Biomaterials, 35(35), 9529–9545.

Li, X., Takashima, M., Yuba, E., Harada, A., & Kono, K., (2014a). PEGylated PAMAM dendrimer-doxorubicin conjugate-hybridized gold nanorod for combined photothermal-chemotherapy. Biomaterials, 35(24), 6576–6584.

Li, Y., He, H., Jia, X., Lu, W. L., Lou, J., & Wei, Y., (2012). A dual-targeting nanocarrier based on poly(amidoamine) dendrimers conjugated with transferrin and tamoxifen for treating brain gliomas. *Biomaterials*, *33*(15), 3899–3908.

Liao, H., Liu, H., Li, Y., Zhang, M., Tomás, H., Shen, M., & Xiangyang, S., (2014). Antitumor efficacy of doxorubicin encapsulated within PEGylated poly(amidoamine) dendrimers. *J. Appl. Polym. Sci.*, *131*(11).

Liauw, S. W., (2013). Molecular mechanisms and clinical use of targeted anticancer drugs. *Aust. Prescr.*, *36*(4), 126–131.

Lim, J., & Simanek, E. E., (2008). Synthesis of water-soluble dendrimers based on melamine bearing 16 paclitaxel groups. *Org. Lett.*, *10*(2), 201–204.

Lim, J., Chouai, A., Lo, S. T., Liu, W., Sun, X., & Simanek, E. E., (2009). Design, Synthesis, Characterization, and Biological Evaluation of Triazine Dendrimers Bearing Paclitaxel Using Ester and Ester/Disulfide Linkages. *Bioconjugate. Chem.*, *20*(11), 2154–2161.

Liu Y., Ng Y., Toh M. R., Chiu G. N., (2015). Lipid-dendrimer hybrid nanosystem as a novel delivery system for paclitaxel to treat ovarian cancer. *J. Control. Release*, *220*(Pt A), 438–446.

Longley, D. B., Harkin, D. P., & Johnston, P. G., (2003). 5-fluorouracil: mechanisms of action and clinical strategies. *Nat. Rev. Cancer*, *3*(5), 330–338.

Malik, N., Evagorou, E. G., & Duncan, R., (1999). Dendrimer-platinate: a novel approach to cancer chemotherapy. *Anticancer. Drugs*, *10*(8), 767–776.

Matai, I., & Gopinath, P., (2016). Hydrophobic myristic acid modified PAMAM dendrimers augment the delivery of tamoxifen to breast cancer cells. *RSC Adv.*, *6*, 24808–24819.

Matai, I., Sachdev, A., & Gopinath, P., (2015b). Self-assembled hybrids of fluorescent carbon dots and PAMAM dendrimers for epirubicin delivery and intracellular imaging. *ACS Appl. Mater. Interfaces.*, *7*(21), 11423–11435.

Matai, I., Sachdeva, A., & Gopinath, P., (2015a). Multicomponent 5-fluorouracil loaded PAMAM stabilized-silver nanocomposites synergistically induce apoptosis in human cancer cells, *Biomater. Sci.*, *3,* 457–468.

Mei, M., Ren, Y., Zhou, X., Yuan, X., Li, F., Jiang, L., Kang, Ch., & Yao, Z., (2009). Suppression of breast cancer cells *in vitro* by polyamidoamine-dendrimer-mediated 5-fluorouracil chemotherapy combined with antisense micro-RNA 21 gene therapy. *J. Appl. Polym. Sci.*, *114*, 3760–3766.

Mintzer, M. A., & Grinstaff, M. W., (2011). Biomedical applications of dendrimers: a tutorial. *Chem. Soc. Rev.*, *40*(1), 173–190.

Morgan, M. T., Carnahan, M. A., Immoos, C. E., Ribeiro, A. A., Finkelstein, S., Lee, S. J., & Grinstaff, M. W., (2003). Dendritic molecular capsules for hydrophobic compounds. *J. Am. Chem. Soc.*, *125*(50), 15485–15489.

Morgan, M. T., Nakanishi, Y., Kroll, D. J., Griset, A. P., Carnahan, M. A., Wathier, M., Oberlies, N. H., Manikumar, G., Wani, M. C., & Grinstaff, M. W., (2006). Dendrimer-encapsulated camptothecins: increased solubility, cellular uptake, and cellular retention affords enhanced anticancer activity *in vitro*. *Cancer Res.*, *66*(24), 11913–11921.

Nam, N. H., & Parang, K., (2003). Current targets for anticancer drug discovery. *Curr. Drug. Targets*, *4*(2), 159–179.

National Cancer Institute (NCI). https://www.cancer.gov/ (accessed Sept 22, 2016).

Neerman, M. F., Chen, H. T., Parrish, A. R., & Simanek, E. E., (2004). Attenuation of drug toxicity using dendrimers based on melamine, candidate vehicles for drug delivery. *Mol. Pharm.*, *1*, 390–393.

Nishiyama, N., Stapert, H. R., Zhang, G. D., Takasu, D., Jiang, D. J., Nagano, T., Aida, T., & Kataoka, K., (2003). Light-harvesting ionic dendrimer porphyrins as new photosensitizers for photodynamic therapy. *Bioconjugate. Chem.*, *14*, 58–66.

Ooya T., Lee J., Park K., (2004). Hydrotropic dendrimers of generations 4 and 5: synthesis, characterization, and hydrotropic solubilization of paclitaxel. *Bioconjug. Chem.*, *15*(6), 1221–1229.

Padilla De Jesús, O. L., Ihre, H. R., Gagne, L., Fréchet, J. M., & Szoka, F. C Jr., (2002). Polyester dendritic systems for drug delivery applications: *in vitro* and *in vivo* evaluation. *Bioconjug. Chem.*, *13*(3), 453–461.

Palecz, B., Buczkowski, A., Piekarski, H., & Kılınçarslan, O., (2016). Thermodynamic interaction between PAMAM G4-NH2, G4-OH, G3. 5-COONa dendrimers and gemcitabine in water solutions. *IJSM*, *3*(1), 21–26.

Papagiannaros, A., Dimas, K., Papaioannou, G., & Demetzos, C., (2005). Doxorubicin-PAMAM dendrimer complex attached to liposomes: cytotoxic studies against human cancer cell lines. *Int. J. Pharm.*, *302*, 29–38.

Pasut, G., Scaramuzza, S., Schiavon, O., Mendichi, R., & Veronese, F. M., (2005). PEG-epirubicin Conjugates with High Drug Loading. *J. Bioact. Compat. Polym.*, *20*(3), 213–230.

Patri, A. K., Kukowska-Latallo, J. F., & Baker, J. R Jr., (2005). Targeted drug delivery with dendrimers: Comparison of the release kinetics of covalently conjugated drug and non-covalent drug inclusion complex. *Adv. Drug Deliv. Rev.*, *57*, 2203–2214.

Patri, A. K., Majoros, I. J., & Baker, J. R. Jr., (2002). Dendritic polymer macromolecular carriers for drug delivery. *Curr. Opin. Chem. Biol.*, *6*, 466–471.

Paul, M. K., & Mukhopadhyay, A. K., (2004). Tyrosine kinase - role and significance in cancer. *Int. J. Med. Sci.*, *1*(2), 101–115.

Payne, S., & Miles, D., (2008). *Mechanisms of Anticancer Drugs*. Scott-Brown's Otorhinolaryngology, Head and Neck Surgery 7Ed. CRC Press, Boca Raton. Chapter 4.

Pazdur, R., Kudelka, A. P., Kavanagh, J. J., Cohen, P. R., & Raber, M. N., (1993). The taxoids: paclitaxel (Taxol) and docetaxel (Taxotere). *Cancer Treat. Rev.*, *19*(4), 351–386.

Peng, C. A., & Hsu, Y. C., (2001). Fluoroalkylated polyethylene glycol as potential surfactant for perfluorocarbon emulsion. *Artif. Cells Blood Substit. Immobil. Biotechnol.*, *29*(6), 483–492.

Piktel, E., Niemirowicz, K., Wątek, M., Wollny, T., Deptuła, P., & Bucki, R., (2016). Recent insights in nanotechnology-based drugs and formulations designed for effective anticancer therapy. *J. Nanobiotechnology*, *14*(1), 39.

Plunkett, W., Huang, P., Xu, Y. Z., Heinemann, V., Grunewald, R., & Gandhi, V., (1995). Gemcitabine: metabolism, mechanisms of action, and self-potentiation. *Semin. Oncol.*, *22*(4 Suppl 11), 3–10.

Quintana, A., Raczka, E., Piehler, L., Lee, I., Myc, A., Majoros, I., Patri, A. K., Thomas, T., Mulé, J., & Baker, J. R. Jr., (2002). Design and function of a dendrimer-based therapeutic nanodevice targeted to tumor cells through the folate receptor. *Pharm. Res.*, *19*(9), 1310–1316.

Reddy, J. A., Dean, D., Kennedy, M. D., & Low, P. S., (1999). Optimization of folate-conjugated liposomal vectors for folate receptor-mediated gene therapy. *J. Pharm. Sciences*, *88*(11), 1112–1118.

Rivankar, S., (2014). An overview of doxorubicin formulations in cancer therapy. *J. Cancer Res. Ther.*, *10*(4), 853–858.

Rowinsky, E. K., (1997). The development and clinical utility of the taxane class of antimicrotubule chemotherapy agents. *Annu. Rev. Med.*, *48*, 353–374.

Ryan, G. M., Kaminskas, L. M., Bulitta, J. B., McIntosh, M. P., Owen, D. J., & Porter, C. J., (2013). PEGylated polylysine dendrimers increase lymphatic exposure to doxorubicin when compared to PEGylated liposomal and solution formulations of doxorubicin. *J. Control. Release*, *172*(1), 128–136.

Sadekar, S., Thiagarajan, G., Bartlett, K., Hubbard, D., Ray, A., McGill, L. D., & Ghandehari, H., (2013). Poly (Amido Amine) dendrimers as absorption enhancers for oral delivery of camptothecin. *Int. J. Pharm.*, *456*(1), 175–185.

Schlessinger, J., (2000). Cell signaling by receptor tyrosine kinases. *Cell*, *103*(2), 211–225.

Shapiro, G. I., & Harper, J. W., (1999). Anticancer drug targets: cell cycle and checkpoint control. *J. Clin. Invest.*, *104*(12), 1645–1653.

She, W., Li, N., Luo, K., Guo, C., Wang, G., Geng, Y., & Gu, Z., (2013). Dendronized heparin-doxorubicin conjugate based nanoparticle as pH-responsive drug delivery system for cancer therapy. *Biomaterials*, *34*(9), 2252–2264.

She, W., Pan, D., Luo, K., He, B., Cheng, G., Zhang, C., & Gu, Z., (2015). PEGylated Dendrimer-Doxorubicin Cojugates as pH-Sensitive Drug Delivery Systems: Synthesis and *In Vitro* Characterization. *J. Biomed. Nanotechnol.*, *11*(6), 964–978.

Shukla, R., Thomas, T. P., Desai, A. M., Kotlyar, A., Park, S. J., & Baker, J. R., (2008). HER2 specific Delivery of Methotrexate by Dendrimer Conjugated anti-HER2 mAb. *Nanotechnology*, *19*(29), 295102.

Sideratou, Z., Kontoyianni, C., Drossopoulou, G. I., & Paleos, C. M., (2010). Synthesis of a folate functionalized PEGylated poly(propylene imine) dendrimer as prospective targeted drug delivery system. *Bioorg. Med. Chem. Lett.*, *20*(22), 6513–6517.

Simanek, E. E., & Lim, J., (2012). Paclitaxel-triazine dendrimer constructs: efficacy, toxicity, and characterization. *Multifunctional Nanoparticles for Drug Delivery Applications*. Nanostructure Science and Technology series. Springer: Boston.

Singh, P., Gupta, U., Asthana, A., & Jain, N. K., (2008). Folate and folate-PEG-PAMAM dendrimers: synthesis, characterization, and targeted anticancer drug delivery potential in tumor bearing mice. *Bioconjug. Chem.*, *19*(11), 2239–2252.

Siriviriyanun, A., Popova, M., Imae, T., Kiew, L. V., Looi, C. Y., Wong, W. F., Lee, H. B., & Chung, L. Y., (2015). Preparation of graphene oxide/dendrimer hybrid carriers for delivery of doxorubicin. *Chem. Eng. J.*, *281*, 771–781.

Sk, U. H., Kambhampati, S. P., Mishra, M. K., Lesniak, W. G., Zhang, F., & Kannan, R. M., (2013). Enhancing the efficacy of Ara-C through conjugation with PAMAM dendrimer and linear PEG: a comparative study, *Biomacromolecules*, *14*(3), 801–810.

Soni, N., Jain, K., Gupta, U., & Jain, N. K., (2015). bControlled delivery of Gemcitabine Hydrochloride using mannosylated poly(propyleneimine) dendrimers. *J. Nanopart. Res.*, *17*, 458.

Soto-Castro, D., Cruz-Morales, J. A., Ramírez Apan, M. T., & Guadarrama, P., (2012). Solubilization and anticancer-activity enhancement of Methotrexate by novel dendrimeric nanodevices synthesized in one-step reaction. *Bioorg. Chem.*, *41–42*, 13–21.

Starpharma. http://www. starpharma. com/ (accessed Oct 22, 2016).

Szulc, A., Appelhans, D., Voit, B., Bryszewska, M., & Klajnert, B., (2012). Characteristics of complexes between poly(propylene imine) dendrimers and nucleotides. *New. J. Chem.*, *36*, 1610–1615.

Szulc, A., Appelhans, D., Voit, B., Bryszewska, M., & Klajnert, B., (2013). Studying complexes between PPI dendrimers and Mant-ATP. *J. Fluoresc.*, *23*(2), 349–356.

Szulc, A., Pulaski, L., Appelhans, D., & Voit, B., (2016). Klajnert-Maculewicz, B. Sugar-modified poly(propylene imine) dendrimers as drug delivery agents for cytarabine to overcome drug resistance. *Int. J. Pharm.*, *513*(1–2), 572–583.

Tacar, O., Sriamornsak, P., & Dass, C. R., (2013). Doxorubicin: an update on anticancer molecular action, toxicity and novel drug delivery systems. *J. Pharm. Pharmacol.*, *65*(2), 157–170.

Taghdisi, S. M., Danesh, N. M., Ramezani, M., Lavaee, P., Jalalian, S. H., Robati, R. Y., & Abnous, K., (2016). Double targeting and aptamer-assisted controlled release delivery of epirubicin to cancer cells by aptamers-based dendrimer *in vitro* and *in vivo*. *Eur. J. Pharm. Biopharm.*, *102*, 152–158.

Tekade, R. K., Dutta, T, Tyagi, A, Bharti, A. C., Das, B. C., & Jain, N. K., (2008). Surface-engineered dendrimers for dual drug delivery: a receptor up-regulation and enhanced cancer targeting strategy. *J. Drug. Target*, *16*, 758–772.

Teow, H. M., Zhou, Z., Najlah, M., Yusof, S. R., Abbott, N. J., & D'Emanuele, A., (2013). Delivery of paclitaxel across cellular barriers using a dendrimer-based nanocarrier. *Int. J. Pharm.*, *441*(1–2), 701–711.

Thorn, C. F., Oshiro, C., Marsh, S., Hernandez-Boussard, T., McLeod, H., Klein, T. E., & Altman, R. B., (2011). Doxorubicin pathways: pharmacodynamics and adverse effects. *Pharmacogenet. Genomics.*, *21*(7), 440–446.

Tomalia, D. A., Baker, H., Dewald, J. R., Hall, M., Kallos, G., Martin, S., Roeck, J., Ryder, J., & Smith, P., (1985). A new class of polymers: starburst dendritic molecules. *Polym. J.*, *17*, 117–132.

Tomalia, D. A., Christensen, J. B., & Boas, U., (2012). *Dendrimers, Dendrons, and Dendritic Polymers. Discovery, Applications, and the Future.* Cambridge University Press, Cambridge.

Tripathi, P. K., Khopade, A. J., Nagaich, S., Shrivastava, S., Jain, S., & Jain, N. K., (2002). Dendrimer grafts for delivery of 5-fluorouracil, *Pharmazie.*, *57*(4), 261–264.

Van der Veldt, A. A., Lubberink, M., Bahce, I., Walraven, M., De Boer, M. P., Greuter, H. N., Hendrikse, N. H., Eriksson, J., Windhorst, A. D., Postmus, P. E., Verheul, H. M., Serné, E. H., Lammertsma, A. A., & Smit, E. F., (2012). Rapid decrease in delivery of chemotherapy to tumors after anti-VEGF therapy: implications for scheduling of anti-angiogenic drugs. *Cancer Cell*, *21*(1), 82–91.

van Dongen, M. A., Rattan, R., Silpe, J., Dougherty, C., Michmerhuizen, N. L., Van Winkle, M., Huang, B., Choi, S. K., Sinniah, K., Orr, B. G., & Banaszak, H. M. M., (2014). Poly(amidoamine) Dendrimer-Methotrexate Conjugates: The Mechanism of Interaction with Folate Binding Protein. *Mol. Pharm.*, *11*(11), 4049–4058.

Venuganti, V. V., & Perumal, O. P., (2008). Effect of poly(amidoamine) (PAMAM) dendrimer on skin permeation of 5-fluorouracil. *Int. J. Pharm.*, *361*(1–2), 230–238.

Wang, F., Bronich, T. K., Kabanov, A. V., Rauh, R. D., & Roovers, J., (2005). Synthesis and evaluation of a star amphiphilic block copolymer from poly(epsilon-caprolactone) and poly(ethylene glycol) as a potential drug delivery carrier. *Bioconjug. Chem.*, *16*(2), 397–405.

Wang, Y., Cao, X., Guo, R., Shen, M., Zhang, M., Zhu, M., & Shi, X., (2011). Targeted delivery of doxorubicin into cancer cells using a folic acid–dendrimer conjugate *Polym. Chem.*, *2*, 1754–1760.

Wei, T., Chen, C., Liu, J., Liu, C., Posocco, P., Liu, X., Cheng, Q., Huo, S., Liang, Z., Fermeglia, M., Pricl, S., Liang, XJ., Rocchi, P., & Peng, L., (2015). Anticancer drug nanomicelles formed by self-assembling amphiphilic dendrimer to combat cancer drug resistance. *Proc. Natl. Acad. Sci. USA*, *112*(10), 2978–2983.

Weiss, R. B., Donehower, R. C., Wiernik, P. H., Ohnuma, T., Gralla, R. J., Trump, D. L., Baker, R., VanEcho, D. A., VonHoff, D. D., & Leyland-Jones, B., (1990). Hypersensitivity reactions from taxol. *J. Clin. Oncol.*, *8*(7), 1263–1268.

Widakowich, C., de Castro, G. Jr., de Azambuja, E., Dinh, P., & Awada, A., (2007). Review: side effects of approved molecular targeted therapies in solid cancers. *Oncologist*, *12*(12), 1443–1455.

Wright, J. C., Prigot, A., & Wright, B. P., (1951). An evaluation of folic acid antagonists in adults with neoplastic diseases. A study of 93 patients with incurable neoplasms. *J. Natl. Med. Assoc.*, *43*, 211–240.

Wu, G., Barth, R. F., Yang, W., Chatterjee, M., Tjarks, W., Ciesielski, M. J., & Fenstermaker, R. A., (2004). Site-specific conjugation of boron-containing dendrimers to anti-EGF receptor monoclonal antibody cetuximab (IMC-C225) and its evaluation as a potential delivery agent for neutron capture therapy. *Bioconjug. Chem.*, *15*(1), 185–194.

Wu, G., Barth, R. F., Yang, W., Lee, R. J., Tjarks, W., Backer, M. V., & Backer, J. M., (2006). Boron containing macromolecules and nanovehicles as delivery agents for neutron capture therapy. *Anticancer Agents Med. Chem.*, *6*, 167–184.

Wu, G., Yang, W., Barth, R. F., Kawabata, S., Swindall, M., Bandyopadhyaya, A. K., Tjarks, W., Khorsandi, B., Blue, T. E., Ferketich, A. K., Yang, M., Christoforidis, G. A., Sferra, T. J., Binns, P. J., Riley, K. J., Ciesielski, M. J., & Fenstermaker, R. A., (2007). Molecular targeting and treatment of an epidermal growth factor receptor-positive glioma using boronated cetuximab. *Clin. Cancer Res.*, *13*(4), 1260–1268.

Wu, W., Driessen, W., & Jiang, X., (2014). Oligo(ethylene glycol)-based thermosensitive dendrimers and their tumor accumulation and penetration. *J. Am. Chem. Soc.*, *136*(8), 3145–3155.

Yabbarov, N. G., Posypanova, G. A., Vorontsov, E. A., Popova, O. N., & Severin, E. S., (2013). Targeted delivery of doxorubicin: drug delivery system based on PAMAM dendrimers. *Biochemistry (Mosc)*, *78*(8), 884–894.

Yalçin, S., Erkan, M., Ünsoy, G., Parsian, M., Kleeff, J., & Gündüz, U., (2014). Effect of gemcitabine and retinoic acid loaded PAMAM dendrimer-coated magnetic nanoparticles on pancreatic cancer and stellate cell lines. *Biomed. Pharmacother*, *68*(6), 737–743.

Yang, R., Mao, Y., Ye, T., Xia, S., Wang, S., & Wang, S., (2016). Study on enhanced lymphatic exposure of polyamidoamin-alkali blue dendrimer for paclitaxel delivery and influence of the osmotic pressure on the lymphatic targeting. *Drug Deliv.*, *23*(7), 2617–2629.

Yang, W., Barth, R. F., Wu, G., Tjarks, W., Binns, P., & Riley, K., (2009). Boron neutron capture therapy of EGFR or EGFRvIII positive gliomas using either boronated monoclonal antibodies or epidermal growth factor as molecular targeting agents. *Appl. Radiat. Isot.*, *67*, 328–331.

Yellepeddi, V. K., Kumar, A., Maher, D. M., Chauhan, S. C., Vangara, K. K., & Palakurthi, S., (2011). Biotinylated PAMAM dendrimers for intracellular delivery of cisplatin to ovarian cancer: role of SMVT. *Anticancer Res.*, *31*(3), 897–906.

Yellepeddi, V. K., Vangara, K. K., & Palakurthi, S. J., (2013). Poly(amido)amine (PAMAM) dendrimer-cisplatin complexes for chemotherapy of cisplatin-resistant ovarian cancer cells. *Nanopart. Res.*, *15*, 1897.

Yellepeddi, V. K., Vangara, K. K., & Palakurthi, S., (2012). *In vivo* Efficacy of PAMAM-Dendrimer-Cisplatin Complexes in SKOV-3 Xenografted Balb/C Nude Mice. *J. Biotechnol. Biomaterial*, S13, 003.

Zamulaeva, I. A., Churyukina, K. A., Matchuk, O., NNikolskaja, E. D., Makarenko, S. A., Zhunina, O. A., Kondrasheva, I. G., & Severin, E. S., (2016). Cytotoxic effects of the combined action of ionizing radiation and doxorubicin conjugates with dendritic polymer and a vector protein to tumor cells *in vitro*. *Radiation and Risk*, *15*(3), 46–56.

Zeng, X., Tao, W., Wang, Z., Zhang, X., Zhu, H., Wu, Y., Gao, Y., Liu, K., Jiang, Y., Huang, L., Mei, L., Feng, S. S., (2015). Docetaxel-Loaded Nanoparticles of Dendritic Amphiphilic Block Copolymer H40-PLA-b-TPGS for Cancer Treatment. *Particle & Particle Systems Characterization, 32*(1), 112–122.

Zhang, C., Pan, D., Luo, K., Li, N., Guo, C., Zheng, X., & Gu, Z., (2014a). Dendrimer-doxorubicin conjugate as enzyme-sensitive and polymeric nanoscale drug delivery vehicle for ovarian cancer therapy. *Polym. Chem.*, *5*, 5227–5235.

Zhang, C., Pan, D., Luo, K., She, W., Guo, C., Yang, Y., & Gu, Z., (2014b). Peptide dendrimer-Doxorubicin conjugate-based nanoparticles as an enzyme-responsive drug delivery system for cancer therapy. *Adv. Healthc. Mater.*, *3*(8), 1299–1308.

Zhang, L., Zhu, S., Qian, L., Pei, Y., Qiu, Y., & Jiang, Y., (2011a). RGD-modified PEG-PAMAM-DOX conjugates: *in vitro* and *in vivo* studies for glioma. *Eur. J. Pharm. Biopharm.*, *79*(2), 232–240.

Zhang, M., Guo, R., Kéri, M., Bányai, I., Zheng, Y., Cao, M., Cao, X., & Sh, I. X., (2014). Impact of dendrimer surface functional groups on the release of doxorubicin from dendrimer carriers. *J. Phys. Chem. B.*, *118*(6), 1696–706.

Zhang, X., Yang, Y., Liang, X., Zeng, X., Liu, Z., Tao, W., Xiao, X., Chen, H., Huang, L., & Mei, L., (2014c). Enhancing therapeutic effects of docetaxel-loaded dendritic copolymer nanoparticles by co-treatment with autophagy inhibitor on breast cancer. *Theranostics.*, *4*(11), 1085–1095.

Zhang, Y., Thomas, T. P., Lee, K. H., Li, M., Zong, H., Desai, A. M., Kotlyar, A., Huang, B., Holl, M. M., & Baker, J. R. Jr., (2011b). Polyvalent saccharide-functionalized generation 3 poly(amidoamine) dendrimer-methotrexate conjugate as a potential anticancer agent. *Bioorg. Med. Chem.*, *19*(8), 2557–2564.

Zhao, J., Zhou, R., Fu, X., Ren, W., Ma, L., Li, R., Zhao, Y., & Guo, L., (2014). Cell-Penetrable Lysine Dendrimers for Anti-Cancer Drug Delivery: Synthesis and Preliminary Biological Evaluation. *Arch. Pharm. Chem. Life Sci.*, *347*, 1–9.

Zhao, P., & Astruc, D., (2012). Docetaxel nanotechnology in anticancer therapy. *Chem. Med. Chem.*, *7*(6), 952–972.

Zhong, Q., & Da Rocha, S. R. P., (2016). Poly (amidoamine) dendrimer–doxorubicin conjugates: *In vitro* characteristics and pseudosolution formulation in pressurized metered-dose inhalers. *Mol. Pharmaceutics*, *13*(3), 1058–1072.

Zhou, Z., D'Emanuele A., & Attwood, D., (2013). Solubility enhancement of paclitaxel using a linear-dendritic block copolymer. *Int. J. Pharm.*, *452*(1–2), 173–179.

Zhou, Z., Ma, X., Murphy, C. J., Jin, E., Sun, Q., Shen, Y., Van Kirk, E. A., & Murdoch, W. J., (2014). Molecularly precise dendrimer-drug conjugates with tunable drug release for cancer therapy. *Angew. Chem. Int. Ed. Engl.*, *53*(41), 10949–10955.

Zhu, S., Hong, M., Tang, G., Qian, L., Lin, J., Jiang, Y., & Pei, Y., (2010). Partly PEGylated polyamidoamine dendrimer for tumor-selective targeting of doxorubicin: the effects of PEGylation degree and drug conjugation style. *Biomaterials*, *31*, 1360–1371.

Zhuo, R. X., Du, B., & Lu, Z. R., (1999). *In vitro* release of 5-fluorouracil with cyclic core dendritic polymer. *J. Control. Release*, *57*(3), 249–257.

Zolotarskaya, O. Y., Xu, L., Valerie, K., & Yang, H., (2015). Click synthesis of a polyamidoamine dendrimer-based camptothecin prodrug. *RSC Adv.*, *5*, 58600–58608.

CHAPTER 11

DENDRIMERIC ARCHITECTURE FOR EFFECTIVE ANTIMICROBIAL THERAPY

RAMADOSS KARTHIKEYAN,[1] ORUGANTI SAI KOUSHIK,[1] and PALANIRAJAN VIJAYARAJ KUMAR[2]

[1]Vignan Pharmacy College, Vadlamudi–522213, A.P., India,
E-mail: rkcognosy@gmail.com

[2]Faculty of Pharmaceutical Sciences, No. 1, Jalan Menara Gading,
UCSI University (South Campus), Taman Connaught, Cheras 56000,
Kuala Lumpur, Malaysia

CONTENTS

ABSTRACT

Agents that are capable of killing pathogenic microorganisms are known as antimicrobial substances. These are of low molecular weight and used for the sterilization of water, as antimicrobial drugs, as food preservatives, and for soil sterilization. Due to their potential to provide quality and safety benefits to many materials, antimicrobials gain interest from both industry and academic research. However, antimicrobial agents with low molecular weight suffer from many disadvantages, such as the antimicrobial ability to short-term and environment toxicity. Antimicrobial functional groups can be introduced into polymer molecules to overcome problems associated with the low molecular weight antimicrobial agents. For enhancing the efficacy of some existing antimicrobial agents, selectivity, increasing their efficiency and prolonging the lifetime of the antimicrobial agents, the use of antimicrobial polymers can be employed. The development of antimicrobial polymers research represents a great a challenge for both the academic world and industry. The polymer research, which presents great modern interest yet has received lacking consideration, is that of the development of polymers with antimicrobial activities, commonly known as polymeric biocides. Biocides immobilized on dendrimers can be more effective if the target sites are cell walls or membranes. It has been shown that small quaternary ammonium compounds exert their antimicrobial action by disintegrating and disrupting the cell membrane, converting functional end groups of dendrimer to ammonium salts, dendrimer biocides can be synthesized. These dendrimer biocides have been shown to be more potent than their small molecule counterparts as they bear high local density active groups. Thus, dendrimer biocides may be very beneficial in terms of activity, reduced toxicity, localization in specific organs and increased duration of action. This chapter will find complete solution for an antimicrobial agent facing problems and advantages of the dendrimeric biocides.

11.1 INTRODUCTION

11.1.1 BACTERIAL INFECTION

Bacteria are common infective agents producing a wide variety of diseases. In developing countries and the origin of massive epidemics bacteria is still an important cause of mortality that spread very easily among the population of developing countries due to the lack of adequate sanitary conditions. Additionally, some bacteria are responsible to produce highly potent toxins that can be considered as biological warfare. The search for therapeutic and prophylactic agents against these toxins is a topic of extreme importance (Aji Alex et al., 2011). The appearance of bacterial resistance is due to the broad use and sometimes abuse of antibiotics that has forced the biomedical researchers to look for new strategies to combat bacterial infections. One of the most attractive alternatives is using antiadhesive molecules in the inhibition of bacterial attachment to target cells during the first stages of the infection. Bacterial adhesion is mediated between lectin-like proteins by specific interactions of carbohydrate-protein at the surface of bacteria and glycoconjugates at the surface of target cells or vice versa. Blocking this interaction should inhibit the attachment of bacteria to the target cell surface and stop the infection. Due to the presence of multivalent interactions, the efficiency of this recognition process is achieved in nature, eventhough carbohydrate-protein interactions are weak. Based on the type of bacterial target, the classification of many different structures is described (Amidi et al., 2006).

11.1.1.1   Bacteria Producing AB_5 Toxins

Gram-negative bacteria are very important group causing thousands of deaths every year by producing toxins. Generally, these toxins consist of six subunits, subunit "A" that is responsible of the infection and a homopentameric subunits "B" (B_5) that are required for the toxin to attach to the cell surface. Inhibition of the attachment of the subunits B_5 should be enough to stop the infection process. In a multivalent way, this B unit has a carbohydrate recognition site that interacts with carbohydrates present at the cell surface (Arsiwala et al., 2013).

11.1.1.2   Cholera Toxin

Cholera toxin is an AB_5 protein secreted by *Vibrio cholerae* causing the disease cholera. This is an infectious intestinal disease characterized by vomiting and severe diarrhea, that if untreated may be life-threatening due to enormous loss of water and electrolytes. The B subunit is able to recognize and interact with ganglioside GM1 at the cell membrane forming a pore for subunit A that blocks GTPase activity of G protein and results in an increase of the synthesis of cAMP. As Cl^- leaves the cells followed by Na^+

and water leads to watery, electrolyte rich diarrhea in the intestine (Awasthi et al., 2005).

11.1.1.3   Shiga and Vero Toxins

Shiga toxins produced by *Shigella dys-enteriae* and Shiga-like toxins also called Vero toxins produced by *Escherichia coli* are AB_5 toxins causing hemorrhagic colitis and watery diarrhea particularly severe in children and elder people (Baba et al., 2007).

11.1.1.4   Heat Labile Enterotoxin

The heat labile enterotoxin of *E. coli* is a cholera-like enterotoxin. It adheres as cholera toxin does to ganglioside GM1 and causes less severe diarrhea due to the same mechanisms. The group of Schengrund has used the same type of experiments to show the activity of oligosaccharide-derivatized dendrimers for cholera toxin and heat labile toxin with similar results based on the structural similarities between cholera toxin and heat labile toxin (Bai et al., 1995).

11.1.1.5   Bacterial Endotoxins

The lipopolysaccharide (LPS) consists of a bilayered component membrane which plays a key role in septic shock present at the outer leaflet of gram-negative bacteria such as Haemophilus influenzae, *Kelbsiella pneumoniae, Bordetellapertussis, Pseudomonas aeruginosa, Chlamydia psittaci, Escherichia coli, Salmonella enterica* and *Legionella pneumophila*. When bacteria multiply, die or lyse, this LPS is liberated from the membrane surface. In many cases of death, they have been recognized LPS as a factor, responsible of toxicity in severe gram-negative bacteria infections. It causes a systemic response that if uncontrolled can lead to septic shock characterized mainly by hypotension, coagulopathy, fever, and organ failure. Lipid A is a glycolipid toxic moiety of LPS considered as a target for the design of drugs against endotoxins. This LPS presents an anionic and amphiphilic nature that is an important feature for the design of compounds able to interact with LPS (Bernfild et al., 1996).

11.1.1.6   Type 1 Fimbriated *Escherichia coli*

Most of the gram-negative bacteria posses type 1 fimbriae, which are responsible for the adherence of *Escherichia coli* to the urinary tract by adhesion organelles causes common urinary tract infection. This adhesion process is governed by the interaction between type 1 fimbriae and mannose conjugates found at the bladder epithelial cell surface (Bareiss et al., 2010).

11.1.1.7   *Streptococcus suis*

Gram-positive bacteria like *Streptococcus suis* is responsible for septicemia, meningitis, and pneumonia in swine, pigs, and other domestic animals. It is responsible for meningitis in humans when being in contact with pigs. In the adhesion process of the bacteria to the host cells this bacterial species contains a galactosyl-1-4-galactose-binding adhesion.. (Bolzinger et al., 2012).

11.1.1.8   *Staphylococcus aureus*

In 1880 and 1882, Staphylococcal disease and its role in sepsis and abscess formation were described for the first time by Ogston. More than 100 years later, *Staphylococcus aureus* remains a dangerous pathogen in humans that can cause ill effects from minor skin infections to life-threatening diseases (Brannon-Peppase et al., 2004).

11.1.1.9   *Actinomyces naeslundii*

A. naeslundii is a gram-positive bacterium that colonizes oral cavities. During this colonization process, *Streptococcus oralis* co-aggregate with *A. naeslundii* through galactose residues present at the surface of *S. oralis* and an adhesion of *A. naeslundii* pili (Brigger et al., 2002).

11.1.2   FUNGAL INFECTION

Candida albicans causes most common infections produced by fungi. Under normal circumstances, *C. albicans* colonizes humans without any harmful effects, although excess may result in candidiasis in skin or mucosa. Systemic candidiasis is often observed in immunocompromised individuals. At the surface of target cells different forms of *C. albicans* are also recognized by different lectins (Cai et al., 2008).

11.1.3 PRION PROTEIN INFECTION

Almost 25 years ago, Stanley Prusiner discovered prion a new infective agent and protein. The prion protein is the product of a normal gene expressed mainly in neural tissue and has several helixes of few sheets in its natural configuration known as PrP^C. This protein adopts an abnormal configuration upon contact with the sheet rich infectious form of the protein known as PrP^{Sc} after scrapie, an old recognized disease of sheep. These types of structures are found in several neurodegenerative disorders such as Creutzfeldt-Jakob disease in humans, bovine spongiform encephalopathy etc. (Cevc et al., 2010).

11.1.4 VIRAL INFECTION

This is the area of application where more efforts have been done respect to develop new anti-infective agents based on dendrimers. Recently, a review describing dendrimers as antivirals has been published. Here, we intend to update the information presented in that review with the most recent publications concerning dendrimers as antiviral drugs. Again, this section will be divided for each different viral agent (Chandak et al., 2008).

11.1.4.1 HIV-1, HIV-2, and SIV

Infection by Human Immunodeficiency Virus (HIV) is a global health problem although, especially dramatic in developing countries in sub-Saharan Africa and Asia where the vast majority of infected patients do not have access to antiretroviral drugs. Recent research in this topic is concentrated mainly on developing microbicides and vaccine development (Chekhonin et al., 2008).

11.1.4.2 Herpes Simplex Virus (HSV) Infection

One of the most prevalent sexual transmitted diseases is Genital human herpes virus infection. HSV-1 and 2 cause mucocutaneous infection, such as herpes labialis and herpes genitalis. Currently, no cure is available, in neurons of the host after primary or initial infection the virus persists for life in a latent form, periodically reactivating (Couvreur et al., 2002).

11.1.4.3 Influenza Virus Infection

Influenza virus is a RNA virus. They are three types of influenza: A, B, and C. Influenza A is the cause of all flu pandemics. Through the interaction of the main envelope glycoprotein: hemagglutinin (HA) this virus adheres to the target cells. HA recognizes sialic acid receptors on the host cell. It is known that monovalent sialic acid was able to prevent influenza A agglutination of chicken erythrocytes (Daneman et al., 2006).

11.1.4.4 Foot-and-Mouth Disease Virus (FMDV) Infection

In farms, virus infects animals through the respiratory tract or skin abrasions. To develop a vaccine against this infection, a very interesting approach was used (Das et al., 2015).

11.1.4.5 Ebola Virus Infection

The Filoviridae family that is responsible of sporadic outbreaks of hemorrhagic fever in Africa characterized by a high death rate. There is no currently any vaccine or specific treatment available for these dangerous agents and only supportive measures can be provided for infected individuals. The envelope of Ebola virus consists of a trimer, highly glycosilated glycoprotein that is recognized both by DC-SIGN and DC-SIGNR/L-SIGN. It has been shown *in vitro* that the presence of these molecules can significantly increase the infectivity facilitating entry in cis and in trans, that is, to susceptible neighboring cells (Davis et al., 2008).

11.2 PROBLEMS ASSOCIATED WITH CONVENTIONAL ANTIMICROBIAL THERAPY

Conventional antimicrobial therapy consists of chemotherapeutic agents to treat the infectious diseases by either killing of the microbes or interfering with their growth. With the commercial production of the first antibiotic penicillin in the late 1940s, use of the antibiotics to treat the infectious diseases increased and to-date many new antibiotics have been developed, ranging from the topical antibiotic ointments to intravenously injected antibiotic solutions. These drugs have proven to be effective in eliminating the microbial infections that arise from minor cuts and scrapes to life threatening infections. An antimicrobial drugs act on the microbes by various mechanisms such as inhibiting cell wall synthesis, inhibiting the protein synthesis, inhibiting the nucleic acid synthesis, inhibiting the metabolic pathways, and by interfering with the membrane integrity. Being a life-saving drug for so many decades, antibiotics do suffer from a range of limitations, which include narrow spectrum of antimicrobial activity, problem regarding the safety and tolerability of the antimicrobial agent, antibiotic-mediated enhancement of microbial virulence properties which may also lead to prolongation of host carrier state and may lead to harmful side effect to the host such as toxicity, or any allergic reaction. Inefficient delivery of the drugs has also been one of the major limitations of conventional antimicrobial therapy (De Jong et al., 2005).

Another major limitation of antimicrobial therapy is the development of bacterial resistance to antibiotics. More than 70% of bacteria causing infections are now resistant to at least one of the drugs most commonly used for the treatment. Some organisms are so reluctant that they can only be treated with the experimental and potentially toxic drugs. These microbes use diverse mechanisms to develop the resistance against the antibiotics such as they may alter the drug target, inactivate enzymes, inhibit efflux transport, or develop alternate metabolic pathways for their growth. One of the serious clinical threats in treating the infections via antibiotics emerged with the development of vancomycin- resistant *Enterococcus* (VRE*)* which showed resistance to many commonly used antibiotics. Another example is that of methicillin resistant *Staphylococcus aureus* (MRSA) strains that have caused great concern due to potential spread of antibiotic resistance. More than 40% of *S. aureus* strains collected from the hospitals were methicillin resistant and some of them were resistant to vancomycin. One of the global and medical challenges in the 21st century is the treatment of vancomycin-resistant microbes because vancomycin is the latest generation of antibiotics and assumed most effective for *S. aureus* infection. Problems with multiple drug resistance are also increasing in noscomial Gram-negative bacteria, which have the capability of developing different mechanisms for antibiotic resistance (DeJong et al., 2008).

The chemotherapy of infections caused by bacteria that inhabit intracellularly presents a number of uncommon challenges. Many bacteria have found the way to produce a "silent" infection inside the cells and to avoid from their bactericidal mechanisms. There is an essential need for new and improved approaches for bacterial destruction. Although the therapeutic efficacy of drugs has been well recognized, inefficient delivery could result in insufficient therapeutic index. It is now clear that a nanotechnology-driven approach using nanoparticles to selectively target and destroy pathogenic bacteria can be successfully implemented. Nanotechnology is one approach to overcome challenges of conventional drug-delivery systems based on the development and fabrication of nanostructures. Some challenges associated with the technology are as it relates to drug effectiveness, toxicity, stability, and pharmacokinetics and drug regulatory control. Localized diseases such as infection and inflammation not only have perforated vasculature but also overexpress some epitopes or receptors that can be used as targets. Thus, nanomedicines can also be actively targeted to these locations. Various types of nanoparticulate systems have been tried as potential drug-delivery systems, containing biodegradable polymeric nanoparticles, polymeric micelles, nanocapsules, nanogels, fullerenes, solid lipid nanoparticles (SLN), nanoliposomes, dendrimers, metal nanoparticles and quantum dots. Nanoparticles have been found useful in the development of systemic, oral, pulmonary, transdermal and other administration routes to study drug targeting, the enhancement of drug bioavailability and protection of drug bioactivity and stability (DeLouise et al., 2012).

In recent years, encapsulation of antimicrobial drugs in nanoparticle systems has emerged as an innovative and promising alternative that enhances therapeutic effectiveness and minimizes the undesirable side effects of drugs. The major goals in designing nanoparticles as delivery systems are to control particle size, surface properties, and release of pharmacologically active agents in order to achieve the site-specific action at the therapeutically optimal rate and dose regimen. This chapter focuses on nanoparticle-based drug-delivery systems and clinical applications to treat a variety of bacterial infectious diseases and their potential applications in the field of medicine and biology (Demoy et al., 1997).

11.2.1 NANOTECHNOLOGY-BASED DRUG DELIVERY SYSTEMS

Perfectly, nanoparticulate drug-delivery system should selectively accumulate in the necessary organ or tissue and at the same time, penetrate target cells to deliver the bioactive agent. It has been proposed that, organ or tissue accumulation could be achieved by the passive or antibody-mediated active targeting, while the intracellular delivery could be mediated by specified ligands or by cell-penetrating peptides. The purpose of drug delivery is to carry out sustained (or slow) and/or controlled drug release and therefore to improve efficacy, safety, and/or patient comfort. Thus, the use of drug-delivery systems has been suggested for passive targeting of infected cells of the mononuclear phagocytic system to enhance the therapeutic index of anti-microbials in the intracellular environment, while minimizing the side effects related with the systemic administration of the antibiotic. These systems propose many advantages in drug delivery, mainly focusing on improved safety and efficacy of the drugs, for example, providing targeted delivery of drugs, improving bioavailability, extending drug or gene effect in target tissue, and improving the stability of therapeutic agents against chemical/ enzymatic degradation. The nanoscale size of these delivery systems is the basis for all these advantages. It is therefore assumed that, DDS with enhanced targeting property is highly promising in increasing the efficiency and efficacy of therapy while at the same time minimizing side effects (Dnyanesh et al., 2003).

11.2.2 CHALLENGES IN TREATING INFECTIOUS DISEASES USING NANOTECHNOLOGY

Use of antibiotics began with commercial production of penicillin in the late 1940s and claimed to be a great success until the 1970–1980s when newer and even stronger antibiotics were additionally improved. Resistance to anti-microbial drugs becomes a threatening problem not only in hospitals but also in communities, resulting in fewer effective drugs available to control infections by "old" well-known bacteria (Drbohlavova et al., 2013).

Carrier systems allow antibiotics to be delivered selectively to phago-cytic cells and to increase their cellular penetration in order to treat intra-cellular infections, particularly in the case of antibiotics active against microorganisms that produce this type of infection but that have a low intracellular penetration capacity. Nevertheless, significant challenges remain for implementation of clinically viable therapies in this field. New challenges in the development of nanotechnology-based drug-delivery sys-tems include: the possibility of scale-up processes that bring innovative therapeutic techniques to the market rapidly, and the possibility of obtain-ing multifunctional systems to carry out several biological and therapeutic requirements. Thus, a drug-delivery system should be multifunctional and possess the ability to switch on and switch off specified functions when urgent. Another important requirement is that different properties of the multifunctional drug-delivery systems are harmonized in an optimal fash-ion. Therefore, design, discovery, and delivery of antimicrobial drugs with improved efficacy and avoidance of resistance are extremely requested (Duncan et al., 2003).

11.2.3   ADVANTAGES OF NANOANTIBIOTICS

The use of NPs as delivery vehicles for antimicrobial agents suggests a new and promising model in the design of effective therapeutics against many pathogenic bacteria. Antimicrobial NPs propose several clinical advantages. First, the surface properties of nanoparticles can be changed for targeted drug delivery for, for example, small molecules, proteins, peptides, and nucleic acids loaded nanoparticles are not known by immune system and efficiently targeted to special tissue types. Second, nanocarriers may over-come solubility or stability issues of the drug and minimize drug-induced side effects. Third, using nanotechnology, it may be possible to achieve co-delivery of two or more drugs or therapeutic modality for combination therapy. Fourth, NP-based antimicrobial drug delivery is promising in over-coming resistance to common antibiotics developed by many pathogenic bacteria. Five, administration of antimicrobial agents using NPs can prog-ress therapeutic index, extend drug circulation (i.e., extended half-life), and achieve controlled drug release, increasing the overall pharmacokinetics. Six, the system can be used for several routes of administration including oral, nasal, parenteral, intra-ocular etc. Thus, antimicrobial NPs are of great interest as they provide a number of benefits over free antimicrobial agents (Egan et al., 2004).

11.3 DISCOVERY OF DENDRIMERS AND DENDRITIC POLYMERS: A BRIEF HISTORICAL PERSPECTIVE

One of the most pervasive topologies observed on our planet is perhaps dendritic architecture. Innumerable examples of these patterns may be found in both abiotic systems as well as in the biological world. In biological systems, these dendritic patterns may be found at dimensional length scales measured in meters (trees), millimeters/centimeters (fungi), or microns (neurons) as illustrated in Figure 11.1. The reasons for such extensive mimicry of these dendritic topologies at virtually all dimensional length scales is not entirely clear. However, one might speculate that these are evolutionary architectures that have been optimized over the past several billion years to provide structures manifesting maximum interfaces for optimum energy extraction/distribution, nutrient extraction/distribution, and information storage/retrieval. The first inspiration for synthesizing such molecular level treelike structures evolved from a lifetime hobby enjoyed by one of the authors as a horticulturist/tree grower. Although perhaps first conceptualized by Flory, the first successful laboratory synthesis of such dendritic complexity did not occur until the late 1970s. It required a significant digression from traditional polymerization strategies with realignment to new perspectives. These perspectives utilized major new synthesis concepts that have led to nearly monodispersed synthetic macromolecules. This was the first time in the history of synthetic polymer science that precise abiotic macromolecules could be synthesized without the use of a biological system. The result was a unique core-shell macromolecular architecture, now recognized as *dendrimers*. The concept of repetitive growth with branching was first reported in 1978 by Buhleier et al. (University of Bonn, Germany) who applied it to the construction of low molecular weight amines. This was followed closely by the parallel and independent development of the divergent, macromolecular synthesis of "true dendrimers" in the Tomalia group (Dow Chemical Co.). The first article using the term "dendrimer" and describing in great detail the preparation of poly(amidoamine) (PAMAM) dendrimers was presented in 1984 at the 1st International Polymer Conference, Society of Polymer Science,

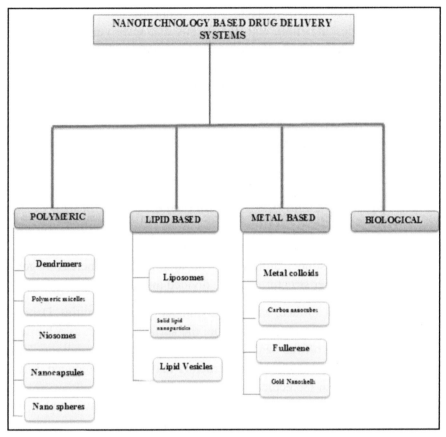

FIGURE 11.1 Different types of nano devices for delivery of antibacterial agents.

Japan (SPSJ). It was then published in 1985, the same year a communication reported the synthesis of arborols (Eijkel et al., 2006).

The divergent methodology based on acrylate monomers was discovered in 1979 and developed in the Dow laboratories during the period of 1979–1985. It did not suffer from the problem of low yields, purity, or purification encountered by Voegtle in his "cascade" synthesis and afforded the first family of well-characterized dendrimers. PAMAM dendrimers with molecular weights ranging from several hundred to over 1 million daltons (i.e., generations 1–13) were prepared in high yields. This original methodology was so successful that today it still constitutes the preferred commercial route to the trademarked Starburst dendrimer family. It is both remarkable and surprising to find that many of these Class IV dendritic structure-controlled macro- molecules (i.e., dendrimers, dendrigrafts, etc.) possess topologies, function, and dimensions that scale very closely to a wide variety of important biological polymers and assemblies. Dendritic polymers, more specifically dendrimers, are expected to play a key role as an "enabling technology" in this challenge during the next century. Justas the first three traditional, synthetic polymer architectures have so success-fully fulfilled the critical material and functional needs for society during the past half-century, it is appropriate to be optimistic about such a role for the "dendritic state" (Fasano et al., 1998).

11.3.1 PHYSICOCHEMICAL PROPERTIES OF DENDRIMERS

As the dendrimer grows, the different compartments of the dendritic struc-ture begin to show distinct features which are amplified with increasing gen-eration. The dendrimer structure may be divided into three parts:

i.   A multivalent surface, with a high number of functionalities. Depen-dent on the dendrimer generation, the surface may act as a border-line shielding off the dendrimer interior from the surroundings. This increasingly "closed" surface structure may result in reduced diffu-sion of solvent molecules into the dendrimer interior.

ii. The outer shell, which have a well-defined microenvironment, to some extent shielded from the surroundings by the dendrimer sur-face. The very high number of functionalities located on the sur-face and the outer shell are well-suited for host–guest interactions and catalysis where the close proximity of the functional motifs is important.

iii. The core, which as the dendrimer generation increases, gets increas-ingly shielded off from the surroundings by the dendritic wedges. The interior of the dendrimer creates a microenvironment which may have very different proper-ties compared to the surroundings. For example as described elsewhere, water-soluble dendrimers with an apolar interior have been constructed to carry hydrophobic drugs in the bloodstream.

11.3.2   TYPES OF DENDRIMERS

11.3.2.1   PAMAM Dendrimer

Poly (amidoamine) dendrimers (PAMAM) are synthesized by the divergent starting ammonia method from ethylene diamine initiator core. PAMAM dendrimers are commercially available, usually as methanol solutions. Starburst dendrimers is applied as a trademark name for a sub-class of PAMAM dendrimers based on a tris-aminoethyleneimine core. The name refers to the star-like pattern observed when looking at the structure of the high-generation dendrimers of this type in two dimensions (Feynman et al., 1960).

11.3.2.2   PAMAMOS Dendrimer

Radially layered poly (amidoamine-organosilicon) dendrimers (PAMAMOS) are inverted unimolecular micelles that consist of hydrophilic, nucleophilic polyamidoamine (PAMAM) interiors and hydrophobic organosilicon (OS) exteriors. These dendrimers are exceptionally useful precursors for the preparation of honeycomb-like networks with nanoscopic PAMAM and OS domains. These are silicone containing first commercial dendrimers (Fix et al., 1996).

11.3.2.3   PPI Dendrimer

PPI-dendrimers stand for "Poly (Propylene Imine)" describing the propylamine spacer moieties in the oldest known dendrimer type developed initially by Vogtle. These dendrimers are generally poly-alkyl amine, having primary amines as end groups; the dendrimer interior consists of numerous of tertiary tris-propylene amines. PPI dendrimers are commercially available up to G5 and have found widespread applications in the field of material science and biology. As an alternative name to PPI, POPAM is sometimes used to describe this class of dendrimers. POPAM stands for Poly (Propylene Amine), which closely resembles PPI (Galindo-Rodriguez et al., 2005).

11.3.2.4   Tecto-Dendrimer

These are composed of a core dendrimer, surrounded by dendrimers of several steps to perform a function necessary for a smart therapeutic nanodevice. Different compounds perform varied functions ranging from diseased cell recognition, diagnosis of disease state, drug delivery, reporting outcomes of therapy (Gao et al., 2008).

11.3.2.5   Multilingual Dendrimers

In these dendrimers, the surface contains multiple copies of a particular functional group (Gaumet et al., 2008).

11.3.2.6   Chiral Dendrimers

The chirality in these dendrimers is based upon the construction of a constitutionally different but chemically similar branch to chiral core (Ghosh et al., 2008).

11.3.2.7   Hybrid Dendrimers Linear Polymers

These are hybrids (block or graft polymers) of dendritic and linear polymers and having properties of both (Gibaud et al., 1996).

11.3.2.8   Amphiphilic Dendrimers

They are built with two segregated chains of which one half is electron donating and other half is electron withdrawing (Gonzalez-Mira et al., 2011).

11.3.2.9   Micellar Dendrimers

These are unicellular micelles of water soluble and hyper branched polyphenylenes (Gowthamarajan et al., 2003).

11.3.2.10 Multiple Antigen Peptide Dendrimers

It is a Dendron like molecular construct upon a polylysine skeleton. Lysine with its alkyl amino side chain serves as a good for the introduction of monomer numerous of branching points. This dendrimer was type of introduce by J.P. Tam in 1988, predominatly found its biological applications, for example, and diagnostic vaccine research (Gradishar et al., 2005).

11.3.2.11 Frechet-Type Dendrimers

It is a more of dendrimer recent type developed by Hawker and Frechet based on poly benzyl ether hyper branched skeleton. These dendrimers usually have carboxylic acid groups as surface groups, serving as a good anchoring point for further surface functionlization and as polar surface groups increase solubility of hydrophobic to this dendrimer type in polar solvents or aqueous media (Grayson et al., 2004).

11.3.3 APPLICATIONS OF DENDRIMERS

11.3.3.1 Solubility Enhancement of Poorly Soluble Drugs

Erythromycin (EM) and tobramycin (TOB) are well-known and widely used antibiotics, belonging to different therapeutic groups. Moreover, they possess different solubility: EM is slightly soluble and TOB is freely soluble in water. PAMAM dendrimers enhanced the pharmacological activity of antifungal drugs by increasing their solubility. PAMAM dendrimers significantly increased the aqueous solubility of EM, despite the increase in the solubility, there was only slight influence on the antibacterial activity of EM and also found that there was no influence of PAMAM on the antibacterial activity of hydrophilic TOB (Hadzijusufovic et al., 2010).

11.3.3.2 Increases Phagocytosis

Pneumococcal virulence factors common to all serotypes, such as choline-binding proteins (CBPs), are promising therapeutic targets in pneumococcal infections. Pneumococcal cultures were exposed to dendrimers containing choline end groups or amino groups as controls, either from the beginning of bacterial growth or at the late exponential phase. Inhibiting CBPs by micro molar concentrations of a choline dendrimer caused the formation of long pneumococcal chains that were readily phagocytosed by microglia. Long bacteria-dendrimer co-incubation resulted in a higher bacterial uptake than short co-incubation. Multivalent dendrimers containing choline end groups are promising antimicrobial agents for the management of pneumococcal diseases (Hamman et al., 2005).

11.3.3.3   Permeability of Bacterial Cell Membranes

To confirm that bacterial cell membranes were disrupted in the presence of PPI dendrimers and AMX, they are treated with propidium iodide (PI). This dye is commonly used for discriminating dead from living cells because it can enter only damaged cells with permeable membrane (Hanaire et al., 2008). PPI dendrimers penetrate through the cell wall more easily than glycodendrimers. C. Bacterial membranes have lower permeabilities of the larger dendrimer analogs. Destroying the cell membranes of microorganisms directly or disrupting multivalent binding interactions between microorganisms and host cells, are the primary mechanisms of antimicrobial action by dendrimers. Cationic dendrimers act via initial electrostatic attraction to the negatively charged bacterium followed by membrane and peptidoglycan disruption. Amoxicillin inhibits bacterial cell wall synthesis; thus, it facilitates contact of the dendrimer with the cell membrane (Illum et al., 2007).

11.3.3.4   Dendrimeric Anti-Microbial Peptide

Antimicrobial drug resistance is a major human health threat. To tackle this problem, Peptide-based dendrimers can be designed to have higher potency than natural antimicrobial peptides with similar chemical structure but varying potency in terms of minimum inhibitory concentration were designed and at the same time they can evade the bacterial defense system (Khafagy et al., 2007).

Therapeutic macromolecules including dendrimers-based drug-delivery systems exploit the pathophysiological patterns of solid tumors, particularly their leaky vasculature, to preferentially extravasate and accumulate in tumor tissue in a process known as the enhanced permeability and retention

(EPR) effect. The amount of dendrimers-based drug-delivery systems that accumulates in tumor tissue is influenced by their size, molecular weight, and surface charge, which affect their residence time in the systemic circulation, transport across the endothelial barrier, and nonspecific recognition and uptake by RES. El-Sayed et al. studied the effect of size, molecular weight, and surface charge on the permeability of fluorescently labeled PAMAM-NH_2 (G0–G4) dendrimers across epithelial and endothelial barriers. The increase in dendrimers size/molecular weight results in a corresponding exponential increase in their extravasation time (τ) across the microvascular endothelium of the cremaster muscle preparation of Syrian hamsters (Kim et al., 2012).

Bio-distribution of Gadolinium-functionalized G2-NH_2 to G10-NH_2 conjugates which showed the influence of dendrimers size/hydrodynamic volume on their transport across the microvascular endothelium *in vivo*. Cationic dendrimers show high nonspecific uptake by the RES particularly in the liver and lungs, which reduces their accumulation in tumor tissue. Upon comparing the biodistribution of cationic G5-NH_2 dendrimers and their neutral counterparts prepared by partial or full acetylation of the surface amine groups in nude mice bearing melanoma and prostate tumors, it showed that both dendrimers displayed a similar distribution profile to all major organs within 1 h after dendrimers injection with particularly high accumulation in the lungs, kidneys, and liver. While the cationic and neutral dendrimers displayed similar biodistribution profiles, cationic dendrimers showed higher net accumulation in each organ due to their favorable electrostatic interaction with the negatively charged epithelial and endothelial cell surface. It is interesting to note that all polylysines, anionic PAMAM-COOH dendrimers, and polyester dendrimers exhibit high distribution to the liver and quick elimination into the urine (Kompella et al., 2001). This biodistribution profile can be attributed to the dendrimer's small hydrodynamic volumes, which results in less than 5% of the initial dose remaining in the systemic circulation 24 h after administration. Attachment of PEG arms to the dendrimer surface increases their size and molecular weight, thus reducing their systemic clearance and improving their biocompatibility. Specifically, attachment of PEG chains with molecular weight up to 20 kDa to the dendrimer's surface groups increases their plasma half-life to 50 h for G3 polyester dendrimers, 75.4 h for polylysine dendrimers, and 100 h for triazine dendrimers. Bhadra et al. showed that the attachment of PEG (5 kDa) chains to 25% of the surface groups of G4-NH_2 dendrimers results in a 3-fold reduction in their hemolytic activity compared to the parent dendrimers (Labouta et al., 2011). These studies clearly indicate the positive effect of surface PEGylation of PAMAM dendrimers by enhancing their plasma residence time and reducing their nonspecific toxicity (Lee et al., 2002).

Cationic antimicrobial peptides constitute an important component of the innate immunity against microbial infections. Recently, there is renewed interest in developing novel approaches for designing peptide-based antibiotics manifested by killing mechanisms that are less likely than conventional antibiotics to develop multidrug resistance. Design elements desirable for therapeutics include activity under physiological conditions (100± 150 mM or high-salt conditions), low toxicity and protcolytic stability. They have designed antimicrobial peptides with unusual structural architectures using rigid scaffoldings such as cyclic peptides highly constrained with a cysteine knot motif on two or three b strands to cluster hydrophobic and charge regions that produce amphipathic structures important for antimicrobial activity. Furthermore, these constraints confer metabolic stability, and impart membranolytic selectivity that minimizes toxicity. Another approach for designing antimicrobial peptides is based on their mechanisms of action. Pathogen-associated motifs include various microbial cell-wall components such as lipopolysaccharide (LPS), peptidoglycans, teichoic acids, mannans, N-formyl peptides, and lipidated peptides. Some well-studied motif-recognizing proteins include LPS-binding protein, soluble and membrane-anchored CD14 and Toll-like LPS receptors as well as mannose-binding protein and the receptors for mannans and manoproteins. Cationic antimicrobial peptides may have also evolved to recognize PAMPs on microbial surfaces. They often possess a broad spectrum of antimicrobial activities against bacteria, fungi or viruses through mechanisms that generally involve the disruption of microbial envelopes (Lee et al., 2010).

11.3.3.5   Membrane Disruption Dendrimer Peptide

Many antimicrobial peptides (AMPs) act by disrupting microbial membranes. These include linear a-helical amphipathic peptides, cyclic peptides and peptoids, foldamers, various amide oligomers, and multivalent lysine dendrimers appended with linear AMPs of various lengths. In all of these cases the multiple positive charges necessary for membrane disruption are brought about by the side chains of basic amino acids. The discovery

of antimicrobial peptide dendrimers such as H1 and bH1 in which positive charges are provided by the multiple amino termini at the dendrimer
periphery. This new type of AMP acts as membrane disrupting agent and
shows remarkably low hemolytic activity. Several of the dendrimers also
showed good activities against *E. coli* and *P. aeruginosa*, in particular the
Dap-branched analogs bH1 and bH2, which were also more than the lysine
branched dendrimers. The peptide dendrimers were generally less active on
gram-negative bacteria, which together with the activity of the dendrimer
stereoisomers and the requirement for a combination of cationic charges and
hydrophobic residues pointed toward membrane disruption as a possible
mechanism of action (Lehr et al., 1994).

11.3.3.6   Nano Conjugates for Antibacterial Therapy

The increased altitudes of control possible over the dendrimers architectural
design are their shape, size, surface functionality and their branching length/
density. The bioactive agents may be condensed into the interior of the dendrimers or physically adsorbed/chemically attached onto the dendrimer surface. It has been shown that Modified dendrimers act as nanodrugs against
bacteria, viruses and tumors. PAMAM dendrimers are potential carriers for
drug delivery due to their unique structure. Penicillin V was used as a model
carboxylic group containing drug to conjugate with full- and half-generation. A drug carrying a carboxylic group (e.g., penicillin V) was coupled to
star polymer via amide and ester bonds. A single-strain bacterium, *Staphylococcus aureus*, was grown up for penicillin-conjugated PEG-PAMAM
star polymer activity test. The bioavailability of modified penicillin after the
ester bond was cleaved (Lemoine et al., 1996).

11.3.3.7   Dual Acting Antibacterial Agents

Besides acting as antimicrobial compounds, dendrimers can be considered
as agents that improve the therapeutic effectiveness of existing antibiotics. A
new approach using amoxicillin (AMX) against reference strains of common
Gram-negative pathogens, alone and in combination with poly (propylene
imine) (PPI) dendrimers, or derivatives. The PPI dendrimers significantly
enhance the antibacterial effect of amoxicillin alone, allowing antibiotic

doses to be reduced. It is important to reduce doses of amoxicillin because it leads to the development of bacterial resistance and environmental pollution (Lenaerts et al., 1984).

11.3.3.8   Against Resistant Strain

PEGylated poly (propylene imine) dendritic architecture was loaded with Ciprofloxacin and targeted to the resistants produce strains of *Staphylococcus aureus* and *Cryptococcus pneumoniae*. The Ciprofloxacin loaded dendrimer has significant antibacterial activity than the plain PPI dendrimer, but standard drug was not shown zone of inhibition upon both microorganisms. The antibacterial activity of system is also relatively safer and hold potential to deliver some other drugs also (Lenhard et al., 2001).

11.3.3.9   Antibacterial Activity

Owing to antibacterial properties of dendrimers, they can be used as alternative water and waste water disinfection with the minimum adverse side effects. It can be used to evaluate the antibacterial activity of predominant bacteria in drinking water resources. Therefore, it is possible to use these nano dendrimers as a safe and effective material for water disinfection in the future. Antimicrobial is a substance that kills or inhibits the growth of microorganisms such as bacteria, fungi, or protozoans. Antimicrobial drugs either kill microbes (microbiocidal) or prevent the growth of microbes (microbiostatic). Disinfectants are antimicrobial substances used on non-living objects or outside the body. Antibacterial and antimicrobial are two similar concepts and sometimes they are used interchangeably. Many antibacterial products for cleaning and hand washing are sold today. PEGylated poly(propylene imine) dendritic architecture was synthesized and loaded with Levofloxacin and targeted to the resistant producing strains of *Klebsiella pneumoniae* and *E.coli*. The Levofloxacin loaded dendrimer has significant antibacterial activity than the plain drug and control. Antibacterial activity of synthesized system is also relatively safer and hold potential to deliver some other drugs also (Lowman et al., 1999).

11.3.3.10   Work of Antibacterial Agents

a.   Interference with cell wall synthesis:
   i.   Lactams: penicillins, cephalosporins, carbapenems, monobactams
   ii.   Glycopeptides: vancomycin, teicoplanin
b.   Protein synthesis inhibition
   i.   Bind to 50S ribosomal subunit: macrolides, chloramphenicol, clindamycin, quinupristindalfopristin, linezolid
   ii.   Bind to 30S ribosomal subunit: aminoglycosides, tetracyclines
   iii.   Bind to bacterial isoleucyl-tRNA synthetase: mupirocin
c.   Interference with nucleic acid synthesis
   i.   Inhibit DNA synthesis: fluoroquinolones
   ii.   Inhibit RNA synthesis: rifampin
d.   Inhibition of metabolic pathway: sulfonamides, folic acid analogs
e.   Disruption of bacterial membrane structure: polymyxins, daptomycin

High molecular surface functional group concentration of dendrimers can dominate antibacterial properties to the interacting molecule. If the end or surface groups of dendrimers are functionalized with biologically active antimicrobial groups, might expect an increase in antimicrobial activity of the dendrimers. Biocides immobilized on dendrimers can be more effective if the target sites are cell walls and/ or membranes. It has been shown that small quaternary ammonium compounds exert their antimicrobial action by disrupting and disintegrating the cell membrane. Dendrimer biocides may be very beneficial in terms of activity, localization in specific organs, reduced toxicity, and increased duration of action. As the bacteria are negatively charged and dendrimer biocides have high positive charge density, electrostatic interactions bring them into contact with each other. High concentrations of dendrimer biocides can lead to complete disintegration of the bacterial membrane causing to a bactericidal effect (Maas et al., 2007).

By adding water soluble functional end groups to dendrimers, water soluble dendrimers are obtained. When these dendrimers interacted with bacteriostatic weak water soluble or insoluble antibiotics, the antibacterial properties antibiotics can be altered, especially improved. Microbiological studies of the quinolones (nadifloxacin and prulifloxacin) showed that strong antimicrobial activities of nadifloxacin and prolifloxacin were still significantly increased in the presence of PAMAM dendrimers and also their water solubility increased. Poly (amido amine) PAMAM (e.g., the generation 3

(G3) dendrimers are the most extensively studied dendrimers. PAMAM dendrimers with a wide variety of functional groups at the periphery are commercially available. Some of the dendrimers having terminal amino groups are shown as they are having low toxicity to eukaryotic cells. Modification of the amino groups of the PAMAM dendrimers with poly (ethylene glycol) (PEG) or lauroly chains further improves the biocompatibility (Maeda et al., 2001).

There is a significant global need for new antibacterial and alternative mechanisms of action given the rise in resistance among bacteria. Of the various known antibacterial agent classes, amphiphilic compounds act through perturbation and disruption of the prokaryotic membrane. It has been hypothesized that amphiphilic anionic dendrimers may exhibit antibacterial activity with minimal eukaryotic cell cytotoxicity, since dendrimers with terminal anionic charges are generally noncytotoxic and have low toxicity in zebrafish whole animal development studies. On the other hand, cationic dendrimers, some of which have antibacterial properties if the positive charge is properly shielded, have repeatedly shown cytotoxicity against a variety of eukaryotic cell lines. Metal containing dendritic nanoparticles called as the Metallodendrimers. They are incorporated with metal atoms. Silver containing compounds and materials have been routinely used to prevent attack of a broad spectrum of microorganisms on prostheses, catheters, vascular grafts, human skin, also used in medicine to reduce infection in burnt treatment, arthroplasty. However, they exhibit low toxicity to mammalian cells. The antibacterial activity of silver nanoparticles increases with the increase in concentration of the active agent. Dispersion of organic/ inorganic hybrid materials could be utilized to form regular thin film coatings with antibacterial effects by using dendritic-polymer templates. The antibacterial activity of the coating films based on the hyper branched core/shell type hybrids and closely associated with the silver ions release of the films. Nano-scaled silver may (1) release silver ions and generate reactive oxygen species (ROS); (2) interact with membrane proteins affecting their correct function; (3) accumulate in the cell membrane affecting membrane permeability; and (4) enter into the cell where it can generate ROS, release silver ions, and affect DNA. Generated ROS may also affect DNA, cell membrane, and membrane proteins, and silver ion release will likely affect DNA and membrane proteins (Mansour et al., 2009).

To inhibit the binding of the B_5 subunit to cell surface GM1 Schengrund group has developed oligosaccharide functionalized dendrimers. The dendritic cores used were tetra (first generation) (1) and octa (propylene imine) (second generation) (2) dendrimers and the first generation of Startburst™ (PAMAM) (3). Pieters et al. have prepared dendrimers with a significant rigid backbone using 3,5-bis(2-aminoethoxy)benzoic acid as repeating unit . Dendrimers of first G1 (4), second G2 (5), and third G3 (6) generations presenting 2, 4, and 8 lactose as sugars. The affinity of these compounds for cholera toxin B subunit was measured using fluorescence spectroscopy. In this study was also pointed out the importance of the size and shape of the spacer used to attach the carbohydrate moiety to the dendritic core. Bernardi and Pieters groups have prepared a dendrimer using the same dendritic core described above functionalized with a GM1 mimic. This system was tested for CTB binding using two techniques, SPR analysis and ELISA assays. ELISA was performed with the aim to evaluate the activity of this octavalent compound. The ELISA wells were coated with the ganglioside (Mathias et al., 2010).

One of the more remarkable example in the rational design of multivalent systems for the inhibition of SLTs was described by named STAR-FISH designed by Bundle et al. Nishikawa et al. have studied the inhibition of infection by shiga toxin-producing *E. coli* 0157:H7 Using carbosilane dendrimers functionalized with the same Gb3 trisaccharides. These dendritic structures named SUPER TWIG are constituted by a core of silicon-carbon bonds that are biologically inert. Polycationic amine-terminated poly(amidoamine) (PAMAM) dendrimers as endotoxin sponge for the therapy of Gram-negative bacterial sepsis. By using BODIPY cadaverine as displacement probe the affinity of these dendrimers for LPS was evaluated using a high-throughput fluorescence displacement method. *In vitro* assays of nitric oxide release in LPS-stimulated murine macrophage were used to analyze the dendrimer activity. To neutralize endotoxin samphiphilic character of these dendrimers partially functionalized with alkyl groups. Lindhorst et al. have developed carbohydrate multivalent systems to inhibit the adhesion of *E. coli* mediated by 1 fimbriae and were made as potential inhibitors based on multivalent systems containing mannopyranosides attached to the scaffold through a thiourea linkage were prepared and their activities were tested. The studies revealed important issues concerning the importance of the mannose orientation on these types of multivalent systems, affinities of these were strongly dependent on the distance between mannose residues; a large distance (> 20 nm) (Mehnert et al., 2012).

The developed small multivalent systems, glycodendrimers and glyco-polymers to inhibit the adhesion of type 1 fimbriateduropathogenic *E. coli*. The activity of these was tested using zone of inhibition growth with bacteria on a solid agar medium. This dendronized system, when it was doped with silver, a clear inhibition was observed. Urbanczyk-Lipkowska et al. have prepared two types of dendrimers based into amino acids (Lysine) to inhibit infection by *E. coli*. Growth was evaluated by Minimal inhibitory concentrations (MIC). Cooper et al. have developed quaternary ammonium functionalized poly(propylene imine) dendrimers as antimicrobials .The biocide activity was evaluated using a bioluminescence method. Magnusson et al. in 1997 prepared a small tetravalent galabiose system to inhibit the hemagglutination by *Streptococcus suis* at nanomolar concentration. Pieters et al. described the preparation of galabiose dendrimers and their inhibition activities against *Streptococcus suis*. To study the inhibition of hemagglutination of human erythrocytes induced by two subtypes of *Streptococcus suis* (P_N and P_O) was performed. The MIC (minimal inhibitory concentration) required for complete inhibition of the agglutination process were measured. The dendrimers described by Cooper et al. and Ur-banczyk-Lipkowska et al. for *E. coli* were also tested against a Gram-positive bacteria *Staphylococcus aureus* NCTC 4163.Davis et al., have developed a newtype of multivalent systems named glycodendriproteins consisting in the functionalization of proteins with glycodendrons and are mimicking glycoproteins avoided the problem of glycoforms. They were tested as potential inhibitors of co-aggregation of *A. naeslundii* and *S. oralis*. Urbanczyk-Lipkowska et al. where dendrimers are used as anti-infective agents in a *C. albicans* infection process. These authors have developed a low molecular mass lysine dendrimer with antimicrobial activity functionalized at the surface by arginine residues. Prusiner et al. have demonstrated that highly branched polyamidoamine PAMAM dendrimers were able to eliminate PrP^{Sc} in a very efficient way. . It seems that the presence of dendrimers make fibrils sensitive to protease Kdegradation. Very recently, Cladera et al. have studied the aggregation of the Alzheimer amyloid peptide A 1-28 and human prion protein PrP185-208 in the presence of PAMAM dendrimers. Higher generations of PAMAM G3-G5 led to a smaller amount of fibrils formed. 1 M of PAMAM G5 was enough to inhibit the fibril formation. Heegaard et al. have used a guanidinium modified dendrimer based on the second generation of poly (propyl-ene imine) (PPI) to destabilize the fibril formation of a peptide fragment of the PrP. Most of the work concerning dendrimers is oriented to developing these microbicides of topical use. HIV inhibition activities and cytotoxicities of these dendrimers were evaluated *in vitro* using CEM-SS cells (human T4-lymphoblastoid cell line) as model. One of the promising compounds developed against HSV is a sulfated polylysine dendrimer named SPL2999. An evolution of this compound is the dendrimer. This dendrimer has been tested as a microbicide candidate against genital herpes in mouse and guinea pig models. A new study concerning the antiviral efficacy, mechanism of action, and toxicity. SPL7013 inhibited virus internalization of both HSV-1 and HSV-2. Also, SPL7013 showed post-exposure activity on HSV infection indicating a therapeutic activity with a pH-independent activity. Again, SPL7013 has been demonstrated as a promising microbicide candidate in STIs (Miyazaki et al., 2007).

Whitesides et al. described in 1999 a pioneer work in the preparation of multivalent systems based on polyacrylamide functionalized with sialoside groups. This glycopolymer strongly inhibited the agglutination of erythrocytes by influenza virus. Baker et al. developed systems based on PAMAM as dendritic core functionalized with sialic acid as described in the original papers and in a recent review. These dendritic structures could inhibit the infection process in a DC-SIGN-dependent manner. Our strategy was oriented to inhibit the entry of Ebola virus blocking the DC-SIGN lectin, a receptor that was described as one of the potential gate of entrance for this virus. These types of glycostructures were able to block the DC-SIGN receptor at the cell surface inhibiting the entrance of the pathogen and therefore, they could be used as microbicides. Most of the structures used as anti-infective agents are based on poly(amido amine) (PAMAM) and poly(propylene imine) (PPI) dendrimers. These are two well-known structures and both of them commercially available up to generation 5 at least. Chemical manipulation of the dendritic surface produces a large number of new structures with interesting antimicrobial. Changes of the functional groups at the surface or even the chemical structure of the core can be manipulated to achieve low toxicities. The dendrimers and dendritic polymers have to play a key role in the nanoscience field. They are very flexible structures to be easily manipulated and modulated by chemists to achieve new properties and pursue future applications in biomedical sciences. In fact, the knowledge provided by all the research summarized above, establishes the bases for new accomplishments in the area of anti-infective agents with a broad spectrum of applications. Having a compound in clinical Phase I means a real step toward achieving the first milestone. We hope that in the next few decades, we could have derivatized dendritic compounds to be used as pharmaceuticals (Moghimi et al., 2001).

11.3.3.11   Dendrimers for Antimicrobial Drug Delivery

Dendrimers are defined as highly ordered and regularly branched globular macromolecules produced by stepwise iterative approaches. The structure of dendrimers consists of three distinct architectural regions: a focal moiety or a core, layers of branched repeat units emerging from the core, and functional end groups on the outer layer of repeat units. In 1978, the first iterative cascade synthetic procedure for branched amines was discovered by Vögtle et al. A few years later, highly branched l-lysine-based dendrimers were patented. In 1984, Tomalia et al. reported the synthesis and characterization of the first family of polyamidoamine (PAMAM) dendrimers, which has become one of the most popular dendrimers since then (Mugumu et al., 2006).

Two synthetic approaches, divergent and convergent approaches, have been developed to synthesize dendritic systems for delivering various types of drugs. The divergent approach initiates the synthesis from a core and emanates outward through a repetition of coupling and activation steps. During the first coupling reaction, the peripheral functional groups of the core react with the complementary reactive groups to form new latent branch points at the coupling sites and increase the number of peripheral functional groups. These latent functional groups are then activated to couple with additional monomers. The activation of the la-tent functional groups can be achieved by removal of protecting groups, coupling with secondary molecules, or reactive functionalization. Large excess of reagents is required to drive the activation step to completion. The final resulted dendrimer products can be separated from the excess reagents by distillation, precipitation or ultrafiltration. Although the divergent approach is ideal for large-scale production, incomplete functionalization or side reactions can occur when the number of generation increases. These flawed dendrimers are usually difficult to be separated from the final products because of structural similarity. In contrast, the convergent approach initiates the synthesis from the periphery and progresses inward. This approach starts with coupling end groups to each branch of the monomer, followed by the activation of a single functional group located at the focal point of the first wedge-shape dendritic fragment or dendron. Higher generation dendron is synthesized by the coupling of the activated dendron to an additional monomer. After repetition of coupling and activation step, a globular dendrimer is formed by attaching a number of dendrons to a polyfunctional core. Dendrimers thus synthesized can be effectively purified. However, synthesis of large dendrimers above the sixth generation is difficult (Mukhopadhyay et al., 1982).

Dendrimers possess several unique properties that make them a good nanoparticle platform for antimicrobial drug delivery. The highly-branched nature of dendrimers provides enormous surface area to size ratio and allows great reactivity with microorganisms *in vivo*. In addition, both hydrophobic and hydrophilic agents can be loaded into dendrimers. Hydrophobic drugs can be loaded inside the cavity in the hydrophobic core, and hydrophilic drugs can be attached to the multivalent surfaces of dendrimers through covalent conjugation or electrostatic interaction. Moreover, by using antimicrobial drugs as a building block, the synthesized dendrimers themselves can become a potent antimicrobial. Dendrimer biocides are such example that contains qua-ternary ammonium salts as functional end groups. Quaternary ammonium compounds (QACs) are antimicrobial agents that disrupt bacterial membranes. Dendrimer biocides have displayed greater antimicrobial activity against target bacteria than small drug molecules because of a high density of active antimicrobials present on the dendrimer surfaces. The polycationic structure of dendrimer biocides facilitates the initial electrostatic adsorption to negatively charged bacteria. The absorption then increases membrane permeability and allows more dendrimers for entering the bacteria, leading to leakage of potassium ions and eventually complete disintegration of the bacterial membrane .PAMAM is one of the most studied dendrimers for antimicrobial delivery because of its higher density of functional groups, which make the dendrimer more hydrophilic and more readily reactive to antimicrobial conjugation. Silver salts loaded PAMAM dendrimers have demonstrated significant antimicrobial activity against *Staphylococcusaureus*, *Pseudomonas aeruginosa*, and *Escherichia coli*. Incorporation of antibacterial agents such as sulfamethoxazole (SMZ) into the ethylenediamine (EDA) core of PAMAM dendrimers has significantly improved the drug's aqueous solubility and antibacterial activity against *E. coli*. Many other antimicrobial drugs have been successfully loaded into dendrimer nanoparticles and have shown improved solubility and therapeutic efficacy (Muller et al., 1993).

Despite the great progresses on nanoparticle-based antimicrobial drug delivery, here we call attention to the need to unite the shared interest between nanoengineers and microbiologists in developing novel nanotechnology targeting a few major unmet challenges of antimicrobial drug delivery. First, acquired microbial drug resistance remains a major challenge for infection treatment. One possible approach is to incorporate more than one antimicrobial drug to a single nanoparticle and then concurrently deliver the drugs to the same microbes. Combinatorial drug therapy is expected to have higher potency as multiple drugs can achieve synergistic effects and overwhelm microbial defense mechanisms. Secondly, premature drug release from the antimicrobial-loaded nanoparticles remains another major challenge, especially for treating systemic and intracellular infections (Muller et al., 2000).

To minimize drug loss before the nanoparticle reach the infectious sites, an infection microenvironment-sensitive drug release nanoparticles can be developed. That is, negligible amount of antimicrobial drugs will be released when the nanoparticles circulate in the blood stream or encounter healthy tissues whereas triggered rapid drug release will occur after the nanoparticles get to the infectious cells or tissues. Potential drug-release triggers include pH value, enzyme and other unique characters of the infection microenvironment (Muller et al., 2006).

Lastly, few of the current nanoparticle-based antimicrobial drug-delivery systems can distinguish microbes or infectious cells from healthy cells due to the lack of the specific targeting ability, although targeted drug delivery has been extensively studied for other disease treatment such as cancers and cardiovascular diseases. It would be beneficial for infection treatment if antimicrobial nanoparticles could be modified with microbe antigen- or infectious cell antigen-specific ligands including antibodies, antibody fragments, aptamers and peptides (Mundargi et al., 2007).

11.4   CONCLUSION

Around the world in many healthcare facilities, bacterial pathogens that express multiple resistance mechanisms are becoming the norm, increasing both human morbidity and financial costs and complicating treatment. To eliminate most intracellular bacteria such as *Brucella* or *Mycobaterium* till now, no antibiotic therapy has been reported. To reduce the disease relapses down to 5-15% still more, a prolonged exposure to combined antibiotics is required. Keeping view in this, drug-delivery scientists are searching for the ideal nanovehicle for the ideal nanodrug-delivery system; one that would dramatically improve in the drug absorption, reduce drug dosage so that the patient can take a smaller dose, and yet have the same benefit, deliver the drug to the right place in the living system, limit or eliminate side effects and increase the local concentration of the drug at the favorite site. Compared with other polymeric particles, colloidal carriers mainly nanoparticles, have appeared more recently as attractive carriers for the delivery of drugs to infected cells. Synthetic biodegradable and biocompatible polymers have been shown to be effective for encapsulating a great variety of antibiotics. Phagocytosis is enhanced powerfully by these polymeric particles and are suitable for antibacterial agents for intracellular delivery. There is no doubt that nanoparticle-based drug-delivery systems will continue to improve treatment to bacterial infections due to the incessant attempts in this field.

KEYWORDS

- **antibacterial therapy**
- **dendrimers**
- **phagocytosis enhancement**
- **resistance strain targeting**
- **solubility enhancement**

REFERENCES

Aji Alex, M. R., Chacko, A. J., Jose, S., & Souto, E. B., (2011). Lopinavir loaded solid lipid nanoparticles (SLN) for intestinal lymphatic targeting. *Eur. J. Pharm. Sci., 42*, 11–18.

Amidi, M., Romeijn, S. G., Borchard, G., Junginger, H. E., Hennink, W. E., Jiskoot, W. 2006. Preparation and characterization of protein-loaded N-trimethyl chitosan nanoparticles as nasal delivery system. *J. Control. Rel., 111*, 107–116.

Arsiwala, A. M., Raval, A. J., & Patravale, V. B., (2013). Nanocoatings on implantable medical devices. *Pharm. Pat. Anal., 2*, 499–512.

Awasthi, V. D., Garcia, D., Goins, B. A., & Phillips, W. T., (2003). Circulation and biodistribution profiles of long-circulating peg-liposomes of various sizes in rabbits. *Int. J. Pharm.*, *253*, 121–132.

Baba, K., Pudavar, H. E., Roy, I., Ohulchanskyy, T. Y., Chen, Y., Pandey, R. K., & Prasad, P. N., (2007). New method for delivering a hydrophobic drug for photodynamic therapy using pure nanocrystal form of the drug. *Mol. Pharm.*, *4*(2), 289–297.

Bai, J. P., Chang, L. L., & Guo, J. H., (1995). Targeting of peptide and protein drugs to specific sites in the oral route. *Crit. Rev. Ther. Drug Carr. Syst.*, *12*, 339–371.

Bareiss, B., Ghorbani, M., Fengfu, Li., Blake, J. A., Scaiano, J. C., Jin Z., Chao D., Merrett, K., Harden, J. L., Diaz-Mitoma, F., & Griffith, M., (2010). Controlled release of acyclovir through bioengineered corneal implants with silica nanoparticle carriers. *Open Tissue Eng. Regen. Med. J.*, *3*, 10–17.

Bernfild, M., Gotte, M., Park, P. W., Reizes, O., Fitzgerald, M. I., Lincecum, J., & Zako, M., (1999). Functions of cell surface heparan sulphate proteoglycans. *Annu. Rev. Biochem.*, *68*, 729–777.

Bolzinger, M. A., Briançon, S., Pelletier, J., & Chevalier, Y., (2012). Penetration of drugs through skin, a complex rate-controlling membrane. *Curr. Opin. Colloid. Interface Sci.*, *17*, 156–165.

Brannon-Peppase, L., & Blanchette, J. Q., (2004). Nanoparticle and targeted systems for cancer therapy. *Adv. Drug Deliv. Rev.*, *56*, 1649–1659.

Brigger, I., Dubernet, C., & Couvreur, P., (2002). Nanoparticles in cancer therapy and diagnosis. *Adv. Drug Deliv. Rev.*, *54*, 631–651.

Cai, W., Gao, T., Hong, H., Sun J., (2008). Applications of gold nanoparticles in cancer nanotechnology. *Nanotechnol Sci Appl.*, 1, 17–32.

Cevc, G., & Vier, I. U., (2010). Nanotechnology and the transdermal route. A state of the art review and critical appraisal. *J. Control. Rel.*, *141*, 277–299.

Chandak, A. R., & Verma, P. R. P., (2008). Development and evaluation of HPMC based matrices for transdermal patches of tramadol. *Clin. Res. Reg. Affairs.*, *25*, 13–30.

Chekhonin, V. P., Gurina, O. I., Ykhova, O. V., Ryabinina, A. E., Tsibulkina, E. A., & Zhirkov, Y. A., (2008). Polyethylene glycol-conjugated immunoliposomes specific for olfactory ensheathing glial cells. *Bull. Exp. Biol. Med.*, *145*, 449–451.

Couvreur, P., Barratt, G., Fattal, E., Legrand, P., & Vauthier, C., (2002). Nanocapsule technology: a review. *Crit. Rev. Ther. Drug Carrier. Syst.*, *19*, 99–134.

Daneman, D., (2006). Type 1 diabetes. *Lancet.*, *367*(9513), 847–858.

Das, P. J., Paul, P., Mukherjee, B., Mazumder, B., Mondal, L., Baishya, R., Chatterjee, M. D., & Dey, S. K., (2015). Pulmonary delivery of voriconazole loaded nanoparticles providing a prolonged drug level in lungs: A promise for treating fungal infection. *Mol. Pharm.*, *12*(8), 2651–2664.

Davis, M. E., Chen, Z. G., & Shin, D. M., (2008). Nanoparticle therapeutics: an emerging treatment modality for cancer. *Nat. Rev. Drug Discov.*, *7*(9), 771–782.

De Jong, W. H., Geertsma, R. E., & Roszek, B., (2005). Possible risks for human health. Report 265001002/2005. National Institute for Public Health and the Environment (RIVM). *Nanotechnology in Medical Applications*. Bilthoven, Netherlands.

DeJong, W. H., & Borm, P. J. A., (2008). Drug delivery and nanoparticles: Applications and hazards. *Int. J. Nano.*, *3*(2), 133–149.

DeLouise, L. A., (2012). Applications of nanotechnology in dermatology. *J. Invest. Dermatol.*, *132*, 964–975.

Demoy, M., Gibaud, S., Andreux, J. P., et al., (1997). Splenic trapping of nanoparticles: complementary approaches for in situ studies. *Pharm. Res.*, *14*, 463–468.

Dnyanesh, N. T., & Vavia, P. R., (2003). Acrylate-based transdermal therapeutic system of nitrendipine. *Drug Dev. Ind. Pharm.*, *29*, 71–78.

Drbohlavova, J., Chomoucka, J., Adam, V., Ryvolova, M., Eckschlager, T., Hubalek, J., & Kizek, R., (2013). Nanocarriers for Anticancer Drugs-New Trends in Nanomedicine. *Curr. Drug Metab.*, *14*(5), 547–564.

Duncan, R., (2003). The dawning era of polymer therapeutics. *Nat. Rev. Drug Disc.*, *2*, 347–360.

Egan, M. E., Pearson, M., Weiner, S. A., Rajendran, V., Rubin, D., Glockner-Pagel, J., Canny, S., Du, K., Lukacs, G. L., & Caplan, M. J., (2004). Curcumin, a major constituent of turmeric, corrects cystic fibrosis defects. *Science*, *304*, 600–602.

Eijkel, J. C. T., & Berg, A. V., (2006). The promise of nanotechnology for separation devices – from a top-down approach to nature-inspired separation devices. *Electrophoresis*, *27*, 677–685.

European Science Foundation, (2005). Policy Briefing (ESF), ESF Scientific Forward Look on Nanomedicine. IREG Strasbourg, France.

Fasano, A., (1998). Modulation of intestinal permeability: an innovative method of oral drug delivery for the treatment of inherited and acquired human diseases. *Mol. Genet. Metab.*, *64*, 12–18.

Ferrari, M., (2005). Cancer nanotechnology: opportunities and challenges. *Nat. Rev. Cancer*, *5*, 161–171.

Feynman, R. P., (1960). There is plenty of room at the bottom. *Engineering & Science Magazine. Volume XXIII No. 5*, California Institute of Technology, Pasadena, USA.

Fix, J. A., (1996). Oral controlled release technology for peptides: status and future prospects. *Pharm. Res.*, *13*, 1760–1764.

Galindo-Rodriguez, S. A., Allemann, E., Fessi, H., & Doelker, E., (2005). Polymeric nanoparticles for oral delivery of drugs and vaccines: a critical evaluation of *in vivo* studies. *Crit. Rev. Ther. Drug Carr. Syst.*, *22*, 419–464.

Gao, L., Zhang, D., & Chen, M., (2008). Drug nanocrystals for the formulation of poorly soluble drugs and its application as a potential drug delivery system. *J. Nanopart. Res.*, *10*(5), 845–862.

Gaumet, M., Vargas, A., Gurny, R., & Delie, F., (2008). Nanoparticles for drug delivery: The need for precision in reporting particle size parameters. *Euro. J. Pharm. Biopharm.*, *69*, 1–9.

Ghosh, P., Han, G., De, M., Kim, C. K., & Rotello, V. M., (2008). Gold nanoparticles in delivery applications. *Adv. Drug Deliv. Rev.*, *60*(11), 1307–1315.

Gibaud, S., Demoy, M., Andreux, J. P., Weingarten, C., Gouritin, B., & Couvreur, P., (1996). Cells involved in the capture of nanoparticles in hematopoietic organs. *J. Pharm. Sci.*, *85*, 944–950.

Gonzalez-Mira, E., Egea, M. A., Souto, E. B., Calpena, A. C., & Garcia, M. L., (2011). Optimizing flurbiprofen-loaded NLC by central composite factorial design for ocular delivery. *Nanotechnology*, *22*, 045101.

Gowthamarajan, K., & Kulkarni, G., (2003). Oral insulin: fact or fiction? *Resonance*, *8*(5), 38–46.

Gradishar, W. J., (2005). Phase III trial of nanoparticle albumin-bound paclitaxel compared with polyethylated castor oil-based paclitaxel in women with breast cancer. *J. Clin. Oncol.*, *23*(31), 7794–7803.

Grayson, A. C., Shawgo, R. S., Johnson, A. M., Flynn, N. T., Li, Y., Cima, M. J., & Langer, R. A., (2004). BioMEMS review: MEMS technology for physiologically integrated devices. *Proc. IEEE*, *92*, 6–21.

Hadzijusufovic, E., Peter, B., Gleixner, K. V., Schuch, K., Pickl, W. F., Thaiwong, T., Yuzbasiyan-Gurkan, V., Mirkina, I., Willmann, M., & Valent, P., (2010). H1-receptor antagonists terfenadine and loratadine inhibit spontaneous growth of neoplastic mast cells. *Exp. Hematol.*, *38*(10), 896–907.

Hamman, J. H., Enslin, G. M., & Kotze, A. F., (2005). Oral delivery of peptide drugs: barriers and developments. *Bio. Drugs.*, *19*, 165–177.

Hanaire, H., Lassmann-Vague, V., Jeandidier, N., Renard, E., Tubiana-Rufi, N., Vambergue, A., Raccah, D., Pinget, M., & Guerci, B., (2008). Treatment of diabetes mellitus using an external insulin pump: The state of the art. *Diabetes Metab.*, *34*, 401–423.

Illum, L., (2007). Nanoparticulate systems for nasal delivery of drugs: A real improvement over simple systems. *J. Pharm. Sci.*, *96*, 473–483.

Khafagy, E. S., Morishita, M., Onuki, Y., & Takayama, K., (2007). Current challenges in noninvasive insulin delivery systems: a comparative review. *Adv. Drug Deliv. Rev.*, *59*(15), 1521–1546.

Kim, T. H., Park, C. W., Kim, H. Y., Chi, M. H., Lee, S. K., Song, Y. M., Jiang, H. H., Lim, S. M., Youn, Y. S., & Lee, K. C., (2012). Low molecular weight (1 kDa) polyethylene glycol conjugation markedly enhances the hypoglycemic effects of intranasally administered exendin-4 in type 2 diabetic db/db mice. *Biol. Pharm. Bull.*, *35*, 1076–1083.

Kompella, U. B., Lee V. H., (2001). Delivery systems for penetration enhancement of peptide and protein drugs: design considerations. *Adv. Drug Deliv. Rev.*, *46*, 211–245.

Labouta, H. I., El-Khordagui, L. K., Krause, T., & Schneider, M., (2011). Mechanism and determinants of nanoparticle penetration through human skin. *Nanoscale*, *3*, 4989–4999.

Lee, H. J., (2002). Protein drug oral delivery: the recent progress. *Arch. Pharm. Res.*, *25*, 572–584.

Lee, R. W., Shenoy, D. B., & Rajiv Sheel, R., (2010). Micellar Nanoparticles: Applications for Topical and Passive Transdermal Drug Delivery. In *Handbook of Non-Invasive Drug Delivery Systems*, Kulkarni, V. S., (ed.), William Andrew, Elsevier, Norwich, New York, pp. 37–58.

Lehr, C. M., (1994). Bioadhesion technologies for the delivery of peptide and protein drugs to the gastrointestinal tract. *Crit. Rev. Ther. Drug Carrier Syst.*, 11, 119–160.

Lemoine, D., Francois, C., Kedzierewicz, F., Preat, V., Hoffman, M., & Maincent, P., (1996). Stability study of nanoparticles of poly (epsilon-caprolactone), poly(DL-lactide) and poly(D, L-lactide-co-glycolide). *Biomaterials*, *17*, 2191–2197.

Lenaerts, V., Nagelkerke, J. F., Van Berkel, T. J., Couvreur, P., Grislain, L., Roland, M., & Speiser, P., (1984). *In vivo* uptake of polyisobutyl cyanoacrylate nanoparticles by rat liver Kupffer, endothelial, and parenchymal cells. *J. Pharm. Sci.*, *73*, 980–982.

Lenhard, M. J., & Reeves, G. D., (2001). Continuous subcutaneous insulin infusion: A comprehensive review of insulin pump therapy. *Arch. Intern. Med.*, *161*, 2293–2300.

Lowman, A. M., Morishita, M., Kajita, M., Nagai, T., & Peppas, N. A., (1999). Oral delivery of insulin using pH-responsive complexation gels. *J. Pharm. Sci.*, *88*(9), 933–937.

Maas, J., Kamm, W., & Hauck, G., (2007). An integrated early formulation strategy-from hit evaluation to preclinical candidate profiling. *Eur. J. Pharm. Biopharm.*, *66*(1), 1–10.

Maeda, H., (2001). The enhanced permeability and retention (EPR) effect in tumor vasculature: the key role of tumor-selective macromolecular drug targeting. *Adv. Enzyme Regul., 41*, 189–207.

Mansour, H. M., Rhee, Y. S., & Wu, X., (2009). Nanomedicine in pulmonary delivery. *Int. J. Nanomed., 4*, 299–319.

Mathias, N. R., & Hussain, M. A., (2010). Non-invasive systemic drug delivery: developability considerations for alternate routes of administration. *J. Pharm. Sci., 99*, 1–20.

Mehnert, W., & Mader, K., (2012). Solid lipid nanoparticles: Production, characterization and applications. *Adv. Drug Deliv. Rev., 64*, 83–101.

Miyazaki, K., & Islam, N., (2007). Nanotechnology systems of innovation – An analysis of industry and academia research activities. *Technovation., 27*, 661–671.

Moghimi, S. M., Hunter, A. C., & Murray, J. C., (2001). Long-circulating and target specific nanoparticles: theory and practice. *Pharmacol. Rev., 53*, 283–318.

Mugumu, H., (2006). Transdermal delivery of Caco3- Nanoparticles Containing Insulin. *Diabetes Technol. Ther., 8*(3), 369–374.

Mukhopadhyay, A., Basu, N., Ghatak, N., & Gujral, P. K., (1982). Anti-inflammatory and irritant activities of curcumin analogues in rats. *Agents Act., 12*, 508–515.

Muller, R. H., & Wallis, K. H., (1993). Surface modification of i. v. injectable biodegradable nanoparticles with poloxamer polymers and poloxamine 908. *Int. J. Pharm., 89*, 25–31.

Muller, R. H., Mader, K., & Gohla, S., (2000). Solid lipid nanoparticles (SLN) for controlled drug delivery-A review of the state of the art. *Eur. J. Pharm. Biopharm., 50*, 161–177.

Muller, R. H., Runge, S., Ravelli, V., Mehnert, W., Thunemann, A. F., & Souto, E. B., (2006). Oral bioavailability of cyclosporine: Solid lipid nanoparticles (SLN) versus drug nanocrystals. *Int. J. Pharm., 317*, 82–89.

Mundargi, R. C., Patil, S. A., Agnihotri, S. A., & Aminabhavi, T. M., (2007). Evaluation and controlled release characteristics of modified xanthan films for transdermal delivery of atenolol. *Drug Dev. Ind. Pharm., 33*, 79–90.

INDEX

Printed and bound by CPI Group (UK) Ltd, Croydon, CR0 4YY

23/10/2024

01777706-0002